THE ARCHAEOLOGY OF NORTH PACIFIC FISHERIES

The Archaeology of
North Pacific Fisheries

Madonna L. Moss and Aubrey Cannon, editors

UNIVERSITY OF ALASKA PRESS
Fairbanks

University of Alaska Press
P.O. Box 756240
Fairbanks, AK 99775-6240

ISBN 978-1-60223-146-7 (paperback); 978-1-60223-147-4 (e-book)

Library of Congress Cataloging-in-Publication Data

The archaeology of North Pacific fisheries / Madonna L. Moss and and Aubrey Cannon, editors.
    p. cm.
Includes bibliographical references and index.
ISBN 978-1-60223-146-7 (pbk. : acid-free paper) — ISBN 978-1-60223-147-4 (electronic book)
1. Fisheries—Northwest Coast of North America—History. 2. Fisheries—North Pacific Ocean—
History. 3. Indians of North America—Northwest Coast of North America—Antiquities.
4. Indians of North America—Northwest Coast of North America—Social life and customs.
5. Ethnoarchaeology—Northwest Coast of North America. 6. Coastal archaeology—Northwest
Coast of North America. 7. Excavations (Archaeology)—Northwest Coast of North America.
8. Fish remains (Archaeology)—Northwest Coast of North America. 9. Northwest Coast of North
America—Antiquities. 10. Northwest Coast of North America—Social life and customs.
 I. Moss, Madonna. II. Cannon, Aubrey.
SH214.4.A73 2011
639.2'2091644—dc22
                    2011003389

Cover art: *Flow of Energy* by Rande Cook; cover design by Jason Gabbert Design, LLC
Interior design and layout by Rachel Fudge

This publication was printed on acid-free paper that meets the minimum requirements for ANSI /
NISO Z39.48–1992 (R2002) (Permanence of Paper for Printed Library Materials).

Support for this publication was provided by the Terris and Katrina Moore Endowment.

Printed in the United States

CONTENTS

Section III: Pacific Cod and Other Gadids: "Cousins" of the Fish That Changed the World

Section IV: Herring and Other Little-Known Fish of the North Pacific Coast

Section V: Conclusion

ILLUSTRATIONS

Figures

Tables

ACKNOWLEDGMENTS

As editors of this book, we are grateful to the authors who submitted their work to be published here. We thank them for their patience, attention to detail, and trust in us.

This publication grew out of a symposium we organized for the 2008 annual meeting of the Society for American Archaeology held in Vancouver, British Columbia. In addition to the contributors to this volume, we also appreciate the participation of Liam Frink, Justin Hays, Iain McKechnie, and Holly McKinney in the symposium. Virginia Butler served as our energetic and incisive discussant to the edification of all present. We appreciate the enthusiasm of the audience, which helped inspire us to work toward publication.

We thank Elisabeth Dabney, acquisitions editor at University of Alaska Press, for seeing this project through its long gestation. It was a pleasure to work with Amy Simpson and the meticulous Sue Mitchell, also with the press. We thank Rachel Fudge for copyediting. We gratefully acknowledge two anonymous reviewers who reviewed the entire manuscript. Their comments helped to improve the overall quality of the volume.

At the University of Oregon, Madonna would like to thank Scott Coltrane and Larry Singell of the College of Arts and Sciences, Russell Tomlin of Academic Affairs, and Carol Silverman of the Department of Anthropology for their support of some time away from teaching that facilitated work on this book. For crucial intellectual and material support over the last three years that contributed in both direct and indirect ways, she thanks Virginia Butler, Chelsea Buell, Scott Byram, Sarah Campbell, Roy Carlson, Molly Casperson, Tom Connolly, Susan Crockford, Jon Erlandson, Daryl Fedje, Sue Jurasin, Brian Kemp, Patty Krier, Gyoung-Ah Lee, Dana Lepofsky, Rob Losey, Al Mackie, Quentin Mackie, Mark McCallum, Iain McKechnie, Alan McMillan, Sandi Morgen, Lesli Morrison, Terry O'Nell, Torrey Rick, Carley Smith, Jane Smith, Martina Steffan, Rebecca Wigen, and Dongya Yang.

To Aubrey: It was simultaneously a privilege and pleasure to collaborate on this effort. I appreciate the breadth and depth of your skills and knowledge, but also your gracious collegiality.

To Jon and Erik Erlandson: Thank you for everything—and for not minding that work on this book kept us from having a Christmas tree in 2010. To Harry Moss: Thank you for teaching me how to fish, and to Rita Moss: Thank you for teaching me how to cook fish. Catching and cooking fish is a large part of what this book is about.—*Madonna*

For Aubrey, this volume is a milestone. From a life growing up next to the Fraser River, in which the salmon, trout, and eulachon of the river and its tributaries played a major role, it has been a long road to this point. He thanks Roy Carlson for initiating his study of fisheries archaeology and Rick Casteel for alerting him early on to the potential and the potential pitfalls of that research. Along the way, many people have shared interests in fish, fishing, and fish bones; their inspiration, guidance, and support have contributed greatly to the directions his work has taken. Foremost, he thanks Deborah Cannon, whose talents and enthusiasm enhanced all aspects of fisheries archaeology and its pursuit. Other individuals also deserve thanks for the key roles they played in enabling his research on the fisheries of the central coast of British Columbia, including Jennifer Carpenter of the Heiltsuk Cultural Education Centre, who facilitated and encouraged new directions and applications in field research; his in-laws, Dan and Stella Yee, whose interest and generous hospitality in providing use of their home as a staging point were keys to the success of field trips; and his research collaborator, Dongya Yang, whose brilliance and enthusiasm helped to shape incredibly productive new directions in the study of ancient fisheries. He also thanks his graduate students, fishy and otherwise, for the interest and enthusiasm to keep the research moving forward. Finally, he acknowledges the generous support of the Canada Foundation for Innovation for providing the funding for the Fisheries Archaeology Research Centre at McMaster University.

To Madonna: It has been both a challenge and a pleasure to work to keep up with you on this project. Your energy, focus, and insight kept the process moving forward at times when I was easily distracted by other responsibilities.—*Aubrey*

On behalf of ourselves and all the contributors, we acknowledge the First Nations of the Pacific Northwest who have shared access to their knowledge, territories, and histories in enabling pursuit of the research presented in this volume. We know we still have much to learn.

Finally, we thank Kwakwaka'wakw artist Rande Cook for permission to use his artwork *Flow of Energy* on the cover of this book. It helps emphasize the importance of fish to our lives and to the health of our planet.

Madonna L. Moss and Aubrey Cannon
Eugene, Oregon, and Hamilton, Ontario, February 2011

CHAPTER 1

The Archaeology of North Pacific Fisheries
An Introduction

Madonna L. Moss, *Department of Anthropology, University of Oregon*
Aubrey Cannon, *Department of Anthropology, McMaster University*

For more than 10,000 years, the First Peoples of the Pacific coast of North America have sustained themselves with a variety of coastal resources, but of these, the remains of fish and shellfish are the most abundant in the region's archaeological sites. Human impacts on fisheries over the last 200 years have left us with a woefully incomplete understanding of the long-term histories of fish species prior to the industrial era. *The Archaeology of North Pacific Fisheries* brings together studies from Alaska to British Columbia to Puget Sound in an effort to start to fill this knowledge gap and to expand our understanding of the histories of the region's fishing cultures. The record derives from the ancestors of the Aleut, Alutiiq, Tlingit, Haida, Tsimshian, Gitksan, Nisga'a, Witsuwit'en, Heiltsuk, Wuikinuxv, Nuu-chah-nulth, and Coast Salish First Nations.

Even though fishing is by convention subsumed under the term *hunter-gatherer*, fishing was of such primary importance to the peoples of the North Pacific region that the less elegant but more accurate term *hunter-fisher-gatherer*, now in widespread use to describe the Mesolithic peoples of Northwest Europe, is more appropriate. Pacific coast societies practiced many different types of fishing, but social groups also maintained control over fishing territories. They were not just fishers but fisheries resource managers, perfecting systems of ecosystem management tailored to their individual circumstances. They successfully harvested a wide range of species, but their management and control of salmon were of special significance. The peoples of the North Pacific mastered the technologies of fish processing and storage, leading them to accumulate significant surpluses. Although the term *food producers* is usually applied only to horticultural or agricultural societies, North Pacific coast societies were clearly food producers, and the keys to that food production were fishing and storage technologies and fisheries harvest and management strategies.

Pacific coast societies invested in infrastructure and altered some aspects of their physical environments to promote fisheries production. Perhaps the most significant aspect of their economic systems was the way they managed harvests through systems of territorial ownership and control, restraining uncontrolled resource use through systems of social relations. Tribes, clans, and households were caretakers of particular watersheds, fish streams, and stretches of ocean shoreline. They managed harvests as trustees who had established long-term relationships not only with resource territories but also with the plant and animal persons with whom they shared their worlds. As Richard Atleo (2005:ix) has explained, proper relationships had to be maintained between all life-forms. He uses salmon as an example:

> Since the salmon and human have common origins they are brothers and sisters of cre-
> ation. Since the assumption of all relationships between all life-forms is a common ances-
> try, protocols become necessary in the exercise of resource management. If the salmon are
> not properly respected and recognized they cannot properly respect and recognize their
> human counterparts of creation. This historical process is neither evolutionary nor devel-
> opmental in the linear sense. Changes are not from simple to complex, as a more modern
> world-view would have it, but from complex to complex, from equal to equal, from one
> life-form to another.

This ideology helps explain the ecological prudence of some Pacific coast societies and the restraint with which they historically managed resources. Another aspect of tradi-tional ecological knowledge involved understanding relationships between species within ecosystems. Although this book focuses on the archaeological record of North Pacific fish-eries, and cannot fully address worldview and philosophy, the ethic described by Atleo (2005) provides a secure foundation for principles of resource management.

Fishing, Fisheries Management, and Fish Processing

Most Pacific coast archaeological sites contain some evidence of fishing. In the typical open shell midden, one may find a harpoon point used on a spear, a fishing gorge, a net weight, or parts of a composite fishhook. Far more abundant will be the remains of the fish themselves, and although salmon are the most widely known, the bones of Pacific cod and other gadids, rockfish, lingcod, greenlings, herring, flatfish, surfperch, and sculpins are common. Halibut, spiny dogfish, skate, eulachon, smelt, ratfish, jack mackerel, tuna, sticklebacks, and pricklebacks are also found (e.g., Cannon 1995; Christensen and Stafford 2005; Crockford 1997; Moss 2004). Fish typically account for the greatest number of bones in the region's archaeological faunal assemblages, usually in the order of 85 to 95 percent. Fishing is represented in town, village, and camp sites across the region.

Isotopic studies of the bones of Shuká Ḵaa show that this man living more than 11,000 years ago on Prince of Wales Island in southeast Alaska had a diet rich in seafood (Kemp et al. 2007). Although we do not know the precise composition of his diet, it likely included fish. At The Dalles on the Columbia River, vast numbers of salmon were caught by site residents over 9300 cal BP (Butler and O'Connor 2004). From Namu, located on the central coast of British Columbia, Cannon and Yang (2006) studied the ancient DNA of salmon bones to better understand the fishery. They found convincing evidence for the long-term reliance on pink salmon going back 7,000 years. Using other seasonal indications from the site, they build a strong case that at least some Namu residents lived at the site year-round and were dependent on a salmon storage economy. Cannon and Yang also argue that storage and sedentism in this case can be decoupled from other indicators of "cultural complexity," including increasing population density and social inequality. Following up on earlier work (Cannon 1991, 1995, 1998), they found additional support for a disruption in the pink salmon fishery about 4,000 years ago that led site residents to compensate for shortfalls with more marginal resources such as ratfish. The cause of this disruption is unknown, but it helps demonstrate how historical contingencies can affect local archaeological records, which are sometimes misread as indicators of long-term evolutionary change. This research does not mean that economies based on salmon as storage staples were established everywhere on the Pacific coast 7,000 or more years ago, but it does shake up the widely held assumption that storage technologies were a late development (e.g., Matson 1992).*

Since the mid-1980s, considerable data have been accumulating from sites found in the water-saturated environments of the intertidal zone, where wood and fiber artifacts and features preserve. These are the remains of intertidal fishing weirs and traps, arrangements of wood stakes driven into muddy substrates (Moss and Erlandson 1998; Moss, Erlandson, and Stuckenrath 1990). Over 1,200 such sites have been found in Alaska, British Columbia, Washington, and Oregon (Moss forthcoming). Of these, fewer than 200 have been radiocarbon dated, with the sites ranging in age from ca. 5500 cal BP to the twentieth century. Weirs and traps made exclusively of stone also number in the hundreds, but because most cannot be directly dated (like the one near the mouth of the Namu River), they have not been intensively studied. These numbers provide evidence of the scale of mass capture of fish on parts of the North Pacific coast.

Ethnographically described technologies include weirs, traps, and nets. A weir is a fence-like structure, set across a river or stream or in an estuarine tidal channel for catching fish. Weirs were built in shallow waters to block the upstream movement of fish or strand them with the outgoing tide. Sometimes stakes were closely spaced; other times, generous spaces between stakes were filled in with minimally altered brush or boughs woven into a framework or with manufactured latticework (rectangular woven screens).

---

* Moss and Erlandson (1995:34) observed that "the focus of Northwest Coast scholars on salmon and storage as harbingers of cultural complexity seems over-emphasized. California Indian groups stored food too, but this is rarely viewed as a key component in the development of cultural complexity along the California coast." This might be thought of as the "even squirrels know how to store nuts" argument. While it is true that storing fish and other seafood requires technological sophistication, the incentive to store food would seem to be hard-wired into mammalian biology, especially for those living in mid- to high latitudes.

A trap might consist of an arrangement of wood stakes, stones, or other elements left in place as an enclosure. These might be large structures embedded in the tidal channel or stream, and some had removable parts. Basketry traps occurred in cylindrical, conical, or globular forms and were often attached to the more stationary weirs or traps described above. Lattice-work sections could be configured into rectangular traps or rolled into a cylinder to form a trap.

When fish were caught in one of the smaller, easily portable traps they could be lifted out of the water. Spears, harpoons, gaff hooks, dip nets, or other devices were used to take fish while a person was positioned on a platform or rock, or from a canoe. Netting was made of nettle or other fibers and graduated needles and gauges were used to make mesh sizes tailored to the size of the targeted fish. Nets were used for eulachon; salmon gill nets or flounder seines were anchored by wooden stakes along the shoreline, while other types of drag nets and reef nets were not set to stakes.

Most intertidal sites with wood stakes have been simply recorded as sites, with some recovery of wooden stakes for radiocarbon dating. One prominent exception is the Little Salt Lake Weir, near Klawock, Alaska, where Langdon (2006, 2007) mapped extensive features with an estimated 100,000 stakes found over a 75 hectare area, dating to the last 2,000 years. Distinctive features included "pavements" with as many as eighty wood stakes per square meter used as barriers or leads to direct fish movements or for people to stand on; "pounds," which are rectangular arrangements of stakes in which fish were impounded; "pairs" of stakes used to brace latticework; and "piles" of stone and stake alignments used as a foundation. Based partly on salmon escapement features, Langdon, Reger, and Campbell (1995) estimated that between 75,000 and 80,000 fish could be caught in the Little Salt Lake Weir annually.

Another spectacular weir and trap complex is located in Comox Harbour, on the east side of Vancouver Island, where Nancy Greene (2010) mapped more than 11,000 wood stakes in overlapping alignments on the surface of the tidal flats. Greene identified two temporally distinct structural designs; the older (ca. 1,000 years old) is a large heart-shaped enclosure with a flattened, or truncated, rear wall, with an opening toward the shoreline, and aligned with the outgoing tide. The younger trap type (ca. 200 years old) has large chevron-winged enclosures with openings toward the shoreline. Both trap types used leads to channel fish into the traps. In their overall configuration, these are similar to the double-lead-and-enclosure wood-stake-and-stone traps near Petersburg, Alaska, so well described by Mobley and McCallum (2001). Such traps may have been used for salmon, herring, and a wide range of other fish.

Beyond technologies, there is an important social dimension to fisheries production. Ethnographies, for example, contain generalizations that strongly gender the labor of fishing: men catch fish and women butcher and process them (e.g., Drucker 1937:232; Emmons 1991:143; de Laguna 1972:384). This generalization conjures up a stereotype of the lone fisherman out in his canoe, with his wife waiting at home, fish knife at the ready. This image is not dissimilar to that of the lone fly fisherman in the remote Western stream, but it is a far cry from the reality of much of Pacific coast fishing. Weir and trap fishing and the use of gill and reef nets were not activities carried out by one person working alone; they required group teamwork. Group fishing could result in large numbers of fish, but processing quantities of fish for long-term storage also required well-organized labor.

Some fish, including halibut or nearshore rockfish and greenling, were caught by individuals with hook and line or gorges, but the capture of schooling taxa such as salmon, herring, eulachon, smelt, and others were group activities.

By scrutinizing ethnographies, we find evidence that women fished on the Pacific coast (Barnett 1955:89; Boas 1921:181–182; Elmendorf and Kroeber 1992:63; de Laguna 1972:386; Olson 1936:29, 32; Singh 1966:58–59; Smith 1940:254, 257, 268; Suttles 1974:114, 188), even in societies that had an ethic precluding women from fishing. Children were also involved in fishing (Byram 2002:169; Smith 1940:268), but who actually caught the fish was only a part of the process—women were often chiefly responsible for gathering and processing the plant materials used in manufacturing the perishable, portable portions of fishing technologies, such as basketry traps, fish baskets, latticework, netting, and nets (Byram 2002:167; Emmons 1991; Gunther 1973:28; Olson 1936:28; Paul 1944). The fabrication of many tools used in fishing is akin to weaving and basketry technologies. In some groups, women were involved in weir construction (Hewes 1940). In other societies, men made some or most of the fishing gear (e.g., Barnett 1937:164; Boas 1921:162; Drucker 1937:232, 1951:16). The rigid division of labor used to characterize Pacific coast fishing breaks down under a close reading of ethnographic sources. Fish butchery is almost always attributed to women, but certainly the drying, smoking, and storage process likely involved work groups made up of a mix of people—women and men, elders and children.

Fish were sliced in different ways depending on the species and the size of the fish, and weather conditions affected stages of fish curing—sunny or windy conditions might allow initial drying to occur outside, whereas rain required drying in a house, smokehouse, or perhaps a rockshelter. As Suttles (1968:63) pointed out long ago, the number of sunny days and the amount of rainfall varies tremendously on the Pacific coast, by latitude and due to orographic factors. Within British Columbia, the city of Victoria, located in the rain shadow of the Olympic Mountains, enjoys over 2,000 hours of sunshine per year, while Prince Rupert receives only half as much. Not all fish preserve equally well. Suttles (1968:63) wrote that his Salish informants said fatter fish (e.g., sockeye salmon) lasted longer than lean species, although elsewhere Suttles (1990:25) stated that sockeye were the hardest fish to keep (see also Romanoff 1985). Processing and storage techniques likely varied as much as fishing technology by season, location, and species, but detailed information is often hard to find. Strategies of fish butchery, cooking, processing, and disposal are of great interest to zooarchaeologists, and while some ethnographies provide important detail (e.g., Boas 1921), many yield only generalizations. Capturing large numbers of fish is pointless, however, without the technological knowledge and organized labor to process them for storage. Stored fish were important winter staples, but a variety of fish products—dried fish, fish oils, fish eggs—were also valued trade items.

Although many fish contributed to the economic foundation of Northwest Coast societies, salmon were especially important. Yet in their review of sixty-three faunal assemblages from the Pacific Northwest, Butler and Campbell (2004) found no evidence for the intensification of salmon relative to other fish over time, nor did they find any evidence of resource depression. Like the work of Cannon and Yang (2006), this challenges earlier views of the economic basis for the so-called emergence of Northwest Coast cultural complexity (e.g., Matson 1992; Matson and Coupland 1995). Ethnographically, we know that across much of the Pacific coast, salmon streams and spawning areas were owned by particular social groups. The antiquity of salmon resource ownership is difficult to document archaeologically but may go back 7,000 years at Namu (Cannon and Yang 2006), where the residents lived year-round, possibly to protect their claim to the river and its salmon. In Chapter 13 of this volume, Moss makes a similar argument for at least 4,000 years of resource ownership at Coffman Cove, Alaska. While many archaeologists invoke the use of salmon as a driver of cultural evolution, another writer has taken a different tack:

> Given that Pacific salmon return to their natal streams to spawn and were beyond the tribes' ability to intercept prior to that time, tribal ownership of streams effectively included secure ownership of native salmon stocks, including the real option to take advantage of new growth opportunities. This gave the tribes the incentive to accumulate the stream-specific knowledge to husband these stocks. Rather than being the fortunate beneficiaries of a naturally rich environment, the compelling conclusion is that the NWC [Northwest Coast] tribes *created* the observed superabundance of salmon through centuries of purposeful husbandry and active management of other resources. In my view, they were not hunter-gatherers content to meet material subsistence needs, they were institutionally sophisticated salmon ranchers who actively sought and proudly achieved prosperity. This view is entirely consistent with the weight of the historical evidence and with the vast body of literature on the economics of property rights. (Johnsen 2004:4–5)

Although the argument for salmon ranching may initially appear radical, it may better capture the essence of at least some North Pacific groups' use of salmon than previously employed concepts. Deliberate stewardship of salmon and other resources would seem the best explanation for Butler and Campbell's (2004) findings.

The Study of Fish Remains on the Pacific Coast: A Brief History

Along the Pacific coast of North America, John E. Fitch of the Los Angeles County Museum of Natural History conducted numerous studies of fish remains from Southern California sites in the 1960s and 1970s. William I. Follett of the California Academy of Sciences worked on California fish remains from the 1940s through the 1980s. Follett also studied fish remains from 49-SIT-244, excavated by Frederica de Laguna (1960) in 1949–1950 near Angoon, Alaska. Fitch and Follett were both biologists, however, and while their work contributed immensely to the recognition that fish remains in archaeological sites were important, such studies were at the time relegated to report appendices and specialized biological journals.

To our knowledge, Richard W. Casteel was the first zooarchaeologist specializing in the study of fish remains to hold a position in a department of anthropology or archaeology in western North America. Casteel received his PhD from the University of California, Davis, in 1972 and went on to be a professor of zooarchaeology in the department of archaeology at Simon Fraser University. His landmark book, *Fish Remains in Archaeology* (1976a), demonstrated how analysts could use fish vertebrae, otoliths, scales, and other elements to identify fish species and seasonality of use, and estimate the body size of individual fish. Casteel argued that fish remains were important sources of data not just to archaeologists but to fisheries biologists, paleoclimatologists, paleontologists, taphonomists, ichthyologists, and ecologists.

Even though *Fish Remains in Archaeology* was written thirty-five years ago, many of Casteel's concerns remain. For example, Casteel (1976a:88–89) pointed out that because some archaeologists had observed that the salmonid skeleton was "relatively soft," this mistakenly led some North American investigators to think that the search for salmon bones in archaeological sites was futile. Bryan (1963:8), then working in Puget Sound, wrote on advice of a fisheries biologist that "salmon bones are cartilaginous and therefore do not preserve." Desautels et al. (1970:320), working on the Aleutian Islands, wrote that salmon and trout "do not possess the high calcification which is present in the bones of the marine species and are therefore so fragile that they may not appear." Casteel corrected both Bryan and Desautels by citing numerous studies (in English and Russian) that reported archaeological salmonid remains. His own work clearly demonstrated that salmonid remains were "abundant constituents of archaeological assemblages in western North America and into Alaska and Canada" (Casteel 1976a:90). Casteel clarified that the semicalcified remains of cartilaginous fish (elasmobranchs) were present in archaeological sites. Yet these fundamental issues of the differential density, survivability, and identifiability of fish remains are ones that persist today, and they are discussed by some of the contributors to this volume.

As recently as the late 1970s it was still possible to hear debate at scholarly conferences over whether Northwest Coast subsistence economies were based primarily on the hunting of sea or land mammals. Publications that carefully documented the relative numbers of bird and mammal bones curiously omitted reference to fish (e.g., Conover 1978; Lyman 1991). Others cited relative proportions of fish in faunal assemblages that could only have included fish as the result of incidental rather than systematic recovery (e.g., Coupland, Stewart, and Patton 2010; Stewart and Stewart 1996). The lack of systematic recovery of fish bones and the small number of comparative collections and analysts with the knowledge and skill to identify fish remains were major impediments to full archaeological recognition of the importance of fish and fishing in Pacific coast economies.

A perennial problem affecting adequate recovery of fish remains on the Pacific coast and elsewhere in the world has been the failure to use screening on a regular basis or, when used, to use mesh small enough to recover the majority of fish remains. Compounding the problems of adequate recovery from shell midden contexts in particular is the near impossibility of seeing fish bones among the sticky black or fine gray matrix that constitutes most sites. The solution, which was to water screen deposits through fine mesh, was first applied in excavations in the late 1960s but became a regular part of archaeological practice starting in the late 1970s. The result was the recovery of tens of thousands of fish remains, though acceptance of quarter-inch (6.35 mm) or even eighth-inch (3.175 mm) screen as the finest mesh in regular use still meant the loss of the vast majority of remains of small fish such as greenling, herring, and eulachon. Water pressure and time and visibility constraints in field screening also compromised full recovery of fish remains, but regular water screening alone resulted in major advances in recognizing the importance and diversity of fishing economies.

Study of fish remains in Pacific coast archaeology was also impeded by the sheer volume of remains and the relatively small number of individuals and facilities available for their analysis. Partly in recognition of the scale of analysis involved in identification of all the fish remains recovered in a typical archaeological excavation, Casteel (1970, 1976b) advocated the use of column and auger sampling for the recovery of fish bones. He showed through direct comparisons that such methods were effective means for recovering a range of shell midden constituents, including fish bones, in relative quantities comparable to those recovered through full-scale excavation. Following Casteel's early work and Stein's (1992) pioneering use of bucket augers in shell midden investigation, Cannon (2000) developed the application of auger sampling to assess inter-site variability among the most ubiquitous of fish remains in shell midden sites on the central coast of British Columbia. This method is now the basis of multisite investigations in various parts of the coast, which are providing a broader basis for understanding variations in fishing economies between locations and over time (see Cannon, Yang, and Speller; Brewster and Martindale; and Caldwell in this volume). Steffen (2006) screened bulk samples over 1 mm mosquito netting to recover tiny calcined fish bones from Early Holocene hearths at the Richardson Island site, in an effort yet to be duplicated.

The final impediment to full archaeological investigation of Pacific coast fisheries was the availability of comparative collections and skilled analysts. Now many universities have adequate collections and full-time faculty and/or technical specialists in the analysis of fish remains. The commercial firm Pacific Identifications Inc., founded and operated by Susan Crockford, Rebecca Wigen, and Gay Frederick and housed at the University of Victoria in British Columbia, retains one of the most comprehensive of extant comparative collections. This firm has been responsible for the analysis of many faunal assemblages from Pacific coast archaeological sites, both for cultural resource management (CRM) firms and for academic archaeologists. A number of currently active zooarchaeologists received their training by working with Pacific Identifications. Further advances in student training have come through opportunities to work with archaeological and comparative collections under the guidance of trained specialists at other universities. Published manuals (D. Cannon 1987) have also made knowledge of fish bone morphology accessible to larger numbers of students. Full analysis of fish bone assemblages from Pacific coast archaeological sites is now standard practice, and it is not surprising that we have seen rapid expansion in knowledge and understanding of fisheries in recent years as a result.

Virginia Butler is a contemporary archaeologist whose work helps illustrate the main developments in the study of North Pacific fish remains over the past twenty years.[*] Trained at the University of Washington, Butler has been a pioneer in the mold of Casteel, having worked through many key methodological, technical, and interpretive issues, some of which we take up in this volume. In her dissertation, Butler (1990) developed criteria to distinguish natural from cultural deposits of salmon bones, suggesting that the salmon bones found in the 9,300-year-old Roadcut site on the Columbia River indicated salmon were an important resource used by Early Holocene peoples (see also Butler 1993).

---

[*] Butler has also made important contributions to the study of fish remains in the Great Basin and Columbia Plateau (e.g., Butler 1996, 2001a, 2004a; Butler and Campbell 2004) and in the Pacific Islands (e.g., Butler 1994, 2001b), although here we focus on her North Pacific research.

Because of potential bias in the Roadcut museum collections, Butler pursued additional field data from the site and was able to establish the cultural use of salmon conclusively (Butler and O'Connor 2004). This work inspired others to use skeletal representation to address butchery and storage practices. Butler's work with the Roadcut materials led her to develop absolute measures of bone density to better gauge its effects on the representation of skeletal elements. Her work on salmon bones (Butler and Chatters 1994) has set the standard for subsequent studies of other species, such as those by her former student, Ross Smith (2008 and this volume). Butler was also the first to pursue analysis of ancient DNA in archaeological salmon bones (Butler and Bowers 1998), which led the way for studies by Cannon and Yang (2006). Butler has conducted meta-analyses of faunal assemblages from large regions, looking for evidence of evolutionary change in the use of animal resources (Butler and Campbell 2004) and indications of paleoenvironmental shifts (Butler 2010). She has also drawn on historical records along with archaeological studies to address contemporary issues in conservation biology and fisheries management (Butler 2004b; Butler and Delacorte 2004). As will become apparent, Butler's work has inspired the work of many other investigators, including the contributors to this volume.

The Genesis of *The Archaeology of North Pacific Fisheries*

This project began in advance of the Society for American Archaeology annual meetings held in Vancouver, British Columbia, in March 2008. In preparing for the meeting during the summer of 2007, Moss was beginning to organize a symposium that would bring together zooarchaeologists studying the remains of Pacific cod and other codfishes in North Pacific coastal sites. Simultaneously, Cannon was organizing a symposium that would bring together zooarchaeologists focusing on salmon across the coast. When we discovered we were inviting some of the same participants, we decided to join forces and consider fish remains more holistically. We agreed that variability in form, focus, and intensity of North Pacific fisheries was becoming increasingly evident, even though the descriptive literature on fisheries resource use is not widely appreciated. Although the importance of salmon is well known to the broader archaeological community as well as the public at large, the timing and extent of salmon use varies tremendously across the coast, and this is not widely recognized. At the same time, the economic and social importance of other fishes—including Pacific cod, herring, eulachon, halibut, other flatfish, rockfish, sculpins, and other taxa—are rarely considered by anyone other than zooarchaeologists.

In the specialized literature that is available, archaeologists attribute spatial and temporal variability to a number of different factors, including environmental variation and change, settlement patterns, technology, transportation, and trade, or to resource depression or control and management. Yet the results of this research have not achieved their full potential. Research on *The Archaeology of North Pacific Fisheries* is central to comprehension of broader archaeological patterns because it helps us understand the economic foundations of North Pacific societies. By examining archaeological variability we can identify patterns and trends in fisheries that also greatly expand the longitudinal record of human use of North Pacific fisheries. Reconstructing this long-term ecological record is absolutely crucial in efforts to understand the dynamic relationships between people and fisheries. In some settings during some time periods, we may find evidence of overexploitation and even of fishing down the food web. In other cases, we may find evidence of conservation and sustainable use. Such historical ecological analyses can ultimately contribute to a broader and more in-depth perspective on the dynamics of aquatic ecosystems, which we think is invaluable in planning for restoration of fisheries that have suffered during the industrial era (Pauly 2009).

In the first of four sections, the contributors to this volume present innovative approaches to long-standing methodological problems. Trevor Orchard and Paul Szpak address the frustration zooarchaeologists have faced due to our inability to identify the most common salmon bone elements in archaeological assemblages—vertebrae—to species based on morphological differences. These authors combine a radiographic technique first used by Cannon (1988) along with metric data to estimate the composition of salmon species in assemblages from Haida Gwaii. Catherine West, Stephen Wischniowski, and Christopher Johnston use Pacific cod otoliths to generate local, high-resolution paleoenvironmental data to understand how the Little Ice Age was manifested in the waters off Kodiak Island frequented by the Alutiiq. This is precisely the kind of study Casteel called for thirty-five years ago in an effort to reconstruct past environments. Ross Smith, Virginia Butler, Shelia Orwoll, and Catherine Wilson-Skogen summarize the results of their important work on fish bone density. Their measurements on Pacific cod, along with Smith's (2008) work on halibut, represent a method that can be used to examine whether or not density-mediated attrition has biased an archaeological assemblage.

The second section of the book presents new studies focused on salmon fisheries. Salmon, the iconic fish of the North Pacific, is a genus under threat from overfishing and, in some areas, climate change (Augerot 2005). As these chapters show, the timing and extent of First Nations salmon use varied widely on the Pacific coast. In their multisite study, Aubrey Cannon, Dongya Yang, and Camilla Speller look at relative abundance of different salmon species—as established by the study of ancient DNA—to better understand a regional picture of the salmon fisheries in Heiltsuk and Wuikinuxv territories on the central coast of British Columbia. In his study, Elroy White interviewed twelve Heiltsuk oral historians to better understand the construction and use of fish traps built by his ancestors. This Heiltsuk perspective reveals how people operated fish traps and how they opened them up after the fishing season. The labor involved in smoke-drying salmon is also described using local knowledge and ethnographic narratives. Paul Prince works on sites along the Skeena River and its tributaries where salmon bones are rarely preserved. Examining other types of clues to salmon resource use, including site locations and stone artifacts, Prince builds a case that frequently overlooked tool types, including cobble spalls, bifaces, small convex scrapers, and microblades, were probably used to clean and process salmon. He suggests that the ancestors of the Tsimshian, Gitksan, Nisga'a, and Witsuwit'en did not require technologies developed on the coast to take full advantage of the abundant salmon in the Nass and Skeena rivers. In a study of southern Haida Gwaii, Trevor Orchard identifies important changes in assemblage composition that he ties to Haida cultural changes. Between 1200 and 800 cal BP, more generalized fishing, with an emphasis on rockfish, appears to have shifted to more specialized economies focused on salmon, particularly chum salmon, known for its storage qualities. Working on the west coast of Vancouver Island, Gregory Monks evaluates faunal assemblages from two sites in Barkley Sound occupied by the Toquaht, a local Nuu-chah-nulth group. Although the ethnographic record suggests a dichotomy of "outside sites" located in exposed locations versus "inside sites" with privileged access to salmon streams, Monks finds a more complicated picture. Although the Toquaht and other Nuu-chah-nulth selected sites for access to key resources, their settlement practices cannot be explained by resource access alone. Monks brings together all faunal data to better contextualize the use of fish.

In the third section, we turn our attention to the underappreciated Pacific cod. Moss's initial chapter provides an overview of this species in comparison to the much more widely renowned Atlantic cod, the "fish that changed the world," as well as the fish that led to the "Fish Event Horizon," identified across much of Western Europe. Mathew Betts, Herbert Maschner, and Don Clark also compare cod fisheries—this time between those used by Sanak Islanders off the Alaska Peninsula in the west to those of the Gulf of Maine in eastern North America. They find that while the Pacific cod populations of the Gulf of Alaska have been resilient over the last 4,500 years, the Gulf of Maine Atlantic cod populations have not. Both sequences show that cod abundances and mean cod size fluctuated significantly prior to commercial exploitation. Studies such as that by Betts, Maschner, and Clark help emphasize how important it is for fisheries managers to understand long-term population histories. Megan Partlow and Robert Kopperl compile data on skeletal elements from a large set of North Pacific sites to identify evidence of cod storage. Although their results are inconclusive, they have developed a method others can use to examine the problem as more data become available. In the last chapter in this section, Moss looks at two assemblages from Coffman Cove in southeast Alaska. While one is dominated by salmon, the other has a much higher proportion of Pacific cod. Different explanations for this pattern are explored, and a case for salmon resource ownership with an antiquity of more than 4,000 years is proposed.

It is a mistake to focus only on salmon or cod, because other fish were—and are—of critical importance in Aboriginal life in the North Pacific region. The final section of the book incorporates more discussion of some of the lesser-known North Pacific fish. Megan Caldwell combines her analysis of fish remains with Nancy Greene's work on the technology of the Comox fish traps to make the case that herring were the focus of fishing in this part of eastern Vancouver Island occupied by the Coast Salish in historic times. Caldwell is following in the footsteps of Monks (1987), who was the first to combine study of a faunal assemblage with fishing technology in his seminal study of Deep Bay. Natalie Brewster and Andrew Martindale try to establish what role fish played in the economy of those who lived on the Dundas Islands, an area historically contested by the Tlingit and the Tsimshian. These authors suggest that the residents of large Dundas villages may have relied to some degree upon salmon transported from the Skeena River fishery over the course of at least 2,000 years. Teresa Trost, Randall Schalk, Mike Wolverton, and Margaret Nelson report an unusual fish assemblage from Puget Sound, displaying another type of local variability. In the final paper of this section, Madonna Moss, Virginia Butler, and Tait Elder review the archaeological record of herring use in southeast Alaska, pointing to ways archaeological data can be used to assess contemporary impacts on herring by today's commercial fishery.

The papers in this volume accomplish a variety of goals. They demonstrate tremendous variation in how different Indigenous peoples in the North Pacific used different constellations of fish species in different settings at different times. The papers also emphasize the overwhelming importance of fish to First Nations. While *The Archaeology of North Pacific Fisheries* reveals new knowledge about ancient lifeways, we maintain that this knowledge is relevant to understanding the long-term histories of fish species. This historical ecological knowledge can contribute to informed management, to sustain both fish and fishing in today's rapidly changing environments.

References Cited

Atleo, E. Richard. "Preface." In *Keeping It Living: Traditions of Plant Use and Cultivation on the Northwest Coast of North America*, D. Deur and N. J. Turner, eds., pp. vii–xi. Seattle: University of Washington Press, 2005.

Augerot, Xanthippe. *Atlas of Pacific Salmon*. Berkeley: University of California Press; Portland, OR: Wild Salmon Center and Ecotrust, 2005.

Barnett, Homer G. "Culture Element Distributions, VII: Oregon Coast." *University of California Anthropological Records* 1(3): 155–204, 1937.

———. *The Coast Salish of British Columbia*. Eugene: University of Oregon Press, 1955.

Boas, Franz. "Ethnology of the Kwakiutl. Parts 1 and 2." In *Thirty-fifth Annual Report of the Bureau of American Ethnology for the Years 1913–1914*. Washington, DC: Government Printing Office, 1921.

Bryan, Alan L. *An Archaeological Survey of Northern Puget Sound*. Pocatello: Idaho State University Museum, Occasional Papers 11, 1963.

Butler, Virginia L. "Distinguishing Natural from Cultural Salmonid Deposits in Pacific Northwest North America." PhD diss., University of Washington, 1990.

———. "Natural vs. Cultural Salmonid Remains: Origin of The Dalles Roadcut Bones, Columbia River, Oregon." *Journal of Archaeological Science* 20 (1993): 1–24.

———. "Fish Feeding Behavior and Fish Capture: The Case for Variation in Lapita Fishing Strategies." *Archaeology in Oceania* 29 (1994): 81–90.

———. "Tui Chub Taphonomy and the Importance of Marsh Resources in the Western Great Basin of North America." *American Antiquity* 61 (1996): 699–717.

———. "Fish Faunal Remains." In *Archaeological Survey and Excavations in the Carson Desert and Stillwater Mountains, Nevada*, R. L. Kelly, ed. Salt Lake City: University of Utah Anthropological Papers 123, 2001a.

———. "Changing Fish Use on Mangaia, Southern Cook Islands: Resource Depression and the Prey Choice Model." *International Journal of Osteoarchaeology* 11 (2001b):88–100.

———. "Fish Remains." In *Marmes Rockshelter: A Final Report on 11,000 Years of Cultural Use*, B.A. Hicks, ed., pp. 320–337. Pullman: Washington State University Press, 2004a.

———. "Where Have All the Native Fish Gone? The Fate of the Fish that Lewis and Clark Encountered on the Lower Columbia River." *Oregon Historical Quarterly* 105 (2004b): 438–463.

———. "Ancient Salmon and the Columbia River." Archaeology Lecture Series, Museum of Natural and Cultural History. University of Oregon, Eugene, October 29, 2010.

Butler, Virginia L., and N. J. Bowers. "Ancient DNA from Salmon Bone: A Preliminary Study." *Ancient Biomolecules* 2 (1998): 17–26.

Butler, Virginia L., and Sarah K. Campbell. "Resource Intensification and Resource Depression in the Pacific Northwest of North America: A Zooarchaeological Review." *Journal of World Prehistory* 18 (2004): 327–405.

Butler, Virginia L., and J. C. Chatters. "The Role of Bone Density in Structuring Prehistoric Salmon Bone Assemblages." *Journal of Archaeological Science* 21 (1994):413–424.

Butler, Virginia L., and M. G. Delacorte. "Doing Zooarchaeology as If It Mattered: Use of Faunal Data to Address Current Issues in Fish Conservation Biology in Owens Valley, California." In *Archaeology and Conservation Biology*, R. Lee Lyman and K. Cannon, eds., pp. 25–44. Salt Lake City: University of Utah Press, 2004.

Butler, Virginia L., and J. E. O'Connor. "9,000 Years of Fishing on the Columbia River." *Quaternary Research* 62 (2004): 1–8.

Byram, R. Scott. "Brush Fences and Basket Traps: The Archaeology and Ethnohistory of Tidewater Weir Fishing in Oregon." PhD diss., University of Oregon, 2002.

Cannon, Aubrey. "Radiographic Age Determination of Pacific Salmon: Species and Seasonal Inferences." *Journal of Field Archaeology* 15 (1988): 103–108.

——. *The Economic Prehistory of Namu: Patterns in Vertebrate Fauna*. Burnaby, BC: Department of Archaeology, Simon Fraser University Publication 19, 1991.

——. "The Ratfish and Marine Resource Deficiencies on the Northwest Coast." *Canadian Journal of Archaeology* 19 (1995): 49–60.

——. "Contingency and Agency in the Growth of Northwest Coast Maritime Economies." *Arctic Anthropology* 35 (1998): 57–67.

——. "Assessing Variability in Northwest Coast Salmon and Herring Fisheries: Bucket-Auger Sampling of Shell Midden Sites on the Central Coast of British Columbia." *Journal of Archaeological Science* 27 (2000): 725–733.

Cannon, Aubrey, and Dongya Y. Yang. "Early Storage and Sedentism on the Pacific Northwest Coast: Ancient DNA Analysis of Salmon Remains from Namu, British Columbia." *American Antiquity* 71 (2006): 123–140.

Cannon, Debbi Yee. *Marine Fish Osteology: A Manual for Archaeologists*. Burnaby, BC: Simon Fraser University, Archaeology Press, 1987.

Casteel, Richard W. "Core and Column Sampling." *American Antiquity* 35 (1970): 465–467.

——. *Fish Remains in Archaeology*. New York: Academic Press, 1976a.

——. "Fish Remains from Glenrose." In *The Glenrose Cannery Site,* R. G. Matson, ed., pp. 82–87. Mercury Series, Archaeological Survey Paper No. 52. Ottawa: National Museum of Man, 1976b.

Christensen, Tina, and Jim Stafford. "Raised Beach Archaeology in Northern Haida Gwaii: Preliminary Results from the Cohoe Creek Site." In *Haida Gwaii: Human History and Environment from the Time of Loon to the Time of the Iron People*, Daryl W. Fedje and Rolf W. Mathewes, eds., pp. 245–273. Vancouver: UBC Press, 2005.

Conover, Kathryn. "Matrix Analyses." In *Studies in Bella Bella Prehistory*, James J. Hester and Sarah M. Nelson, eds., pp. 67–99. Burnaby, BC: Department of Archaeology, Simon Fraser University Publication 5, 1978.

Coupland, Gary, Kathlyn Stewart, and Katherine Patton. "Do You Never Get Tired of Salmon? Evidence for Extreme Salmon Specialization at Prince Rupert Harbour, British Columbia." *Journal of Anthropological Archaeology* 29 (2010): 189–207.

Crockford, Susan Janet. "Archeological Evidence of Large Northern Bluefin Tuna, *Thunnus thynnus,* in Coastal Waters of British Columbia and Northern Washington." *Fishery Bulletin* 95 (1997): 11–24.

Desautels, R. J., A. J. McCurdy, J. D. Flynn, and R. R. Ellis (eds.). *Archaeological Report, Amchitka Island, Alaska, 1969–1970*. Los Angeles: Archaeological Research, Inc., 1970.

Drucker, Philip. *The Tolowa and Their Southwest Oregon Kin*. University of California Publications in American Archaeology and Ethnology 36(4): 221–300, 1937.

———. *The Northern and Central Nootkan Tribes*. Washington, DC: Smithsonian Institution Bureau of American Ethnology Bulletin 144, 1951.

Elmendorf, William W., and A. L. Kroeber. *The Structure of Twana Culture*. Pullman: Washington State University Press, 1992.

Emmons, George T. *The Tlingit Indians*. Frederica de Laguna, ed. Seattle: University of Washington Press; Vancouver: Douglas and McIntyre; New York: American Museum of Natural History, 1991.

Greene, Nancy. "Comox Harbour Fish Trap Site. WARP (Wetland Archaeological Research Project) Web Report." http://newswarp.info/wp-content/uploads/2010/03/WARP-web-report.pdf, 2010.

Gunther, Erna. *Ethnobotany of Western Washington: The Knowledge and Use of Indigenous Plants by Native Americans*. Seattle: University of Washington Press, 1973.

Hewes, Gordon W. Field Notes from Interviews with Indian People of Several Northwest California Tribes, on the Topic of Traditional Fishing Techniques. MS on file, Berkeley: Bancroft Library, University of California, 1940.

Johnsen, D. Bruce. "A Culturally Correct Proposal to Privatize the British Columbia Salmon Fishery." George Mason University School of Law, Working Paper Series, Paper 8. The Berkeley Electronic Press, http://law.bepress.com/gmulwps/gmule/art8, 2004.

Kemp, Brian M., Ripan S. Malhi, John McDonough, Deborah A. Bolnick, Jason A. Eshleman, Olga Rickards, Cristina Martinez-Labarga, John R. Johnson, Joseph G. Lorenz, E. James Dixon, Terence E. Fifield, Timothy H. Heaton, Rosita Worl, and David Glenn Smith. "Genetic Analysis of Early Holocene Skeletal Remains from Alaska and Its Implications for the Settlement of the Americas." *American Journal of Physical Anthropology* 132 (2007): 605–621.

de Laguna, Frederica. *The Story of a Tlingit Community: A Problem in the Relationship Between Archeological, Ethnological and Historical Methods*. Washington, DC: Smithsonian Institution Bureau of American Ethnology Bulletin 172, 1960.

———. *Under Mount Saint Elias: The History and Culture of the Yakutat Tlingit*. Washington, DC: Smithsonian Contributions to Anthropology, vol. 7, 1972.

Langdon, Stephen J. "Tidal Pulse Fishing: Selective Traditional Tlingit Salmon Fishing Techniques on the West Coast of the Prince of Wales Archipelago." In *Traditional Ecological Knowledge and Natural Resource Management*, Charles R. Menzies, ed., pp. 21–46. Lincoln: University of Nebraska Press, 2006.

———. "Sustaining a Relationship: Inquiry into the Emergence of a Logic of Engagement with Salmon among the Southern Tlingits." In *Native Americans and the Environment: Perspectives on the Ecological Indian*, Michael E. Harkin and David Rich Lewis, eds., pp. 233–273. Lincoln: University of Nebraska Press, 2007.

Langdon, Stephen J., Douglas Reger, and Neil Campbell. "Pavements, Pairs, Pound, Piles and Puzzles: Investigating the Intertidal Fishing Structures in Little Salt Lake, Prince of Wales Island, Southeast Alaska." Paper presented at Hidden Dimensions: The Cultural Significance of Wetland Archaeology conference, University of British Columbia, Vancouver, April, 1995.

Lyman, R. Lee. *Prehistory of the Oregon Coast*. San Diego: Academic Press, 1991.

Matson, R. G. "The Evolution of Northwest Coast Subsistence." In *Long-Term Subsistence Change in Prehistoric North America*, D. R. Croes, R. A. Hawkins, and B. L. Isaac, eds., pp. 367–428. Research in Economic Anthropology Supplement 6. Greenwich, CT: JAI Press, 1992.

Matson, R. G., and Gary Coupland. *Prehistory of the Northwest Coast*. San Diego: Academic Press, 1995.

Mobley, Charles M., and W. Mark McCallum. "Prehistoric Intertidal Fish Traps from Central Southeast Alaska." *Canadian Journal of Archaeology* 25 (2001): 28–52.

Monks, Greg G. "Prey as Bait: The Deep Bay Example." *Canadian Journal of Archaeology* 11 (1987): 119–142.

Moss, Madonna L. *Archaeological Investigation of Cape Addington Rockshelter: Human Occupation of the Rugged Seacoast on the Outer Prince of Wales Archipelago, Alaska*. University of Oregon Anthropological Paper No. 63. Eugene: Museum of Natural History, 2004.

———. "Fishing Traps and Weirs on the Northwest Coast of North America: New Approaches and New Insights." In *Oxford Handbook of Wetland Archaeology*, Francesco Menotti and Aidan O'Sullivan, eds. Oxford: Oxford University Press, forthcoming.

Moss, Madonna L., and Jon M. Erlandson. "Reflections on North American Pacific Coast Prehistory." *Journal of World Prehistory* 9 (1995): 1–45.

———. "A Comparative Chronology of Northwest Coast Fishing Features." In *Hidden Dimensions: The Cultural Significance of Wetland Archaeology*, K. Bernick, ed., pp. 180–198. Vancouver: UBC Press, 1998.

Moss, Madonna L., Jon M. Erlandson, and Robert Stuckenrath. "Wood Stake Weirs and Salmon Fishing on the Northwest Coast: Evidence from Southeast Alaska." *Canadian Journal of Archaeology* 14 (1990): 143–158.

Olson, Ronald L. *The Quinault Indians*. University of Washington Publications in Anthropology 6 (1936): 1–190.

Paul, Frances. *Spruce Root Basketry of the Alaska Tlingit*. Washington, DC: Bureau of Indian Affairs, 1944.

Pauly, Daniel. "Aquacalypse Now." *The New Republic,* September 28, 2009, http://www.tnr.com/article/environment-energy/aquacalypse-now.

Romanoff, Steven. "Fraser Lillooet Salmon Fishing." *Northwest Anthropological Research Notes* 19 (1985): 119–160.

Singh, Ram Raj Prasad. *Aboriginal Economic System of the Olympic Peninsula Indians, Western Washington*. Sacramento, CA: Sacramento Anthropological Society Papers 4, 1966.

Smith, Marian. *The Puyallup-Nisqually*. New York: Columbia University Contributions to Anthropology 32, 1940.

Smith, Ross. "Structural Bone Density of Pacific Cod (*Gadus macrocephalus*) and Halibut (*Hippoglossus stenolepis*): Taphonomic and Archaeological Implications." Master's thesis, Portland State University, 2008.

Steffen, Martina Lianne. "Early Holocene Hearth Features and Burnt Faunal Assemblages at the Richardson Island Archaeological Site, Haida Gwaii, British Columbia." Master's thesis, University of Victoria, 2006.

Stein, Julie K. "Sediment Analysis of the British Camp Shell Midden." In *Deciphering a Shell Midden*, J. K. Stein, ed., pp. 135–162. San Diego: Academic Press, 1992.

Stewart, Frances L., and Kathlyn M. Stewart. "The Boardwalk and Grassy Bay Sites: Patterns of Seasonality and Subsistence on the Northern Northwest Coast, B.C." *Canadian Journal of Archaeology* 20 (1996): 39–60.

Suttles, Wayne. "Coping with Abundance: Subsistence on the Northwest Coast." In *Man the Hunter*, R. B. Lee and I. DeVore, eds., pp. 56–68. Chicago: Aldine, 1968.

———. *The Economic Life of the Coast Salish of Haro and Rosario Straits*. New York: Garland, 1974.

———. "Environment." In *Northwest Coast*, Handbook of North American Indians, vol. 7, W. Suttles, ed., pp. 16–29. Washington, D.C.: Smithsonian Institution, 1990.

CHAPTER 2

Identification of Salmon Species from Archaeological Remains on the Northwest Coast

Trevor J. Orchard, *Department of Anthropology, University of Toronto*
Paul Szpak, *Department of Anthropology, University of Western Ontario*

Fish were universally an important resource to Northwest Coast First Nations ethnographically (e.g., Donald 2003; Suttles 1990), and a wide variety of fish species are represented in archaeological assemblages from the culture area (e.g., Butler and Campbell 2004; Hanson 1991; Orchard and Clark 2005). The most commonly utilized taxa prehistorically and ethnographically tend also to be those species that have been prominent in modern commercial fisheries, namely salmon (*Oncorhynchus* spp.), Pacific cod (*Gadus macrocephalus*) and other gadids (Gadidae), herring (*Clupea pallasii*), and halibut (*Hippoglossus stenolepis*). Salmon have been particularly prominent in discussions of Northwest Coast culture history (Butler and Campbell 2006; Coupland, Stewart, and Patton 2010; Schalk 1977), and in some areas they have been a major component of traditional diets for as many as 9,000 years (Butler and O'Connor 2004; Cannon 1991; Cannon and Yang 2006). Although the Northwest Coast is thus often characterized as dominated by salmon fishing (Butler and Campbell 2006; Donald 2003; but see Monks 1987), the region is perhaps better characterized by considerable variability in the local use of fish species (Cannon 2000; Orchard 2007, 2009). Even in areas where salmon was a predominant resource ethnographically, it is increasingly evident that there is considerable variability in the timing of the adoption of salmon-focused economies. Documenting and understanding this variability in both the adoption of salmon-focused economies and the use of other fish taxa is critical for understanding variability in Northwest Coast cultural developments, and more generally for understanding the rise of the sedentary, socially complex cultures documented ethnographically (e.g., Ames 2003; Lepofsky et al. 2005; Prentiss et al. 2007). Insights into long-term patterns of fish use are also becoming increasingly important as baselines for understanding more recent historic patterns in fisheries and fish populations (Jackson et al. 2001; McKechnie 2007; Pauly et al. 2002).

Salmon pose a particular problem for archaeological analysis. A thorough understanding of the timing, technology, and logistics of prehistoric salmon fishing depends largely on the species of salmon involved (Kew 1992). Unfortunately, as with other sets of closely related animal species (Bochenski 2008), salmon bones are notoriously difficult to identify to species, particularly when the bulk of salmon assemblages tends to consist of vertebrae (Butler and Bowers 1998; Cannon 1988).

Pacific Salmon Life Histories

The various species of Pacific salmon typically spawn at different ages and at different adult sizes (Table 2.1; Hart 1973; Healey 1986; Quinn 2004). In coastal British Columbia, pink salmon (*Oncorhynchus gorbuscha*), with only very rare exceptions, spawn at two years of age and at mean weights of roughly 2.2 kg. Coho salmon (*O. kisutch*) spawn at ages ranging from two to six years, with the vast majority (95 percent) spawning during their third

Table 2.1. Summary data on the sizes and life history characteristics of Pacific salmon

Species	Length at maturity	Weight at maturity*	Age at spawning	Freshwater	Oceanic
Pink (*O. gorbuscha*)	48.5 to 57.2 cm (♂) 46.0 to 56.1 cm (♀) Mean (♀): 52 cm	Typical: 1.4–2.3 kg (3–5 lbs) Mean: 2.2 kg (4.8 lbs) Max: 5.5 kg (12 lbs)	2 years	Rapid migration to ocean following emergence of fry	Reside in ocean 18 months
Chum (*O. keta*)	51.6 to 88.9 cm (♂) 55.5 to 77.6 cm (♀) Mean (♀): 68 cm	Typical: 4.5–6.8 kg (10–15 lbs) Mean: 5.4 kg (11.7 lbs) Max: 15 kg (33 lbs)	2–7 years; usually 3–5 years	Rapid migration to ocean following emergence of fry	Reside in ocean 2–7 years (usually 3–5)
Coho (*O. kisutch*)	21.0 to 72.7 cm (♂) 27.8 to 70.0 cm (♀) Mean (♀): 64 cm	Typical: 2.7–5.4 kg (6–12 lbs) Mean: 4.1 kg (9 lbs) Max: 14 kg (31 lbs)	2–6 years; usually 3 years	Typically reside in streams for 1 or 2 years	Typically spend 1–2 years at sea
Sockeye (*O. nerka*)	30.0 to 73.7 cm (♂) 46.6 to 72.4 cm (♀) Mean (♀): 55 cm	Typical: 2.3–3.6 kg (5–8 lbs) Mean: 2.7 kg (6.0 lbs) Max: 6.8 kg (15 lbs)	3–8 years; usually 4 or 5 years	Typically reside in lakes or rivers for 1–3 years	Reside in ocean 1–4 years
Chinook (*O. tshawytscha*)	10.2 to 115.0 cm Mean (♀): 87 cm	Typical: 4.5–6.8 kg (10–15 lbs) Mean: 6.8 kg (15 lbs) Max: 57 kg (125 lbs)	3–8 years; usually 4 or 5 years	Rapid migration or 1 year spent in rivers	Reside in ocean 1–5 years

*With the exception of pink salmon, mean weights are from Kew (1992) for Fraser River stocks specifically. These provide some general indications of average sizes.
Source: Compiled from Hart (1973), Healey (1986), Kew (1992), and Quinn (2004).

year of life, and are slightly larger than pinks, averaging 4.1 kg at maturity. Chum salmon (*O. keta*) spawn at ages ranging from two to seven years, with the vast majority spawning at three to five years of age, and are larger still, averaging 5.4 kg at maturity. Finally, sockeye salmon (*O. nerka*) and chinook salmon (*O. tshawytscha*) generally spawn at four or five years of age, though both species can spawn at ages ranging from three to eight years. Sockeye are relatively small, spawning at a mean size of roughly 2.7 kg, while chinook are the largest of the Pacific salmon, reaching a mean size of 6.8 kg at maturity. Importantly, the life histories of these five species of Pacific salmon consist of variable amounts of time spent in freshwater and saltwater environments. Pink and chum salmon spend the majority of their lives in saltwater, chinook may spend their first year in freshwater, coho typically spend one to two years in freshwater, and sockeye typically reside in freshwater for the first one to three years of life. In fact, some sockeye populations, known as "kokanee," are permanently resident in freshwater environments (Quinn 2004:15).

Two other species of salmon are also represented along the Pacific coast. The steelhead (*O. mykiss*) and the coastal cutthroat trout (*O. clarkii*) are somewhat less common than the five species discussed above, and they do not form the massive spawning runs typical of the other Pacific salmon species. While specimens of steelhead and cutthroat may, and undoubtedly have, turned up in archaeological assemblages in low numbers, they will not be considered further here.

Species Identification of Salmon Assemblages: Previous Approaches

The variable spawning ages of the different species of Pacific salmon led Cannon (1988) to propose a method of ageing salmon remains based on examination of growth lines or annuli in vertebrae. Cannon's experiments indicated that these annuli were easily visible through radiographic analysis, and relatively large numbers of vertebrae could be examined quickly and easily on X-ray plates. Given the tendency of the various salmon species to spawn at different ages, the resulting age profiles could be used to infer the probable species compositions of the assemblages under consideration. Surprisingly, relatively few researchers have employed this technique subsequent to Cannon's (1988) initial article, though the application of this approach has increased slightly in recent years. Berry (2000; Hayden 1997) employed radiography to infer the species of salmon represented in assemblages from the Keatley Creek site. Notably, Berry's interpretations have been challenged by more recent ancient DNA (aDNA) analysis of salmon remains from that site (Speller, Yang, and Hayden 2005). More recently, Coupland and colleagues (Coupland, Colten, and Conlogue 2002) and Trost (2005) utilized the radiographic technique to examine the composition of salmon assemblages from sites in the Prince Rupert Harbour area and in the Gulf of Georgia, respectively. Most recently, Clark (2007) employed radiographic analysis of salmon vertebrae from sites in the Gulf Islands of the southern Gulf of Georgia to explore the rise of reef netting in that area. These applications of Cannon's (1988) radiographic technique highlight the potential utility of this approach, and the relative ease with which it can be applied.

In recent years, aDNA analysis has also been increasingly used to identify salmon species from archaeological vertebrae (Butler and Bowers 1998; Cannon and Yang 2006;

Speller, Yang, and Hayden 2005; Trost 2005; Yang, Cannon, and Saunders 2004). This approach has proven very successful, with archaeological salmon vertebrae generally showing excellent DNA preservation and facilitating the secure species identification of small samples of vertebrae from a number of sites. While aDNA analysis is the most accurate means of speciating salmon bones, it is prohibitively expensive in many cases. The increasing study of aDNA, however, has led to a refinement in the more cost-effective method of radiography, and it has also highlighted the potential of using metric characteristics of salmon vertebrae to provide additional information on the species composition of archaeological assemblages. Specifically, Cannon and colleagues have compared the results of aDNA analysis of salmon vertebrae from Namu to the results obtained through radiographic analysis and to measurements of the transverse diameter of the vertebrae (Cannon and Yang 2006; Yang, Cannon, and Saunders 2004; also see Trost 2005). Two particular observations arising from these comparisons are of interest here. First, aDNA analysis of Namu salmon vertebrae revealed that pink salmon was the most common species represented in the assemblage, contributing roughly 42 percent of the analyzed vertebrae (Cannon and Yang 2006:128). In contrast, previous analyses based entirely on radiography had concluded that pink salmon contributed less than 5 percent of the salmon vertebrae at Namu (Cannon 1988:107). Pink salmon generally spawn during their second year, and thus their vertebrae should show only a single winter annulus. However, Cannon and Yang (2006:128) found that both a modern comparative specimen of pink salmon and many archaeological vertebrae identified by aDNA analysis as pink salmon showed two growth lines. This is in accordance with biological studies that indicate that pink salmon often exhibit a supplementary growth check line on their scales that can be confused with a winter annulus (Bilton and Ricker 1965). Second, through the comparison of the metric traits of vertebrae to the species determination of those vertebrae through aDNA analysis, Cannon and Yang (2006:132) found that vertebrae with a transverse diameter of 10.5 mm or greater were all identified as either chum or chinook salmon, while vertebrae with a transverse diameter of 8.0 mm or less were all identified as either pink or sockeye salmon. Vertebrae falling between 8.0 and 10.5 mm derived from the full range of species except chinook (Cannon and Yang 2006:132). These results have been expanded by subsequent data from Namu that indicate that vertebrae less than or equal to 8.0 mm are overwhelmingly dominated by pink and sockeye salmon, but also contain small numbers of chum and coho (Aubrey Cannon, pers. comm., May 2010).

Species Identification of Salmon Assemblages: Moving Forward

The recent applications of Cannon's (1988) radiographic technique, combined with the insights arising from applications of aDNA analysis (Cannon and Yang 2006) outlined above, provide a basis for employing a more refined technique combining radiography and metric analysis. While Cannon and Yang (2006), for example, suggest that the confusion arising from the tendency of pink salmon vertebrae to display two radiographic annuli reduces the utility of the radiographic method, we argue that the realization that pink salmon can exhibit one or two apparent growth annuli under radiographic analysis can improve the radiographic technique and allow more accurate assessment of the possible range of species present in an archaeological assemblage. Similarly, the apparently consistent patterning in metric characteristics (i.e., transverse vertebral diameter) exhibited by the different salmon species identified by Cannon and Yang (2006) provides a means for providing better estimates of the composition of large assemblages of salmon vertebrae (e.g., Trost 2005).

Consideration of a limited number of modern comparative specimens generally supports the metric observations of Cannon and Yang (2006). Transverse vertebral width was measured for all vertebrae from twenty individual Pacific salmon from the zooarchaeological comparative collections of McMaster University, the University of Toronto, and the University of Victoria (Table 2.2). The sample sizes for some species, particularly chum (one individual, sixty-one total vertebrae), are small, and the comparative samples as a whole likely do not encompass the full range of sizes of each species (Table 2.1). Measurements of additional comparative specimens in the future will help to refine this approach. Nevertheless, these data generally support the size categories suggested by Cannon and Yang (2006). For example, only specimens of chum and chinook include vertebrae greater than 10.5 mm, while neither chum nor chinook include vertebrae that fall below 8.0 mm in width. Only two aspects of these comparative data are different from the results presented by Cannon and Yang (2006). First, chinook salmon represent a relatively wide range of sizes, and need to be included in the 8.0 mm to 10.5 mm size category. Cannon and Yang's (2006) failure to include chinook in this category resulted from the very limited number of chinook present in their sample (only a single vertebra measuring 18.0 mm). Similarly, coho are well represented in the 8.0 mm or less size category among the comparative specimens measured for this paper, while the Namu coho sample was again too small to represent this diversity (Cannon and Yang 2006). As indicated above, subsequent analysis of Namu samples has identified coho in the small size category (Aubrey Cannon, 2010, pers. comm.). Notably, small numbers of chum vertebrae also fall into the small size category in these more recent Namu analyses, though the proportion of chum vertebrae less than or equal to 8.0 mm is very small (Aubrey Cannon, 2010, pers. comm.). Overall, the suggestion of a tripartite division of salmon vertebrae by size is supported.

In Cannon's (1988) original application of the radiographic technique, he used a portable dental X-ray machine to generate radiographs of salmon vertebrae on dental X-ray films.

Table 2.2. Metric data from modern comparative salmon specimens of spawning age

Species	Individuals (n)	Vertebrae (n)	Mean transverse diameter (mm)	Standard deviation	Range (mm)
Chinook	7	437	12.46	1.12	9.82–15.37
Chum	1	61	10.09	0.64	8.84–10.86
Coho	5	303	8.30	0.84	6.08–9.78
Pink	3	179	6.86	1.35	4.55–8.94
Sockeye	4	242	7.89	0.54	6.55–8.96

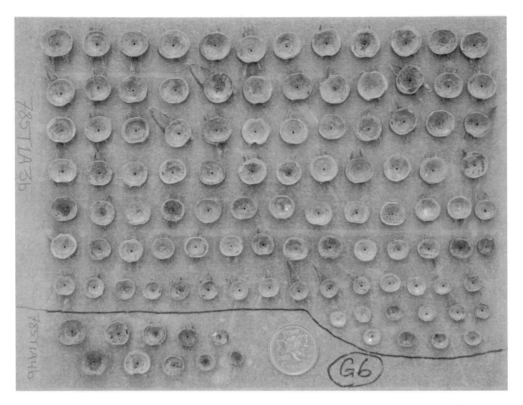

Figure 2.1. Example of a sheet of cardstock (13.5 x 18.5 cm) with vertebrae attached using double-sided tape.

While this approach was straightforward and facilitated the radiography of relatively large numbers of vertebrae with some ease, advances in radiographic technologies have further simplified the process. In particular, the development of digital radiography machines allows for the rapid processing of large numbers of vertebrae without the expense or time required for the purchasing and processing of radiographic films. In a case study of the approach outlined here, for example, Orchard (this volume) imaged a large sample of vertebrae with the assistance of Gord Mawdsley using a General Electric Senographe 2000D digital mammography machine at Sunnybrook Health Sciences Centre in Toronto (also see Clark 2007). Vertebrae were prepared by fixing them to 13.5 cm by 18.5 cm sheets of card stock using double-sided tape. As many as 100 to 120 vertebrae could be affixed to a single sheet using this method (Figure 2.1). Completed sheets of vertebrae were placed in sealed plastic Ziploc bags to further secure the vertebrae to the sheets. Sheets of vertebrae in their plastic bags were directly placed in the digital mammography machine and exposed to a radiographic output of 24 kV and 56.0 mA. High-resolution digital images (Figure 2.2) were generated for each sheet and were outputted to a CD-ROM for later analysis. The use of such a digital radiography machine allowed a large number of vertebrae to be imaged very quickly and easily. The total sample of 1,908 vertebrae, on twenty sheets of card stock, was imaged in less than an hour. The digital images were viewed and analyzed using a computer program called QuickQC, developed internally within the Sunnybrook and Women's College Health

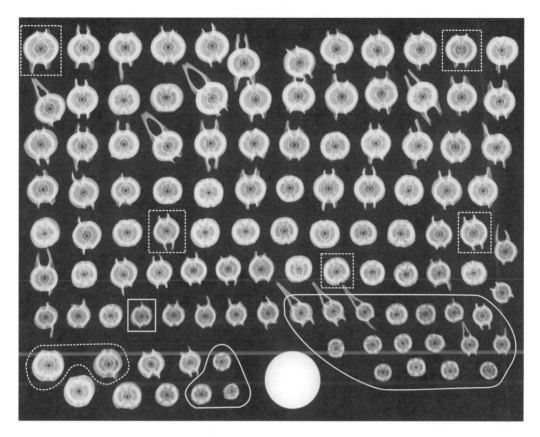

Figure 2.2. Digital radiograph of imaged vertebrae showing the denser growth annuli. Vertebrae show a combination of four-year-old fish (three annuli; vertebrae enclosed in dashed lines), two-year-old fish (one annulus; vertebrae enclosed in solid lines), and three-year-old fish (two annuli; all remaining vertebrae).

Sciences Centre. This program facilitated the adjustment of image contrasts to improve the visibility of radio-opaque growth lines in the vertebrae.

Combining the expectations for salmon vertebrae characteristics based on metric and radiographic approaches provides a means of categorizing archaeological salmon vertebrae to better assess the species composition of an assemblage. As summarized in Table 2.3, the combination of vertebral width (transverse diameter) and annuli as revealed through radiography provides twenty-one potential combinations of vertebral traits. These combinations in many cases allow the attribution of vertebrae to a single species while at worst narrowing down the possible species. Given the relative ease and generally low cost with which large numbers of salmon vertebrae can be digitally radiographed and measured, this approach can provide a good estimate of the various salmon species in an archaeological assemblage.

An additional, and as yet largely unexplored, line of evidence for the species composition of salmon bone assemblages comes from the analysis of stable isotopes of carbon ($\delta^{13}C$) and nitrogen ($\delta^{15}N$). Together, these values can provide insight into the diet of the animals through chemical analysis of their tissues (e.g., Cannon, Schwarcz, and Knyf 1999; DeNiro and Epstein 1978, 1981; Pollard et al. 2007). Stable carbon isotope ratios can be used to assess the relative contributions of isotopically distinct sources of primary

Table 2.3. Summary of metric and radiographic vertebral categories useful in the consideration of salmon assemblage compositions

Radiographic annuli	Transverse diameter (mm)		
	≤ 8	8–10.5	≥ 10.5
1	Pink; (coho); (chum)	Pink; (coho); (chum)	(chum)
2	Pink; Coho; (sockeye); (chum)	Pink; Coho; Chum; (chinook); (sockeye)	Chum; (chinook)
3	Sockeye; (coho); (chum)	Chum; Sockeye; Chinook; (coho)	Chum; Chinook
4	Sockeye; (coho); (chum)	Chum; Sockeye; Chinook; (coho)	Chum; Chinook
5	(coho); (sockeye); (chum)	(coho); (chum); (chinook); (sockeye)	(chum); (chinook)
6	(sockeye); (chum)	(chum); (chinook); (sockeye)	(chum); (chinook)
7	(sockeye)	(chinook); (sockeye)	(chinook)

Note: More likely categories are capitalized; less likely categories are in parentheses.

production (e.g., macrophyte algae and seagrass versus phytoplankton) to an animal's diet, while stable nitrogen isotopes indicate the trophic level of those animals within the food web, with a regular enrichment in ^{15}N at each successive trophic level (e.g., Fredriksen 2003; Kaehler, Pakhomov, and McQuaid 2000). Isotopic values of primary producers, and the animals consuming them, are distinct in freshwater and oceanic environments, with the latter characterized by higher δ^{15}N and δ^{13}C values (Schoeninger and DeNiro 1984). In terms of general dietary trends, five species of Pacific salmon can be broadly divided into two major subgroups (Healey 1986). The first, including sockeye, chum, and pink, primarily consume amphipods, euphausiids, and copepods and spend the marine component of their lives pelagically in the open ocean. The second, containing coho and chinook, are primarily piscivorous and spend much of their marine lives in coastal and continental shelf waters.

These variable diets are reflected in typical nitrogen and carbon stable isotope values. Satterfield and Finney (2002) record three groups of salmon based on carbon and nitrogen isotope values (also see Welch and Parsons 1993). The group with the lowest values for both δ^{13}C and δ^{15}N contains pink, chum, and sockeye salmon (Satterfield and Finney 2002), corresponding well with their consumption of low trophic-level (zooplankton) foods. An intermediate group with slightly higher values for both δ^{13}C and δ^{15}N consists primarily of coho salmon, though with some chum and sockeye (Satterfield and Finney 2002). Finally, chinook salmon are completely distinct from the other four species, with the highest values for both δ^{13}C and δ^{15}N (Satterfield and Finney 2002), because this species consumes relatively higher proportions of fish. Based on these data, chinook salmon are clearly distinguishable from the other four species due to their foraging behavior, while coho salmon show moderate separation from the other species. The data presented by Satterfield and Finney (2002), however, are not directly applicable to species identification in archaeological contexts because they are derived from muscle tissue, which remodels very rapidly (Hobson and Clark 1992). Muscle tissue only records the oceanic portion of an adult salmon's life, while the same would not be true for bone collagen extracted from archaeological specimens. This is particularly relevant since salmon are anadromous, as discussed below. Satterfield and Finney's (2002) data do demonstrate that chinook salmon are distinct in both $\delta^{13}C$ and $\delta^{15}N$ values. Given the relatively lengthy amount of time this species spends in the ocean, this elevation in isotopic values should also be present in bone collagen. This is consistent with bone collagen data presented by Schoeninger and DeNiro (1984), though this was based only on single specimens of chinook and pink salmon.

In addition to these dietary differences, Pacific salmon species spend variable portions of their life cycles in fresh and marine waters. Since bone collagen remodels slowly and reflects dietary intake over several years or the entire lifetime of an animal (Hobson 1990), isotopic values of salmon bone collagen reflect a mix of both oceanic and freshwater dietary input (Schoeninger and DeNiro 1984). Because the different salmon species spend variable amounts of time in freshwater, the degree to which this is reflected in stable isotope values of bone collagen should be species specific. Coho and sockeye salmon, which spend significantly more time in freshwater as juveniles than do chum, pink, or chinook, should exhibit $\delta^{13}C$ values indicative of the consumption of more substantial amounts of freshwater protein. Ben-David (1996) demonstrates that juvenile coho salmon exhibit significantly more depleted $\delta^{13}C$ values than do adults (~5 percent), while juvenile chum and pink salmon tend to be very similar to adults (difference of 0.91 percent and 1.12 percent, respectively). Based on this and because sockeye spend even longer periods in freshwater than do coho, it is likely that a similar or even greater difference in stable carbon isotopes exists between adults and juveniles of this species. The isotopic data obtained thus far for modern salmon, fragmentary though they are, suggests a pattern in isotopic composition for $\delta^{13}C$ of bone collagen of sockeye < coho < pink ≈ chum < chinook, and pink < sockeye ≈ chum ≤ coho < chinook for $\delta^{15}N$. Although more detailed analyses of the isotopic composition of bone collagen of Pacific salmon would be immensely useful, the species-specific variability in $\delta^{13}C$ and $\delta^{15}N$ values discussed here demonstrates that stable isotope analysis has the potential to contribute to the species identification of archaeological salmon remains, particularly when combined with radiography and metric analyses (Orchard, this volume).

Conclusions

While ancient DNA analysis provides the most accurate means of speciating archaeological salmon bones, a combination of radiographic and metric analyses provides a much more cost-effective approach. Digital radiography simplifies the process, reduces the cost, and allows for the contrast of radiographs to be fine-tuned during analysis to aid in visual identification of annuli. Combining the counting of radiographic annuli with the recording of the size (transverse diameter) of salmon vertebrae provides a series of discrete categories, each corresponding to at most four and in many cases only one or two possible species of Pacific salmon. This combined technique can thus provide more reliable identification of species in an archaeological assemblage (Orchard, this volume). The analysis of carbon (δ^{13}C) and nitrogen (δ^{15}N) isotopes in archaeological salmon bone samples is another promising technique. Given the importance of salmon to virtually all First Nations groups on the Northwest Coast, and given the variable timing and nature of the seasonal spawning aggregations of the different species of Pacific salmon, the ability to identify the species represented in archaeological salmon bone assemblages in the region facilitates a greater understanding of the timing, technology, process, and location of prehistoric subsistence activities. Furthermore, the determination of the relative abundance of the various species of salmon in archaeological sites that span the temporal and geographic range of human occupation on the Northwest Coast will provide data that are of considerable use in understanding the long-term histories of these species from biological, ecological, and commercial perspectives (e.g., Jackson et al. 2001; Pauly et al. 2002).

Acknowledgments

We would like to thank Aubrey Cannon and Madonna Moss for the invitation to participate in the North Pacific Fisheries session at the 2008 SAA Conference. Financial support during the preparation of this paper was provided by the Social Science and Humanities Research Council of Canada. Gord Mawdsley at Sunnybrook Health Sciences Centre in Toronto provided access to the digital radiography equipment and helped with the radiographic analysis of salmon vertebrae. Terence Clark also provided assistance with the salmon vertebrae radiography. Meghan Burchell, Aubrey Cannon, Terence Clark, and Madonna Moss provided very useful comments on earlier drafts of this paper.

References Cited

Ames, Kenneth M. "The Northwest Coast." *Evolutionary Anthropology* 12 (2003): 19–33.

Ben-David, Merav. "Seasonal Diets of Mink and Martens: Effects of Spatial and Temporal Changes in Resource Abundance." PhD diss., University of Alaska Fairbanks. Ann Arbor, MI: University Microfilms, 1996.

Berry, Kevin. "Prehistoric Salmon Utilization at Keatley Creek." In *The Ancient Past of Keatley Creek,* Volume II: *Socioeconomy,* Brian Hayden, ed., pp. 135–142. Burnaby, BC: Simon Fraser University, Archaeology Press, 2000.

Bilton, H. T., and W. E. Ricker. "Supplementary Checks on the Scales of Pink Salmon (*Oncorhynchus gorbuscha*) and Chum Salmon (*O. keta*)." *Journal of the Fisheries Research Board of Canada* 22 (1965): 1477–1489.

Bochenski, Zbigniew M. "Identification of Skeletal Remains of Closely Related Species: The Pitfalls and Solutions." *Journal of Archaeological Science* 35 (2008): 1247–1250.

Butler, Virginia L., and Nancy J. Bowers. "Ancient DNA from Salmon Bone: A Preliminary Study." *Ancient Biomolecules* 2 (1998): 17–26.

Butler, Virginia L., and Sarah K. Campbell. "Resource Intensification and Resource Depression in the Pacific Northwest of North America: A Zooarchaeological Review." *Journal of World Prehistory* 18 (2004): 327–405.

———. "Northwest Coast and Plateau Animals." In *Environment, Origins, and Population*, Handbook of North American Indians, vol. 3, Douglas H. Ubelaker, ed., pp. 263–273. Washington, DC: Smithsonian Institution, 2006.

Butler, Virginia L., and Jim E. O'Connor. "9000 Years of Salmon Fishing on the Columbia River, North America." *Quaternary Research* 62 (2004): 1–8.

Cannon, Aubrey. "Radiographic Age Determination of Pacific Salmon: Species and Seasonal Inferences." *Journal of Field Archaeology* 15 (1988): 103–108.

———. *The Economic Prehistory of Namu*. Burnaby, BC: Department of Archaeology, Simon Fraser University, 1991.

———. "Faunal Remains as Economic Indicators on the Pacific Northwest Coast." In *Animal Bones, Human Societies*, Peter Rowley-Conwy, ed., pp. 49–57. Oxford: Oxbow Books, 2000.

Cannon, Aubrey, Henry P. Schwarcz, and Martin Knyf. "Marine-based Subsistence Trends and the Stable Isotope Analysis of Dog Bones from Namu, British Columbia." *Journal of Archaeological Science* 26 (1999): 399–407.

Cannon, Aubrey, and Dongya Y. Yang. "Early Storage and Sedentism on the Pacific Northwest Coast: Ancient DNA Analysis of Salmon Remains from Namu, British Columbia." *American Antiquity* 71 (2006): 123–140.

Clark, Terence. "Radiography and Reef Nets: Prehistoric Evidence for the Antiquity and Distribution of Reef Netting on the Northwest Coast." Paper presented at annual meeting of the Society for American Archaeology, April 2007, Austin, TX.

Coupland, Gary, Roger Colten, and Jerry Conlogue. "Household Production of Salmon on the Northwest Coast of North America: Radiographic Evidence from the McNichol Creek Site." Paper presented at the International Council for Archaeozoology (ICAZ) Conference, August 2002, Durham, UK.

Coupland, Gary, Kathlyn Stewart, and Katherine Patton. "Do You Never Get Tired of Salmon? Evidence for Extreme Salmon Specialization at Prince Rupert Harbour, British Columbia." *Journal of Anthropological Archaeology* 29 (2010): 189–207.

DeNiro, Michael J., and Samuel Epstein. "Influence of Diet on the Distribution of Carbon Isotopes in Animals." *Geochimica et Cosmochimica Acta* 42 (1978): 495–506.

———. "Influence of Diet on the Distribution of Nitrogen Isotopes in Animals." *Geochimica et Cosmochimica Acta* 45 (1981): 341–351.

Donald, Leland. "The Northwest Coast as a Study Area: Natural, Prehistoric, and Ethnographic Issues." In *Emerging from the Mist: Studies in Northwest Coast Culture History*, R. G. Matson, Gary Coupland, and Quentin Mackie, eds., pp. 289–327. Vancouver: UBC Press, 2003.

Fredriksen, Stein. "Food Web Studies in a Norwegian Kelp Forest Based on Stable Isotope (δ^{13}C and δ^{15}N) Analysis." *Marine Ecology Progress Series* 260 (2003): 71–81.

Hanson, Diane K. "Late Prehistoric Subsistence in the Strait of Georgia Region of the Northwest Coast." PhD diss., Simon Fraser University. Ann Arbor, MI: University Microfilms, 1991.

Hart, J. L. *Pacific Fishes of Canada*. Ottawa: Fisheries Research Board of Canada Bulletin 180, 1973.

Hayden, Brian. *The Pithouses of Keatley Creek: Complex Hunter-Gatherers of the Northwest Plateau*. Fort Worth, TX: Harcourt Brace College Publishers, 1997.

Healey, M. C. "Optimum Size and Age at Maturity in Pacific Salmon and Effects of Size-Selective Fisheries." In *Salmonid Age at Maturity*, D. J. Meerburg, ed., pp. 39–52. Canadian Special Publications in Fisheries and Aquatic Sciences 89, 1986.

Hobson, Keith A. "Stable Isotope Analysis of Marbled Murrelets: Evidence for Freshwater Feeding and Determination of Trophic Level." *The Condor* 92 (1990): 897–903.

Hobson, Keith A., and Robert G. Clark. "Assessing Avian Diets Using Stable Isotopes I: Turnover of ^{13}C in Tissues." *The Condor* 94 (1992): 181–188.

Jackson, Jeremy B. C., Michael X. Kirby, Wolfgang H. Berger, Karen A. Bjorndal, Louis W. Botsford, Bruce J. Bourque, Roger H. Bradbury, Richard Cooke, Jon Erlandson, James A. Estes, Terence P. Hughes, Susan Kidwell, Carina B. Lange, Hunter S. Lenihan, John M. Pandolfi, Charles H. Peterson, Robert S. Steneck, Mia J. Tegner, and Robert R. Warner. "Historical Overfishing and the Recent Collapse of Coastal Ecosystems." *Science* 293 (2001): 629–638.

Kaehler, S., E. A. Pakhomov, and C. D. McQuaid. "Trophic Structure of the Marine Food Web at the Prince Edward Islands (Southern Ocean) Determined by δ^{13}C and δ^{15}N Analysis." *Marine Ecology Progress Series* 208 (2000): 13–20.

Kew, Michael. "Salmon Availability, Technology, and Cultural Adaptation in the Fraser River Watershed." In *A Complex Culture of the British Columbia Plateau*, Brian Hayden, ed., pp. 177–221. Vancouver: UBC Press, 1992.

Lepofsky, Dana, Ken Lertzman, Douglas Hallett, and Rolf Mathewes. "Climate Change and Culture Change on the Southern Coast of British Columbia 2400–1200 Cal. BP: An Hypothesis." *American Antiquity* 70 (2005): 267–293.

McKechnie, Iain. "Investigating the Complexities of Sustainable Fishing at a Prehistoric Village in Western Vancouver Island, British Columbia, Canada." *Journal for Nature Conservation* 15 (2007): 208–222.

Monks, Gregory G. "Prey as Bait: The Deep Bay Example." *Canadian Journal of Archaeology* 11 (1987): 119–142.

Orchard, Trevor J. "Otters and Urchins: Continuity and Change in Haida Economy during the Late Holocene and Maritime Fur Trade Periods." PhD diss., University of Toronto. Ann Arbor, MI: University Microfilms, 2007.

———. *Otters and Urchins: Continuity and Change in Haida Economy during the Late Holocene and Maritime Fur Trade Periods*. BAR International Series 2027. Oxford: Archaeopress, 2009.

Orchard, Trevor J., and Terence Clark. "Multidimensional Scaling of Northwest Coast Faunal Assemblages: A Case Study from Southern Haida Gwaii, British Columbia." *Canadian Journal of Archaeology* 29 (2005): 88–112.

Pauly, Daniel, Villy Christensen, Sylvie Guénette, Tony J. Pitcher, U. Rashid Sumaila, Carl J. Walters, R. Watson, and Dirk Zeller. "Towards Sustainability in World Fisheries." *Nature* 418 (2002): 689–695.

Pollard, A. Mark, Catherine M. Batt, Benjamin Stern, and Suzanne M. M. Young. *Analytical Chemistry in Archaeology*. Cambridge: Cambridge University Press, 2007.

Prentiss, Anna Marie, Natasha Lyons, Lucille E. Harris, Melisse R. P. Burns, and Terrence M. Godin. "The Emergence of Status Inequality in Intermediate Scale Societies: A Demographic and Socio-economic History of the Keatley Creek Site, British Columbia." *Journal of Anthropological Archaeology* 26 (2007): 299–327.

Quinn, Thomas P. *The Behaviour and Ecology of Pacific Salmon and Trout*. Vancouver: UBC Press, 2004.

Satterfield, Franklin R., IV, and Bruce P. Finney. "Stable Isotope Analysis of Pacific Salmon: Insight into Trophic Status and Oceanographic Conditions over the Last 30 Years." *Progress in Oceanography* 53 (2002): 231–246.

Schalk, Randall F. "The Structure of an Anadromous Fish Resource." In *For Theory Building in Archaeology*, L. R. Binford, ed., pp. 207–249. New York: Academic Press, 1977.

Schoeninger, Margaret J., and Michael J. DeNiro. "Nitrogen and Carbon Isotopic Composition of Bone Collagen from Marine and Terrestrial Animals." *Geochimica et Cosmochimica Acta* 48 (1984): 625–639.

Speller, Camilla F., Dongya Y. Yang, and Brian Hayden. "Ancient DNA Investigation of Prehistoric Salmon Resource Utilization at Keatley Creek, British Columbia, Canada." *Journal of Archaeological Science* 32 (2005): 1378–1389.

Suttles, Wayne, ed. *Northwest Coast,* Handbook of North American Indians, vol. 7. Washington, DC: Smithsonian Institution, 1990.

Trost, Teresa. "Forgotten Waters: A Zooarchaeological Analysis of the Cove Cliff Site (DhRr-18), Indian Arm, British Columbia." Master's thesis, Simon Fraser University, 2005.

Welch, David W., and Timothy R. Parsons. "$\delta^{13}C$–$\delta^{15}N$ Values as Indicators of Trophic Position and Competitive Overlap for Pacific Salmon (*Oncorhynchus* spp.)." *Fisheries Oceanography* 2 (1993): 11–23.

Yang, Dongya Y., Aubrey Cannon, and Shelley R. Saunders. "DNA Species Identification of Archaeological Salmon Bone from the Pacific Northwest Coast of North America." *Journal of Archaeological Science* 31 (2004): 619–631.

CHAPTER 3

Little Ice Age Climate
Gadus macrocephalus *Otoliths as a Measure of Local Variability*

Catherine F. West, *Department of Anthropology, University of Washington*
Stephen Wischniowski, *International Pacific Halibut Commission, Seattle*
Christopher Johnston, *Alaska Fisheries Science Center, National Marine Fisheries Service, NOAA, Seattle*

The Late Holocene is a time of rapid global climate change punctuated by warming and cooling events that have been widespread throughout the northern hemisphere. Paleoenvironmental reconstruction has long been used in archaeology to interpret the effects of such climate changes on prehistoric humans, and ecological research has become increasingly relevant as archaeologists begin to make connections between ancient and modern climate changes. As a result, reconstructions of Late Holocene climate and environment are widely applied to address hypotheses about human-environmental interactions and the effects of climate change on prehistoric populations.

Despite the global nature of these events, their timing and magnitude is debated, and research indicates that climate variation is rarely regionally consistent (Mayewski et al. 2004). The effect of regional climate and environmental change on local areas varies widely, and geography plays a significant role in paleoenvironmental reconstruction. Therefore, details about the local effects of these changes are vital, and applying nonlocal paleoenvironmental reconstructions to archaeological interpretation is potentially problematic.

While archaeologists commonly employ broad temporal and spatial paleoenvironmental reconstructions, it is possible to get more detailed and localized environmental information directly from the calcium carbonate structures preserved in the archaeological record (Wurster and Patterson 2001; Wurster, Patterson, and Cheatham 1999). Calcium carbonate structures, such as fish otoliths (or fish ear bones), preserved in archaeological sites are a source of both local and high-resolution paleoenvironmental data. The result is that otoliths provide direct connections among climate, archaeological sites, and prehistoric fish populations, all within a distinct geographic area.

Here, we present the results of an otolith sampling study that uses archaeological collections from Kodiak Island, Alaska. Pacific cod (*Gadus macrocephalus*) otoliths are sampled to reconstruct local changes in the paleoenvironment during the Little Ice Age (LIA). This method is used here to test whether paleoenvironmental records for a specific geographic location—in this case, Kodiak Island—conform to climate patterns recorded for the wider Gulf of Alaska region.

Background on Otoliths

All fish have three pair of otoliths (sagittae, asteriscae, and lapillae) used for acoustic perception and balance. Otoliths, or "ear stones," are composed of calcium carbonate in the mineral form aragonite and grow in incremental bands that are produced in daily, seasonal, and yearly cycles (Figure 3.1). Since the discovery that otolith aragonite precipitates in oxygen isotope equilibrium with water depending on temperature (Devereux 1967), salinity (Eldson and Gillanders 2002), and biological processes (Kalish 1989), otoliths have been used for several purposes: in age and growth studies (Pannella 1971), as a modern and prehistoric temperature proxy (Andrus et al. 2002; Andrus, Crowe, and Romanek 2002), and to study the dynamics of both contemporary and ancient fish populations (Daverat et al. 2005; Ivany, Patterson, and Lohmann 2000).

Otoliths provide an effective measure of paleotemperature because the oxygen deposited in the aragonitic structure can be used to estimate the temperature of the water where the fish was living (Campana 1999; Devereux 1967). Oxygen is present in several stable isotopes, and the most abundant are ^{16}O (99.7 percent) and ^{18}O (0.2 percent) (Faure 1986). The ratio of these isotopes in a system, such as the ocean, is dependent on several factors, including temperature. Therefore, the ratio of ^{18}O to ^{16}O deposited in a fish's growing otoliths reflects the ratio present in the water, as a function of temperature. The otoliths can be sampled to measure the amount of ^{18}O and ^{16}O relative to an international standard for carbonates (in this case, Vienna Pee Dee belemnite [VPDB], International Atomic Energy Agency, Vienna). The resulting oxygen isotope abundances are calculated relative to this standard to produce a delta (δ) value (Faure 1986; Ostermann and Curry 2000):

$$\delta = \left(\frac{R_{sample} - R_{standard}}{R_{standard}} \right) \times 1000 \, (^o/_{oo})$$

The $\delta^{18}O_{VPDB}$ values are expressed as parts per thousand (‰), and a higher value indicates that the sample is enriched in ^{18}O relative to VPDB while a lower value is depleted in ^{18}O.

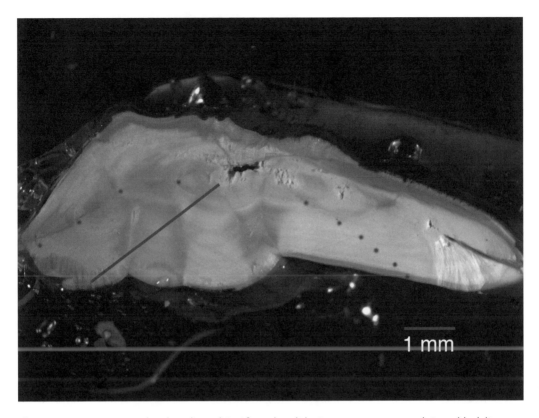

Figure 3.1. Cross-sectioned archaeological Pacific cod otolith. Dots represent annual rings, black line represents sampled transect. This fish was aged to five-plus years old. Photograph by C. Johnston; Alutiiq Museum catalog number AM193F.11.

Variability in the $\delta^{18}O$ values in a carbonate structure may be interpreted as changes in temperature and environmental conditions recorded by a living organism.

Although temperature is one influence on the ratio of oxygen isotopes present in a sample, precise temperature reconstructions are limited by other factors. "Vital effects" or metabolic processes do not appear to significantly influence the deposition of oxygen isotopes in otoliths (Høie, Otterlei, and Folkvord 2004; Thorrold et al. 1997); however, variation in isotope fractionation among fish species may influence the calculation of temperature based on $\delta^{18}O$ values (Høie, Otterlei, and Folkvord 2004; Radtke et al. 1996). As well, water salinity is known to affect oxygen isotope ratios, and the combined influence of glacial activity, mixing, and freshwater input on ocean salinity likely affects the otolith oxygen isotope record (Eldson and Gillanders 2002; Faure 1986; Xiong and Royer 1984). There is little information about how salinity may have varied during the Little Ice Age in the Gulf of Alaska, however, and the salinity tolerance of Pacific cod is not well known. Finally, a lack of information about prehistoric water conditions continues to fuel debate about how to accurately reconstruct paleotemperatures from oxygen isotope data (Campana 1999; Pentecost 2005). As a result of these limitations, the research presented here examines general trends, rather than attempting to reconstruct the actual temperature of the paleoenvironment.

Case Study: The Kodiak Archipelago

To perform the stable oxygen isotope analysis, otoliths were sampled from the Karluk-1 archaeological site located on Kodiak Island. Kodiak is just south of the Alaska Peninsula in the Gulf of Alaska (Figure 3.2), where the temperate, productive marine environment and a complex coastline provide rich biological resources and protected habitats for the island's human occupants. The earliest evidence of human occupation in the Kodiak Archipelago is 7500 cal BP, when small groups of maritime-oriented hunter-gatherers likely colonized Kodiak by boat (Fitzhugh 2003; Steffian, Saltonstall, and Kopperl 2006). By the time of Russian contact in the eighteenth century AD, Kodiak Islanders had developed into large groups of socially and politically complex hunter-gatherers, who lived in well-established villages throughout the archipelago (Clark 1979, 1984, 1998; Fitzhugh 2003; Knecht 1995).

The best evidence of Kodiak's late prehistoric culture, or the Koniag phase, was preserved in the Karluk-1 archaeological site on the southwest side of Kodiak Island at the mouth of the Karluk River. Karluk-1 was excavated in the mid-1980s by a team from Bryn Mawr College and the Kodiak Area Native Association, led by Richard Jordan and Richard Knecht (Crowell, Steffian, and Pullar 2001; Jordan and Knecht 1988; Knecht 1995). This 4 m deep excavation uncovered a series of ten stratified house floors and three middens that date to the Koniag phase, when Kodiak's hunter-gatherers had developed complex social, religious, and political systems (Jordan and Knecht 1988; Knecht 1995).

While the excavators previously dated the Karluk-1 site using charcoal and wood planks (Jordan and Knecht 1988; Knecht 1995), new radiocarbon samples were taken on short-lived botanical materials to create a tight chronology for comparing the otolith data with other lines of paleoenvironmental evidence (Table 3.1; West 2011). Recent accelerator mass spectrometry (AMS) dates indicate the deposits date from 550 to 500 cal BP, or cal AD 1450 to 1400, until Russian contact in the eighteenth century. To align the various datasets, ages are presented in Table 3.1 as calibrated years AD, calibrated years BP at the 2-sigma range, and the median date, using the calibration methods outlined by Stuiver and Reimer (1993).

The Karluk-1 deposits existed in an anaerobic environment, which preserved an extraordinary array of organic artifacts and faunal remains, including Pacific cod otoliths (Jordan and Knecht 1988; Knecht 1995; West 2009). Pacific cod are a migratory, demersal fish, and they move from the nearshore environment to the edge of the continental shelf during their brief winter migration (ADFG 1985; Mecklenburg, Mecklenburg, and Thorsteinson 2002; OCSEAP 1986). Because Pacific cod move only as far as the edge of the narrow continental shelf during this migration, their otoliths provide a relatively local measure of coastal conditions through time. Further, recent tagging data collected by the Alaska Department of Fish and Game in Kodiak appear to indicate that Pacific cod living near the coast of Kodiak Island may stay near shore year-round (D. Urban, 2005, pers. comm.).

Given the fish's life history and habitat requirements, people on Kodiak likely caught cod year-round or in the spring/early summer with hook and line (ADFG 1985; Hrdlicka 1944; Partlow 2000; Saltonstall and Steffian 2006). Because this is a demersal fish, one or more hooks were secured to the ocean floor using line weights and a rig spreader, and fish may have been caught singly or two or three at a time (Hrdlicka 1944; Knecht 1995). While

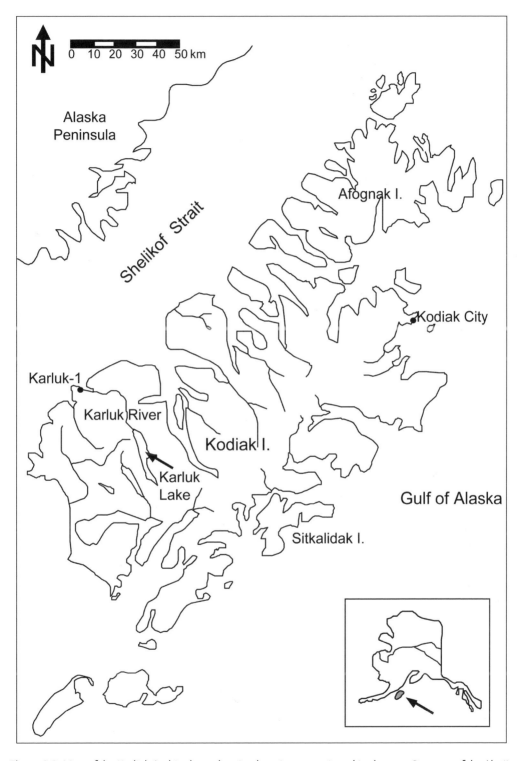

Figure 3.2. Map of the Kodiak Archipelago showing locations mentioned in the text. Courtesy of the Alutiiq Museum.

Table 3.1. Number of otoliths sampled by stratigraphic level

Level	AMS date (BP)	NOSAMS accession #	Cal BP range (2s)	Median date	Dated material	No. otoliths sampled
1	130±30	OS-58218	280–10	130	Wood/twig	1
2	190±25	OS-58223	260–140	180	Wood/twig	1
3	295±30	OS-58180	460–290	380	Seed (*Rubus* sp.)	2
4	315±30	OS-58210	460–300	390	Seed (*Rubus* sp.)	2
6	260±30	OS-61045	430–150	300	Wood/bark	1
7	250±30	OS-61047	430–150	300	Wood/bark	2
8	405±30	OS-58213	520–320	480	Seed (*Rubus* sp.)	3
UBM	400±30	OS-58063	510–330	470	Wood/twig	1
10	480±35	OS-61136	550–490	520	Seed (*Rubus* sp.)	1
10B	500±30	OS-58181	550–500	530	Seed (*Sambucus* sp.)	1

Note: Dates were produced by NOSAMS and are listed here with 2 standard deviation error terms. They were calibrated using CALIB version 5.0 (Stuiver and Reimer 1993). For full discussion of dating, see West (2009a, 2011).

there is relatively little information about the prehistoric cod fishery on Kodiak, limited historical observations and archaeological data indicate that people consumed these fish immediately and may have preserved them for storage by either drying or smoking (Foster 2006; Gideon 1989; Hays 2007; Hrdlicka 1944; Kopperl 2003; Partlow 2000; Saltonstall and Steffian 2006; Steffian, Saltonstall, and Kopperl 2006), but see Partlow and Kopperl (this volume) for a recent assessment of this topic.

Methods

The otoliths used in this study were sampled from the faunal remains collected at the Karluk-1 site (West 2009a, 2009b). Each bulk faunal sample was screened through eighth-inch mesh and then sorted to isolate the fish bone. Fifty-one otoliths, both whole and fragmentary, were pulled from the fish bone portion and identified using the protocol outlined by Morrow (1979), the photographic catalog compiled by Harvey et al. (2000), as well as comparison with reference specimens. All otoliths were identified to the family Gadidae, and the majority were narrowed to the Pacific cod (*Gadus macrocephalus*) species.

Fifteen otoliths representing nine house floors and one midden were selected for sampling, based on availability and condition (Table 3.1). Each otolith was graded on its preservation, and those whole otoliths with the least amount of chipping and flaking were chosen as candidates for sectioning and micromilling. Whole otoliths were chosen for two reasons: (1) so the fish could be aged accurately and (2) so the temperature reconstruction would represent each fish's full lifetime across all seasons.

Pacific cod otoliths are composed of calcium carbonate, typically in the form of aragonite. However, calcium carbonate can also take on the structural form of calcite or vaterite as a result of significant changes in temperature, burial, or through the influence of groundwater (Faure 1986). Calcite and vaterite may distort the paleoenvironmental analyses (Campana 1999; Faure 1986), so the mineral structure of a subsample of otoliths was examined through X-ray diffraction to establish that they are aragonite. The selected otoliths were then cleaned ultrasonically in Type II water, secured to glass slides, and sectioned using a Buehler diamond saw fit with a 3.5-inch blade. The otoliths were sectioned to reveal the growth rings, or annuli, for aging (Figure 3.1; Kimura and Anderl 2005). The growth rings were also used to differentiate between summer and winter growth, visible in each annulus as light and dark bands. It may be possible to estimate the season of capture using the last ring on the edge of the otolith, but this was not done for two reasons: (1) there is variability in the edge growth in Pacific cod otoliths that makes this estimation difficult in modern samples and (2) post-depositional degradation of the archaeological otoliths blurs the last season of growth and increases the uncertainty of determining the season of capture.

Using a micromill drill, the otoliths were sampled in two ways to produce 20–100 μg of powder for analysis: (1) by individual summer growth rings in years one through four and (2) in transects across growth rings to get average values across the fishes' lifetimes (Dettman and Lohmann 1995; Wurster and Patterson 2001; Wurster, Patterson, and Cheatham 1999). The isotopic composition of the carbonate samples was measured in a Finnigan MAT253 mass spectrometer system with a Kiel III carbonate device (Ostermann and Curry 2000).

Results and Discussion

The results of the otolith analysis indicate changes in the marine environment during the last 500 years. The data presented here are mean values based on transects taken across the growth rings of the otoliths in each stratigraphic layer; the transects represent an average of five to six years of growth in each fish (West 2009). The $\delta^{18}O_{VPDB}$ values indicate the Pacific cod otoliths recorded variable environmental conditions over this time period (Figure 3.3). The $\delta^{18}O$ values range from 0.78 percent to 1.48 percent with an error term of ±0.1 percent (Ostermann and Curry 2000; D. Ostermann, 2005, pers. comm.). In three periods of time, $\delta^{18}O$ values are higher, meaning ocean conditions along this portion of the continental shelf were relatively cooler: at the onset of occupation 500 ± 30 BP or 530 cal BP, at 405 ± 30 BP or 320 cal BP, and finally after 130 ± 30 BP or 130 cal BP. The clearest warming period is seen 315 ± 30 BP or 390 cal BP.

Recall that the otolith data are used here to test whether paleoenvironmental records for Kodiak Island conform to climate patterns recorded for the wider Gulf of Alaska region. The best recorded and most widely studied Late Holocene climate change in this period is the Little Ice Age (LIA), when climate throughout the northern hemisphere cooled rapidly from approximately 650–100 cal BP (Mann et al. 1998; Mayewski et al. 2004). Global datasets indicate that the LIA was punctuated by both warming and cooling events, primarily

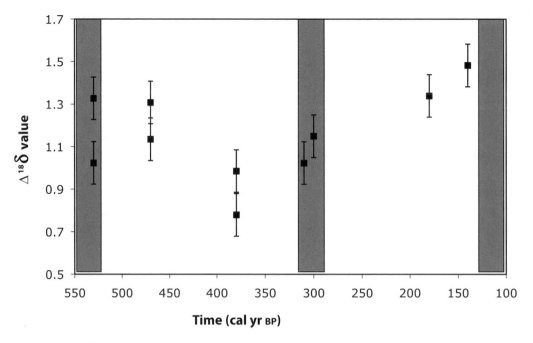

Figure 3.3. Results of the otolith analysis through time, based on the mean value of the transects taken across otoliths in each stratigraphic layer. The x-axis represents time, and the y-axis represents the stable isotope analysis: higher δ18O values indicate cooler temperatures; lower δ18O values indicate warmer temperatures. Each data point has an error term of ±1‰; dates are based on median ages produced by CALIB version 5.0 (Stuiver and Reimer 1993). The shaded bars represent the Little Ice Age cool periods suggested for the Gulf of Alaska in the early fifteenth, mid-seventeenth, and late nineteenth centuries (Barclay, Wiles, and Calkin 1999; Calkin, Wiles, and Barclay 2001; Mann et al. 1998; Wiles, D'Arrigo, and Jacoby 1996).

in response to changes in insolation (Grove 1988; Mayewski et al. 2004). In the Gulf of Alaska, it is generally accepted that the LIA occurred for a relatively short time between approximately 650 and 100 cal BP, with several major periods of cooling: in the early fifteenth century (~550 cal BP), in the mid-seventeenth century (~300 cal BP), and in the late nineteenth century (~100–200 cal BP) (Barclay, Wiles, and Calkin 1999; Calkin, Wiles, and Barclay 2001; Mann et al. 1998; Wiles, D'Arrigo, and Jacoby 1996).

To examine whether changes seen in the otolith record represent local variations of larger, broad-scale trends in the Little Ice Age Gulf of Alaska, these data were compared to other temporally fine-grained datasets: glacier and tree ring records. As expected, the otolith dataset is slightly different from the other records, but it does correspond to the major cyclical cooling phases recorded for the Little Ice Age during the last 600 years. As seen in Figure 3.3, the colder conditions recorded by the otoliths in the earliest stratigraphic layers correspond loosely to the cooling LIA phase in the fifteenth century, and the cold conditions recorded by the otoliths in the youngest layers correspond to the nineteenth-century LIA cooling phase.

A more detailed comparison is possible, based on calibrated radiocarbon dates from the glacial record, tree rings, otolith studies, and the hypothesis that major phases of cooling occurred in the early fifteenth, mid-seventeenth, and late nineteenth centuries (Barclay, Wiles, and Calkin 1999; Calkin, Wiles, and Barclay 2001; Mann et al. 1998; Wiles, D'Arrigo,

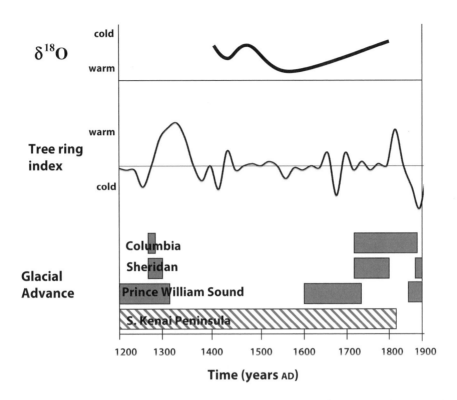

Figure 3.4. Otolith δ^{18}O record for Kodiak Island, tree ring chronology for Prince William Sound, and glacial history for Prince William Sound and the Kenai Peninsula. Shaded bars represent periods of glacial
advance, separated by periods of glacial retreat; hatch-marked bar represents complex period of glacial
advance and retreat in the southern Kenai Mountains (Mann et al. 1998). Redrawn from Barclay, Wiles,
and Calkin (1999), Calkin, Wiles, and Barclay (2001), and Mann et al. (1998).

and Jacoby 1996). As shown in Figure 3.4, glaciers advanced in the southern Kenai Mountains 1390 to 1520 cal AD (560–430 cal BP) and cooler conditions were recorded by tree rings
in Prince William Sound approximately 1400 cal AD (550 cal BP). This is reflected in the
otolith record, which produced relatively higher δ^{18}O values during this period. The otoliths
recovered from the period between 1500 and 1680 cal AD (450–270 cal BP) in house floors
three to seven show slightly lower δ^{18}O values, and a warm period is seen in the Sheridan,
Columbia, and western Prince William Sound glacial records before approximately 1600 cal
AD (350 cal BP). These slightly warmer conditions are reflected in the tree ring record just
after 1600 cal AD (350 cal BP). Finally, the Columbia Glacier, Sheridan Glacier, and Prince
William Sound and southern Kenai Mountains glaciers advanced between 1700 and 1800
cal AD (250–150 cal BP), indicating increasingly cooler conditions that match the high δ^{18}O
values seen in the otoliths after 1730 cal AD (220 cal BP).

The purpose of this comparison is to test whether local paleoenvironmental records
based on archaeological otoliths conform to climate patterns recorded for the wider Gulf of
Alaska in the Little Ice Age. The dataset produced by using otoliths from the Karluk-1 site on
Kodiak Island indicates that while periods of cooling and warming recorded by Pacific cod
during the Little Ice Age correspond loosely to broader cooling and warming trends throughout the Gulf, the timing of these trends appears to be specific to southwestern Kodiak.

There are several possible explanations for the differences seen in the Pacific cod otoliths when compared to other paleoenvironmental records. The first is methodological: this study compares several datasets based on calibrated radiocarbon dates. Given the error terms associated with radiocarbon dating, there is inherent variability in comparing the different sets of dates (Reimer et al. 2004; Stuiver and Reimer 1993). This limitation reinforces the advantage of using carbonates for paleoenvironmental reconstruction that are collected directly from the archaeological sites being analyzed (where possible), so that the radiocarbon dates associated with the paleoenvironmental and archaeological records will be identical.

The second potential limitation is biological: Pacific cod are mobile and will move in response to changes in ocean temperature when these changes could be fatal or threaten reproductive success (ADFG 1985; OCSEAP 1986). Therefore, the otoliths will produce data only within the animal's temperature tolerance, and more dramatic changes in climate recorded by glaciers and tree rings may be missed. This problem may be resolved by comparing the fish otolith records to data from archaeological shellfish—clams and mussels are sessile animals, and their carbonate structures are commonly used in paleoenvironmental studies.

Despite these limitations, given the broad similarities among the glacial, tree ring, and otolith records shown here, the variation is likely due to the localized effects of climate change. As demonstrated in the wide variety of Holocene paleoenvironmental studies produced for the Gulf of Alaska, there is dramatic variability in the environmental changes recorded by glaciers, tree rings, sea level, pollen, and carbonates (Anderson, Edwards, and Brubaker 2004; Barclay, Wiles, and Calkin 1999; Calkin, Wiles, and Barclay 2001; Heusser 1985; Heusser, Heusser, and Peteet 1985; Jordan and Maschner 2000; Josenhans et al. 1995; Kaufman, Porter, and Gillespie 2004; Mann et al. 1998; Wiles, D'Arrigo, and Jacoby 1996). Therefore, in a broad geographic context, the Karluk-1 otolith results further illustrate this variability.

Conclusions

Carbonate structures derived from materials in archaeological sites are widely employed for paleoenvironmental reconstruction throughout the world, and they can provide data pertinent to questions of archaeological interpretation in the Gulf of Alaska region. In this case, archaeological otoliths recovered from the late prehistoric Karluk-1 site on Kodiak Island were used to reconstruct ocean conditions during the Little Ice Age. The results of this analysis indicate that Pacific cod experienced variable ocean conditions during the LIA, which were shown to correspond to broad trends seen in LIA cooling and warming during the period from 530±30 cal BP to 140±130 cal BP. However, more detailed comparison of the otolith $\delta^{18}O$ values to other lines of paleoenvironmental data reveals that local trends in environmental change were variable throughout the Gulf of Alaska during the Late Holocene.

These results are important because they reinforce that relying on wide-ranging climate reconstructions in archaeological interpretation may be problematic. Rather, detailed analysis of local environmental and climate variability is necessary when making connections between archaeological and paleoenvironmental data. For example, West (2009a, 2009b) addresses the effects of climate change and resource availability on Kodiak Island's late prehistoric fisheries. The results of this otolith study are used to test whether changing fish use through time at the Karluk-1 site can be explained by changes in local marine conditions (West 2009). Preliminary results of this study indicate that the environmental changes recorded in the otoliths and the relative abundance of Pacific cod in the Karluk-1 zooarchaeological remains both change cyclically, which may reflect changes in climate conditions and cod availability (West 2009). Given the short occupation at Karluk-1, this relationship would not be visible in the context of broad, regional paleoenvironmental reconstruction.

The close connection between archaeological and paleoenvironmental data available in archaeological otoliths has the potential to strengthen interpretations about prehistoric human-environmental interactions and to explain the potentially important influences of climate change on human cultural development, subsistence, and resource availability. Holocene environmental and climate conditions in the North Pacific Ocean have been variable and volatile, and documenting this variability is central to understanding the long-term history of fisheries in this region. Detailed fish and environmental histories for the North Pacific Ocean have the potential to contribute to contemporary understandings of fish biogeography, migration patterns, and the effects of current climate changes on fish populations. As fisheries scientists learn more about the relationship between changes in modern fish populations and climate in the North Pacific, detailed paleoenvironmental reconstruction becomes increasingly relevant to archaeological interpretation, biological studies, and fisheries management.

Acknowledgments

The authors would like to thank Madonna Moss and Aubrey Cannon for organizing this volume. This research is supported by the Prince William Sound Oil Spill Recovery Institute, the National Science Foundation, the Alaska Anthropological Association, and the University of Washington Quaternary Research Center. Thanks also to Koniag, Inc., Alutiiq Museum and Archaeological Repository, University of Wisconsin Zoological Museum, University of Washington Fish Collection, National Ocean Sciences AMS facility, and Dorinda Ostermann, Benjamin Walther, and Simon Thorrold at the Woods Hole Oceanographic Institute Micropaleontology Lab.

References Cited

Alaska Department of Fish and Game (ADFG). "Pacific Cod (*Gadus macrocephalus*)." In *Alaska Habitat Management Guide, Southcentral Region*, vol. 1: *Life Histories and Habitat Requirements of Fish and Wildlife*. Juneau: State of Alaska Department of Fish and Game, Division of Habitat, 1985.

Anderson, P. M., M. E. Edwards, and L. B. Brubaker. "Results and Paleoclimate Implications of 35 Years of Paleoecological Research in Alaska." In *The Quaternary Period in the United States*, A. R. Gillespie, S. C. Porter, and B. F. Atwater, eds., pp. 427–440. Vol. 1 of *Developments in Quaternary Science*. New York: Elsevier Press, 2004.

Andrus, C. F. T., D. E. Crowe, and C. S. Romanek. "Oxygen Isotope Record of the 1997–1998 El Nino in Peruvian Sea Catfish (*Galeichthys peruvianus*) Otoliths." *Paleoceanography* 17 (2002): 1053–1063.

Andrus, C. F. T., D. E. Crowe, D. H. Sandweiss, E. J. Reitz, and C. S. Romanek. "Otolith $\delta^{18}O$ Record of Mid-Holocene Sea Surface Temperatures in Peru." *Science* 295 (2002): 1508–1511.

Barclay, D. J., G. C. Wiles, and P. E. Calkin. "A 1119-Year Tree-Ring-Width Chronology from Western Prince William Sound, Southern Alaska." *Holocene* 9 (1999): 79–84.

Calkin, P. E., G. C. Wiles, and D. J. Barclay. "Holocene Coastal Glaciation of Alaska." *Quaternary Science Reviews* 20 (2001): 449–461.

Campana, S. E. "Chemistry and Composition of Fish Otoliths: Pathways, Mechanisms, and Applications." *Marine Ecology Progress Series* 188 (1999): 263–297.

Clark, D. W. *Ocean Bay: An Early North Pacific Maritime Culture.* Archaeological Survey of Canada Mercury Series 86. Ottawa: Canadian Museum of Man, 1979.

———. "Prehistory of the Pacific Eskimo Region." In *Arctic,* Handbook of North American Indians, vol. 5, D. Damas, ed., pp. 136–148. Washington, DC: Smithsonian Institution, 1984.

———. "Kodiak Island: The Later Cultures." *Arctic Anthropology* 35 (1998): 172–186.

Crowell, A. L., A. F. Steffian, and G. L. Pullar. *Looking Both Ways: Heritage and Identity of the Alutiiq People.* Fairbanks: University of Alaska Press, 2001.

Daverat, F., J. Tomas, M. Lahaye, M. Palmer, and P. Elise. "Tracking Continental Habitat Shifts of Eels Using Otolith Sr/Ca Ratios: Validation and Application to the Coastal, Estuarine, and Riverine Eels of the Gironde-Garonne-Dordogne Watershed." *Marine and Freshwater Research* 56 (2005): 619–627.

Dettman, D. L., and K. C. Lohmann. "Microsampling Carbonates for Stable Isotope and Minor Element Analysis: Physical Separation of Samples on a 20 Micrometer Scale." *Journal of Sedimentary Research, Section A: Sedimentary Petrology and Processes* 65 (1995): 566–569.

Devereux, I. "Temperature Measurements from Oxygen Isotope Ratios of Fish Otoliths." *Science* 155 (1967): 1684–1685.

Eldson, T. S., and B. M. Gillanders. "Interactive Effects of Temperature and Salinity on Otolith Chemistry: Challenges for Determining Environmental Histories of Fish." *Canadian Journal of Fisheries and Aquatic Sciences* 59 (2002): 1796–1808.

Faure, G. *Principles of Isotope Geology.* 2nd ed. New York: John Wiley and Sons, 1986.

Fitzhugh, J. B. *The Evolution of Complex Hunter-Gatherers: Archaeological Evidence from the North Pacific.* New York: Kluwer Academic-Plenum, 2003.

Foster, C. W. "Analysis of Faunal Remains." In *The Archaeology of Horseshoe Cove: Excavations at KOD-415, Uganik Island, Kodiak Archipelago, Alaska,* P. G. Saltonstall and A. F. Steffian, eds., pp. 130–134. *Occasional Papers in Alaskan Field Archaeology* 1. Anchorage: Bureau of Indian Affairs, 2006.

Gideon, H. *The Round the World Voyage of Hieromonk Gideon 1803–1809.* Richard A. Pierce, ed., Lydia T. Black, trans. Kingston, ON: Limestone Press, 1989.

Grove, J. M. *The Little Ice Age.* New York: Methuen, 1988.

Harvey, J. T., T. R. Loughlin, M. A. Perez, and D. S. Oxman. "Relationship between Fish Size and Otolith Length for 63 Species of Fishes from the Eastern North Pacific Ocean." NOAA Technical Report NMFS 150, 2000.

Hays, J. M. "The Horseshoe Cove Site: An Example of Early Kachemak Subsistence Strategies from Faunal Remains in the Kodiak Archipelago." Master's thesis, University of Alaska Anchorage, 2007.

Heusser, C. J. "Quaternary Pollen Records from the Pacific Northwest Coast: Aleutians to the Oregon-California Boundary." In *Pollen Records of Late-Quaternary North American Sediment,*

Vaughn M. Bryant, Jr., and Richard G. Holloway, eds., pp. 141–165. Dallas: American Association of Stratigraphic Palynologists Foundation, 1985.

Heusser, C. J., L. E. Heusser, and D. M. Peteet. "Late-Quaternary Climatic Change on the American North Pacific Coast." *Nature* 315 (1985): 485–487.

Hrdlicka, A. *The Anthropology of Kodiak Island.* Philadelphia: Wistar Institute of Anatomy and Biology, 1944.

Høie, H., E. Otterlei, and A. Folkvord. "Temperature-dependent Fractionation of Stable Oxygen Isotopes in Otoliths of Juvenile Cod (*Gadus morhua* L.)." *ICES Journal of Marine Science* 61 (2004): 243–251.

Ivany, L. C., W. P. Patterson, and K. C. Lohmann. "Cooler Winters as a Possible Cause of Mass Extinctions at the Eocene/Oligocene Boundary." *Nature* 407 (2000): 887–890.

Jordan, J. W., and H. D. G. Maschner. "Coastal Paleogeography and Human Occupation of the Western Alaska Peninsula." *Geoarchaeology* 15 (2000): 385–414.

Jordan, R. H., and R. A. Knecht. "Archaeological Research on Western Kodiak Island, Alaska: The Development of Koniag Culture." In *The Late Prehistoric Development of Alaska's Native People*, R. D. Shaw, R. K. Harritt, and D. E. Dumond, eds., pp. 356–453. Anchorage: Alaska Anthropological Association Monograph No. 4, 1988.

Josenhans, H. W., D. W. Fedje, K. W. Conway, and J. V. Barrie. "Post Glacial Sea Levels on the Western Canadian Continental Shelf: Evidence for Rapid Change, Extensive Subaerial Exposure, and Early Human Habitation." *Marine Geology* 125 (1995): 73–94.

Kalish, J. M. "Otolith Microchemistry: Validation of the Effects of Physiology, Age, and Environment on Otolith Composition." *Journal of Experimental Marine Biology and Ecology* 132 (1989): 151–178.

Kaufman, D. S., S. C. Porter, and A. R. Gillespie. "Quaternary Alpine Glaciation in Alaska, the Pacific Northwest, Sierra Nevada, and Hawaii." In *The Quaternary Period in the United States*, A. R. Gillespie, S. C. Porter, and B. F. Atwater, eds., pp. 77–103. *Developments in Quaternary Science* 1. New York: Elsevier Press, 2004.

Kimura, D. K., and D. M. Anderl. "Quality Control of Age Data at the Alaska Fisheries Science Center." *Marine and Freshwater Research* 56 (2005): 783–789.

Knecht, R. A. "The Late Prehistory of the Alutiiq People: Culture Change on the Kodiak Archipelago from 1200–1750 A.D." PhD diss., Bryn Mawr College, 1995.

Kopperl, R. E. "Cultural Complexity and Resource Intensification on Kodiak Island, Alaska." PhD diss., University of Washington, 2003.

Mann, D. H., A. L. Crowell, T. D. Hamilton, and B. P. Finney. "Holocene Geologic and Climatic History around the Gulf of Alaska." *Arctic Anthropology* 35 (1998): 112–131.

Mayewski, P. A., E. E. Rohling, J. C. Stager, W. Karlén, K. A. Maasch, L. D. Meeker, E. A. Meyerson, F. Gasse, S. van Kreveld, K. Holmgren, J. Lee-Thorp, G. Rosqvist, F. Rack, M. Staubwasser, R. R. Schneider, and E. J. Steig. "Holocene Climate Variability." *Quaternary Research* 62 (2004): 243–255.

Mecklenburg, C. W., T. A. Mecklenburg, and L. K. Thorsteinson. *Fishes of Alaska.* Bethesda, MD: American Fisheries Society, 2002.

Morrow, J. E. "Preliminary Keys to Otoliths of Some Adult Fishes of the Gulf of Alaska, Bering Sea, and Beaufort Sea." NOAA Technical Report NMFS Circular 420. Washington, DC: Department of Commerce, 1979.

Ostermann, D. R., and W. B. Curry. "Calibration of Stable Isotopic Data: An Enriched O Standard Used for Source Gas Mixing Detection and Correction." *Paleoceanography* 15 (2000): 353–360.

Outer Continental Shelf Environmental Assessment Program (OCSEAP). "Marine Fishes: Resources and Environments." In *The Gulf of Alaska: Physical Environment and Biological Resources*, D. W. Hood and S. T. Zimmerman, eds., pp. 399–415. Anchorage: National Oceanic and Atmospheric Administration, 1986.

Pannella, G. "Fish Otoliths: Daily Growth Layers and Periodical Patterns." *Science* 173 (1971): 1124–1127.

Partlow, M. A. "Salmon Intensification and Changing Household Organization in the Kodiak Archipelago." PhD diss., University of Wisconsin–Madison, 2000.

Pentecost, A. *Travertine*. Berlin: Springer-Verlag, 2005.

Radtke, R. L., W. Showers, E. Moksness, and P. Lenz. "Environmental Information Stored in Otoliths: Insights from Stable Isotopes." *Marine Biology* 127 (1996): 161–170.

Reimer, P. J., M. G. L. Baillie, E. Bard, A. Bayliss, J. W. Beck, C. Bertrand, P. G. Blackwell, C. E. Buck, G. Burr, K. B. Cutler, P. E. Damon, R. L. Edwards, R. G. Fairbanks, M. Friedrich, T. P. Guilderson, K. A. Hughen, B. Kromer, F. G. McCormac, S. Manning, C. Bronk Ramsey, R. W. Reimer, S. Remmele, J. R. Southon, M. Stuiver, S. Talamo, F. W. Taylor, J. van der Plicht, and C. E. Weyhenmeyer. "IntCal04 Terrestrial Radiocarbon Age Calibration, 0–26 Cal Kyr BP." *Radiocarbon* 46 (2004): 1029–1058.

Saltonstall, P. G., and A. F. Steffian. *The Archaeology of Horseshoe Cove: Excavations at KOD-415, Uganik Island, Kodiak Archipelago, Alaska*. Occasional Papers in Alaskan Field Archaeology 1. Anchorage: Bureau of Indian Affairs, 2006.

Steffian, A. F., P. G. Saltonstall, and R. E. Kopperl. "Expanding the Kachemak: Surplus Production and the Development of Multi-Season Storage in Alaska's Kodiak Archipelago." *Arctic Anthropology* 43 (2006): 93–129.

Stuiver, M., and P. J. Reimer. "Extended ^{14}C Data Base and Revised CALIB 3.0 ^{14}C Age Calibration Program." *Radiocarbon* 35 (1993): 215–230.

Thorrold, S. R., S. E. Campana, C. M. Jones, and P. K. Swart. "Factors Determining δ^{13}C and δ^{18}O Fractionation in Aragonitic Otoliths of Marine Fish." *Geochimica et Cosmochimica Acta* 61 (1997): 2909–2919.

West, C. F. "Human Dietary Response to Climate Change and Resource Availability." PhD diss., University of Washington, 2009a.

West, C. F. "Kodiak Island's Prehistoric Fisheries: Human Dietary Response to Climate Change and Resource Availability." *Journal of Island and Coastal Archaeology* 4 (2009b): 223-239.

West, C. F. "A Revised Radiocarbon Sequence for Karluk-1 and the Implications for Kodiak Island Prehistory." In *Arctic Anthropology*, forthcoming in 2011.

Wiles, G. C., R. D. D'Arrigo, and G. C. Jacoby. "Temperature Changes Along the Gulf of Alaska and the Pacific Northwest Coast Modeled from Coastal Tree Rings." *Canadian Journal of Forest Research* 26 (1996): 474–481.

Wurster, C. M., and W. P. Patterson. "Late Holocene Climate Change for the Eastern Interior United States: Evidence from High Resolution δ^{18}O Values of Sagittal Otoliths." *Palaeogeography, Palaeoclimatology, Palaeoecology* 170 (2001): 81–100.

Wurster, C. M., W. P. Patterson, and M. M. Cheatham. "Advances in Micromilling Techniques: A New Apparatus for Acquiring High-Resolution Oxygen and Carbon Stable Isotope Values and Major/Minor Elemental Ratios from Accretionary Carbonate." *Computers and Geoscience* 25 (1999): 1159–1166.

Xiong, Q., and T. C. Royer. "Coastal Temperature and Salinity in the Northern Gulf of Alaska." *Journal of Geophysical Research* 89 (1984): 8061–8066.

CHAPTER 4

Pacific Cod and Salmon Structural Bone Density

Implications for Interpreting Butchering Patterns in North Pacific Archaeofaunas

Ross E. Smith, *Northwest Archaeological Associates, Inc., Seattle*
Virginia L. Butler, *Department of Anthropology, Portland State University*
Shelia Orwoll, *Oregon Health and Sciences University, Bone Mineral Unit*
Catherine Wilson-Skogen, *Oregon Health and Sciences University, Bone Mineral Unit*

Fish skeletal element representation is one line of evidence that has been used to document the intensive use and surplus production of dried fish for export and trade along the North Atlantic coast during the medieval period (Barrett 1997; Barrett, Locker, and Roberts 2004; Barrett, Nicholson, and Ceron-Carrasco 1999; Colley 1984; Perdikaris and McGovern 2009). Ethnohistorical accounts of Native American fish use along the North Pacific coast describing the abundance of salmon (*Oncorhynchus* sp.) and intensive strategies for procuring, processing, and storing salmon inspired many researchers to document the prehistoric development of such practices using salmon remains from archaeological assemblages. Some researchers have sought direct evidence of salmon storage from analyses of remains associated with storage features and patterns of body part representation from numerous sites along the Northwest Coast (Bernick 1983; Boehm 1973; Coupland, Colten, and Case 2003; Croes 2003; Grier 2003; Matson 1992; Matson and Coupland 1995; Wigen 2003, 2005). Recent faunal analyses (e.g., Bowers and Moss 2001; Kopperl 2003; McKechnie 2005; Orchard 2001; Partlow 2000; see chapters in Sections III and IV of this volume) demonstrate that nonsalmonid fish taxa, such as Pacific cod (*Gadus macrocephalus*), were also significant components of North Pacific Native people's subsistence; in some cases their importance stemmed from their potential as a stored resource. However, the extent and duration of cod fishing, variations in processing methods, and the extent cod use changed over time relative to other fish taxa are currently unclear. This chapter illustrates how combining analyses of body part representation and structural bone density can contribute to documenting fish processing techniques and expand our understanding of North Pacific fisheries.

Using body part representation to infer fish preparation for storage has a deep history in archaeology. In 1875, H. E. Sauvage proposed a simple behavioral model to explain uneven representation of salmon skeletal parts based on his analysis of Pleistocene fish remains in French caves:

> It is an interesting fact that among the numerous salmon remains from the caves, which we have examined, we have not met with an entire skeleton, having seen only portions of the vertebral column, as if only the edible portions were taken home to the caves. The bones of the head of the salmon, had they been there, would have been as well preserved as those of the small cyprinoids [chub, bream] which we find in the same deposits. These cyprinoids, on the contrary, which constituted what we may call the every-day fishing of the Aborigines, are recognised in all parts of their skeleton. They were evidently caught near the abode, and furnished fresh food; whilst the salmon went to form a food reserve. (Sauvage 1875:223)

In this passage Sauvage assumes that salmon and cyprinoids should be equally well preserved in cave deposits and uses the disproportionate abundance of salmon postcranial remains to make three inferences: (1) salmon were processed in a manner different from the cyprinoids; (2) processing resulted in salmon cranial and postcranial remains being deposited in separate locations; and (3) the pattern of salmon element representation indicated use of stored salmon at these sites. Nearly one hundred years later a similar model was independently developed by Boehm (1973) to explain the disproportionate representation of salmon postcranial remains at the St. Mungo Cannery site in British Columbia. Based on ethnographic accounts of Upper Stalo (now Stó:lō) salmon processing, Boehm (1973:95) proposed that (1) if salmon heads were removed before the postcranial portions were preserved, then archaeologists could expect to find disproportionate percentages of salmon cranial and postcranial remains (i.e., a large number of vertebrae and few cranial elements) at sites where stored salmon were consumed; (2) if the preparation and consumption of dried or smoked salmon increased through time, then the overrepresentation of salmon postcranial remains should increase through time; and (3) the disproportionate abundance of salmon postcranial remains could be a seasonal indicator of winter occupation of archaeological sites, based on the logic that mainly stored salmon would be consumed at such locales. Other researchers subsequently used Boehm's model to infer prehistoric salmon storage, site function, and seasonality from the relative abundance of salmon postcranial remains at archaeological sites along the Northeast Pacific coast (e.g., Bernick 1983; Coupland, Colten, and Case 2003; Croes 2003; Grier 2003; Matson 1992; Matson and Coupland 1995; Wigen 2003, 2005). Most researchers who have employed Boehm's model, however, limited their analysis to an assemblage from a single site and have not considered the potential effects of taphonomic factors structuring element representation in archaeological contexts.

According to Thomas and Mayer (1983), archaeologists must consider the location of a site in the context of the subsistence and settlement system when interpreting patterns in archaeofaunal samples. Since the attributes of a faunal assemblage are the product of the processing, transport, and disposal activities conducted at a location within a larger subsistence-settlement system, inferences based on the archaeofaunal assemblage from one site, or one type of site, may not be representative of the whole system (Binford 1978; Lyman 1994; Thomas and Mayer 1983:368–369). Boehm's model was often used to infer that sites containing disproportionately high numbers of vertebrae represent residential bases where preserved and stored salmon were consumed. If this inference was accurate, one would expect that head parts should dominate faunal samples from contemporaneous logistical camps if, in fact, heads were cut off and deposited at these field-processing locations (Butler 1990). While this regional approach, comparing body part representation at residential villages versus logistical processing camps to infer past salmon processing patterns, has been used in the Aleutian Islands (Hoffman, Czederpiltz, and Partlow 2000), most studies using fish bones to identify storage on the North Pacific coast have focused on single site records. Furthermore, when salmon cranial or postcranial remains are disproportionately represented in North Pacific faunal assemblages, this pattern is often uncritically viewed as evidence of specific processing activities. Other taphonomic factors, however, such as bone density likely affect salmon element preservation and should be considered.

Butler and Chatters (1994) used methods developed to measure mammal bone density (Lyman 1984) to quantify the structural bone density (bone mineral content/unit volume) of various elements in the salmon skeleton, which they then compared to salmon element representation in three archaeological sites in the Pacific Northwest. The logic behind the comparison is this: Element bone density is a proxy measure of preservation potential. The higher the density, the more likely the element will endure the rigors of postdepositional impacts (i.e., trampling, scavenger gnawing, chemical breakdown, and weathering). They found that bone density values of most salmon cranial elements were lower than vertebrae and reasoned that vertebrae should resist destruction better than cranial remains when exposed to destructive processes. Significant correlations between salmon element representation and salmon bone density values were found at two sites, suggesting that density-mediated bone attrition accounted for cranial element representation in those locales, while processing behavior better accounted for the relative abundance of postcranial remains at a third site, where the correlation was low. This study concluded that skeletal element representation patterns (namely, low frequency of cranial elements), cited by Boehm and others as evidence for salmon butchering, transport, and storage practices, could be explained more simply by differential preservation of the skeleton.

The commonly employed approach to inferring salmon storage and processing patterns using high numbers of salmon vertebrae and relatively low numbers of cranial remains is overly simplistic, usually narrowly focused on one part of the subsistence-settlement system, and does not consider variation between sites. While salmon bone density data have been available for more than ten years, and ethnohistorical sources contain descriptions of salmon head preservation methods (e.g., Drucker 1951:63; Emmons 1991:143; see also Moss 1989), archaeologists working on the Northwest Coast continue to cite the simple behavioral model and infer salmon storage from element representation without considering whether salmon element density could account for body part representation. Expanding our understanding of North Pacific Native fisheries requires researchers to broaden the scope of analysis to include nonsalmonid fishes and consider how taphonomy may affect the attributes of North Pacific fish archaeofaunal assemblages. Recent analyses of Pacific cod (*Gadus macrocephalus*) structural bone density were undertaken to generate a similar set of data that could be used to assess the role of bone density or preservation in accounting for cod body part representation. Unlike salmon, where vertebrae have higher density than cranial elements, density is similar across the Pacific cod skeleton, suggesting that density or preservation per se cannot explain uneven representation of cod body parts in archaeological sites. Archaeofaunal records from two sites in southeast Alaska are evaluated in light of these new bone density data.

Pacific Cod Use in the North Pacific Region

Nonsalmonid fishes, such as Pacific cod, were important sources of protein, oils, and fats and were key to Northwest Coast economies (Ames 1994); in some archaeofaunas, remains of nonsalmonid fishes comprise a larger proportion than salmon (Bowers and Moss 2001; Butler and Campbell 2004; Croes 1992; Hanson 1991; Huelsbeck 1994; McKechnie 2005). According to ethnohistorical descriptions of Native subsistence, Pacific cod were highly regarded resources in Native communities from the Aleutian Islands to the coast of Washington (Boas 1921; Drucker 1951; Emmons 1991; Jewitt 1988; Jochelson 1933; de Laguna 1972; Swan 1870). Their importance was due to their consistent seasonal availability (cod were caught in the winter and early spring when other food sources were scarce), their oil and fat content, and the length of time that they could be stored in dried or smoked form (but see Partlow and Kopperl, this volume, for another perspective on storage). At a basic level physiological differences in the accessibility of meat and fats in cod carcasses affected butchering and transport decision-making. In most accounts of cod processing, separating cod meat from the carcass required disarticulation or cutting certain bones to separate body parts (Partlow and Kopperl, this volume). In the process, certain elements remained attached to cod meat fillets. If Pacific cod were butchered and processed at logistical camps separate from the residential bases where dried fillets were consumed, then logistical camps and residential bases should differ in the representation of whole or partial carcasses in an assemblage.

Like salmon, variations in cod cranial/postcranial body part representation in Pacific Northwest archaeological deposits have been used to examine processing, preservation, and transport strategies (Bowers and Moss 2001; Partlow and Kopperl 2007). And as with salmon, few studies have directly addressed the potential influence of density-mediated attrition on fish element representation in archaeological sites; therefore, ambiguity persists as to whether archaeological patterns of fish body part representation resulted from human behavior or differential bone preservation. Until now, the lack of quantitative data describing the biomechanical properties of Pacific cod skeletal elements has prevented researchers from adequately testing for the effects of taphonomic processes such as density-mediated attrition.

Pacific Cod Structural Bone Density

Smith (2008) determined the structural bone density of twelve Pacific cod elements (whole and partial sections), from five individual fish, representing all parts of the skeleton. Bone mineral content was measured using dual energy X-ray absorptiometry (DEXA) at the Oregon Health and Science University (OHSU), Bone and Mineral Unit, Portland. The volume of each element was then measured using hydrostatic weighing and joined with the bone mineral content (BMC) to calculate bone volume density (BVD). To allow for comparison with salmon bone data, Smith used similar analytic protocols, instrumentation, and elements as Butler and Chatters (1994). The DEXA instrument housed at OHSU, however, was different from the system used by Butler and Chatters. Therefore, prior to measuring the cod elements, the authors reanalyzed a sample of nine chinook salmon (*Oncorhynchus tshawytscha*) altas vertebrae originally measured by Butler and Chatters (1994) and compared the datasets using Spearman's Rank Order Correlation Coefficient (r_s). While small differences were present between the 1991 and 2006 scan results, the two datasets were highly and significantly correlated ($r_s = 0.954$, $p<0.001$), indicating that the BMC measurements produced by the OHSU DEXA instrument were comparable to those produced by the DEXA instrument used by Butler and Chatters (1994).

Comparison of density values from the Pacific cod and salmon elements shows some striking differences (Figure 4.1). With the exception of the basipterygium, Pacific cod elements are between two and five times denser than salmon elements. All things being equal, cod skeletal elements should resist destruction better than salmon remains. Such intertaxonomic variation has important implications for interpreting taxonomic representation in North Pacific fish faunal assemblages that will be considered in a future publication. For now, we focus on another important difference between cod and salmon: density in vertebrae versus cranial elements. In salmon, vertebrae have consistently higher densities than cranial elements, whereas with Pacific cod, vertebrae densities fall within the range of values determined for the head. While the density of one cranial element, the hyomandibula, is lower than the standard deviations of the vertebrae measures, the remaining cranial elements fall within the standard deviation of, or exhibit higher densities than, the vertebrae.

Such differences in the distribution of density values across Pacific cod and salmon skeletons mean that for salmon skeletons, vertebrae, which have higher densities, are more likely to be preserved than cranial elements. Thus, in interpreting body part representation for salmon bone assemblages, this preservation overprint must be examined. For Pacific cod, however, the consistency in density across the skeleton indicates that density or preservation cannot explain variation in abundance of cranial elements versus vertebrae.[*]

Below we review the Pacific cod record from the North Point (49-SUM-25) and Cape Addington (49-CRG-188) sites (Figure 4.2) to consider the implications of the new Pacific cod bone density data in inferring butchering and processing from body part representation. The sites were both occupied during the last 2,500 years, a period in which

---

[*] Besides density as measured by bone mineral content, element shape may affect preservation and resistance to destruction. For example, Morales (1984) suggested that fish cranial elements tend to be laminar and more readily broken than spool-shaped vertebrae. We currently lack a rigorous way to test for the "shape" effect. However, given that most fish vertebrae have a spool-like shape, the bias would be consistent across taxa. Therefore, differences in element representation across taxa could not be easily linked to this shape difference per se.

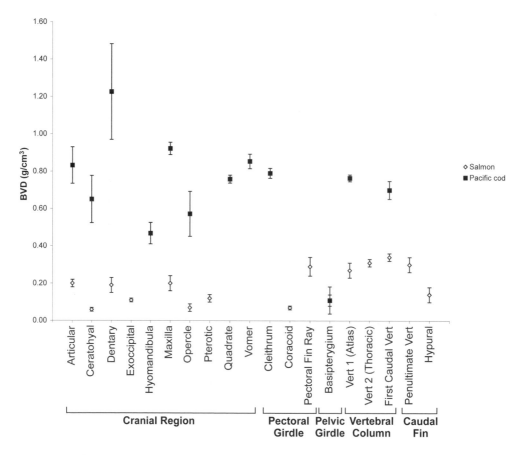

Figure 4.1. Pacific cod and salmon element BVD mean and standard deviation.

archaeological evidence suggests Native subsistence economies along the Northwest Coast were reliant on stored food (Ames and Maschner 1999). The North Point site has been interpreted as a residential base while the Cape Addington Rockshelter site is thought to represent a logistical camp. The assemblages were collected, analyzed, and reported by the same researcher using a consistent methodology.

North Point (49-SUM-25)

The North Point site is located along the mainland coast north of Petersburg, Alaska, at the confluence of Stephens Passage and Frederick Sound. A diverse array of terrestrial fauna, resident shorebirds, and numerous remains of Pacific cod led Bowers and Moss (2001:170) to conclude that North Point was a residential base for subsistence procurement activities in the local terrestrial uplands and nearshore marine environments during the late winter and early spring between 2760 and 2505 BP (Bowers and Moss 2001:162). Excavated sediments were passed through quarter-inch mesh and the majority of fish faunal remains were recovered from midden deposits (Bowers and Moss 2001:166, Table 3).

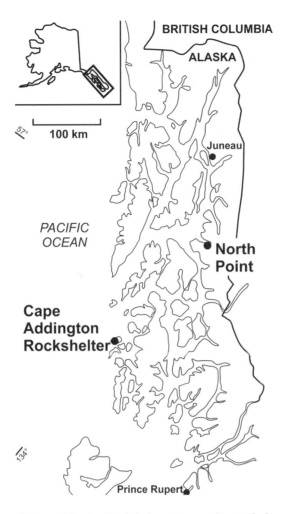

Figure 4.2. North Point and Cape Addington Rockshelter sites, southeast Alaska.

While salmon remains were recovered, small sample sizes preclude quantitative analysis of body part representation, and the focus here is on Pacific cod.

In the course of the faunal analysis, Bowers and Moss (2001) noted that Pacific cod vertebrae were far more abundant than cranial remains; the estimated number of cod individuals based on the vertebrae (MNI = 35) was considerably higher than the number of cod crania (MNI = 20, based on basioccipitals). Citing ethnographic accounts of Tlingit and Kwakiutl cod processing and cooking, the authors proposed three possible explanations to account for the disproportionate ratio of cod cranial to postcranial elements: (1) cod were butchered offsite and their heads were discarded offsite, (2) cooking methods (e.g., boiling, steaming, or roasting) resulted in the differential destruction of cranial elements, and (3) cod were butchered at North Point, the cod heads were removed from the site for further processing, and the remains were discarded offsite (Bowers and Moss 2001:172).

The overall pattern of cod skeletal element representation is presented in Figure 4.3; it illustrates the patterning that Bowers and Moss had noted; the assemblage is dominated

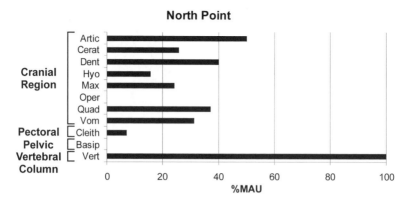

Figure 4.3. Pacific cod element representation (%MAU) at North Point.

by vertebrae with many fewer cranial elements (and pelvic and pectoral fin elements) than would be expected if entire skeletons were present. The Pacific cod density data indicate that all of the elements have a similar likelihood of preserving, thus density (and preservation factors per se) cannot explain the striking contrast in cranial versus vertebrae representation. Undoubtedly density and destruction play some role in site formation and element representation, and help explain the varying representation of cranial elements (i.e., the absence or low representation of elements such as the opercle, ceratohyal, and hyomandibula, which have the lowest densities of the cranial elements). On the other hand, density simply cannot explain the scarcity of heads and prominence of the trunk in North Point deposits. We suggest that the underrepresentation of Pacific cod cranial remains is better explained through off-site disposal of cod heads. In this case, cod body part representation supports the assertion that cod cranial and postcranial body parts were treated differently; it is possible that the North Point assemblage contains the remains of stored cod consumed at this site.

Cape Addington Rockshelter (49-CRG-188)

The Cape Addington Rockshelter is located on Noyes Island, along the outer coastline of the Prince of Wales Archipelago in southeast Alaska. Madonna Moss, assisted by a team of archaeologists, tested this site in 1996 and 1997 (Moss 2004). The dry environment and ample air circulation within the rockshelter led Moss to suggest that the site provided an ideal environment for preserving fish for long-term storage (Moss 2004:160). Ethnohistorical sources from this region report that cod were fished during the spring and summer and that these fish were processed for long-term storage (Emmons 1991; Newton and Moss 2005; Oberg 1973). Based on the relative abundances and diversity of taxa represented in the Cape Addington faunal assemblage, Moss (2004:211) concluded that the rockshelter was a seasonally occupied logistical camp where locally procured resources were processed and likely preserved.

Archaeological deposits within the rockshelter were excavated using arbitrary levels and following natural stratigraphic boundaries. Faunal remains were recovered in the field using quarter-inch mesh and bulk samples were screened using quarter-inch and eighth-inch mesh during lab processing. A majority of the identified cod specimens from

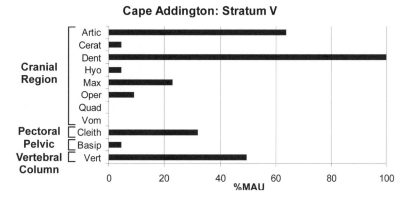

Figure 4.4. Pacific cod element representation (%MAU) from Stratum V, Cape Addington Rockshelter.

the natural strata in the Cape Addington Rockshelter were recovered from Stratum V (NISP = 341), which yielded dates from 1880 ± 70 to 1710 ± 70 BP (Moss 2004:60, Table 5–2). Like North Point, the number of cranial and postcranial elements revealed a disproportionate ratio of cod heads and trunks. At Cape Addington, however, cod heads are disproportionately abundant (MAU = 11 based on the dentary) while postcranial portions of the cod carcass are underrepresented (MAU = 5 calculated from the vertebrae).

Given the similarities between cod cranial and postcranial element densities, density-mediated destruction cannot be used to explain cod element representation in the Cape Addington Stratum V assemblage (Figure 4.4). We are at a loss to explain the highly uneven representation of cranial elements (e.g., scarcity of ceratoyhal, hyomandibula, opercle, quadrate, and vomer; prominence of articular and dentary); sampling and in situ destruction of elements are possible factors. We think it highly unlikely that only parts of the crania were deposited in the rockshelter. Overall, the high frequency of crania (indicated by the dentary) show that cod crania tended to be separated from the postcranial carcass and left in the rockshelter while postcranial portions of cod carcasses tended to be removed and deposited elsewhere. Based on ethnohistorical descriptions of cod butchery from the Northwest Coast, the overrepresentation of cod heads at Cape Addington is consistent with the expected pattern of body part representation produced by processing for storage.

Conclusions

Future analyses of North Pacific archaeofaunal data should examine fish body part representation to determine how different taxa were used and whether specific body parts are over- or underrepresented in archaeofaunal assemblages. The potential role of density-mediated element attrition in structuring the characteristics of these assemblages, however, must be addressed before body part representation is taken as evidence of past human behavior. While salmon body part representation (i.e., heads versus trunks) may be affected by density differences, our new Pacific cod bone density records demonstrate that density per se does not control Pacific cod body part representation. Pacific cod elements from the cranial and postcranial regions exhibit similar densities, and thus density-mediated element attrition should be consistent across the skeleton. Partlow and Kopperl (2007, this volume) postulate that faunal assemblages at residential sites where stored cod were consumed should be dominated by postcranial elements, particularly those from the pectoral girdle (i.e., cleithra), and that assemblages from logistical camps where cod were processed should be dominated by cranial elements and exhibit significantly fewer postcranial remains; the patterns of Pacific cod body part representation identified at the North Point and Cape Addington sites conform to these expectations.

Our work also highlights the differences in bone density between Pacific salmon and cod, suggesting that cod elements are more likely to survive destructive physical and chemical processes than salmon elements. At sites where the remains of cod and salmon were discarded, measures of taxonomic abundance may be biased in favor of taxa possessing denser skeletal elements. Future analyses will further explore how intertaxonomic structural and bone density differences affect taxonomic representation in archaeofaunal assemblages.

References Cited

Ames, Kenneth M. "The Northwest Coast: Complex Hunter-Gatherers, Ecology, and Social Evolution." *Annual Review of Anthropology* 23 (1994): 209–229.

Ames, Kenneth M., and Herbert D. G. Maschner. *Peoples of the Northwest Coast: Their Archaeology and Prehistory.* London: Thames and Hudson, 1999.

Barrett, James H. "Fish Trade in Norse Orkney and Caithness: A Zooarchaeological Approach." *Antiquity* 71 (1997): 616–638.

Barrett, James H., Alison M. Locker, and Callum M. Roberts. "The Origins of Intensive Marine Fishing in Medieval Europe: The English Evidence." *Proceedings of the Royal Society of London, Series B.* 271 (2004): 2417–2421.

Barrett, James H., Rebecca A. Nicholson, and Ruby Ceron-Carrasco. "Archaeo-ichthyological Evidence for Long-Term Socioeconomic Trends in Northern Scotland: 3500 BC to AD 1500." *Journal of Archaeological Science* 26 (1999): 353–388.

Bernick, Kathryn M. *A Site Catchment Analysis of the Little Qualicum River Site, DiSc1: A Wet Site on the East Coast of Vancouver Island, B.C.* Mercury Series, Archaeological Survey Paper No. 118. Ottawa: National Museum of Man, 1983.

Binford, Lewis R. *Nunamiut Ethnoarchaeology.* New York: Academic Press, 1978.

Boas, Franz. *Ethnology of the Kwakiutl.* Washington, DC: Government Printing Office, 1921.

Boehm, S. Gay. "Cultural and Non-cultural Variation in the Artifact and Faunal Samples from the St. Mungo Cannery Site, B.C. (DgRr2)." Master's thesis, University of Victoria, 1973.

Bowers, Peter M., and Madonna L. Moss. "The North Point Wet Site and the Subsistence Importance of Pacific Cod on the Northern Northwest Coast." In *People and Wildlife in Northern North America: Essays in Honor of R. Dale Guthrie,* S. Craig Gerlach and Maribeth S. Murray, eds., pp. 159–177. BAR International Series. Oxford: Archaeopress, 2001.

Butler, Virginia L. "Distinguishing Natural from Cultural Salmonid Deposits in Pacific Northwest North America." PhD diss., University of Washington, 1990.

Butler, Virginia L., and Sarah K. Campbell. "Resource Intensification and Resource Depression in the Pacific Northwest of North America: A Zooarchaeological Review." *Journal of World Prehistory* 18 (2004): 327–405.

Butler, Virginia L., and James C. Chatters. "The Role of Bone Density in Structuring Prehistoric Salmon Bone Assemblages." *Journal of Archaeological Science* 21 (1994): 413–424.

Colley, Sarah M. "Some Methodological Problems in the Interpretation of Fish Remains from Archaeological Sites in Orkney." In *2nd Fish Osteoarchaeology Meeting,* N. Desse-Berset, ed., pp. 117–131. Notes et Monographies Techniques 16. Paris: Centre de la Recherche Scientifique, 1984.

Coupland, Gary, Roger H. Colten, and Rebecca Case. "Preliminary Analysis of Socioeconomic Organization at the McNichol Creek Site, British Columbia." In *Emerging from the Mist: Studies in Northwest Coast Culture History,* R. G. Matson, Gary Coupland, and Quentin Mackie, eds., pp. 152–169. Vancouver: UBC Press, 2003.

Croes, Dale R. "Exploring Prehistoric Subsistence Change on the Northwest Coast." In *Long-Term Subsistence Change in Prehistoric North America,* Dale R. Croes, Rebecca A. Hawkins, and Barry L. Isaac, eds., pp. 337–366. Research in Economic Anthropology, Supplement 6. Greenwich, CT: JAI Press, 1992.

———. "Northwest Coast Wet-Site Artifacts: A Key to Understanding Resource Procurement, Storage, Management and Exchange." In *Emerging from the Mist: Studies in Northwest Coast Culture History,* R. G. Matson, Gary Coupland, and Quentin Mackie, eds., pp. 51–75. Vancouver: UBC Press, 2003.

Drucker, Philip. *The North and Central Nootkan Tribes.* Washington, DC: Smithsonian Institution Bureau of American Ethnology Bulletin 144, 1951.

Emmons, George T. *The Tlingit Indians.* Frederica de Laguna, ed. Seattle: University of Washington Press; Vancouver: Douglas and McIntyre; New York: American Museum of Natural History, 1991.

Grier, Colin. "Dimensions of Regional Interaction in the Prehistoric Gulf of Georgia." In *Emerging from the Mist: Studies in Northwest Coast Culture History,* R. G. Matson, Gary Coupland, and Quentin Mackie, eds., pp. 170–187. Vancouver: UBC Press, 2003.

Hanson, Diane K. "Late Prehistoric Subsistence in the Strait of Georgia Region of the Northwest Coast." PhD diss., Simon Fraser University, 1991.

Hoffman, Brian W., Jessica M. C. Czederpiltz, and Megan A. Partlow. "Heads or Tails: The Zooarchaeology of Aleut Salmon Storage on Unimak Island, Alaska." *Journal of Archaeological Science* 27 (2000): 699–708.

Huelsbeck, David R. "Mammals and Fish in the Subsistence Economy of Ozette." In *Ozette Archaeological Research Reports,* Volume II: *Fauna,* Stephan R. Samuels, ed., pp. 17–91. Pullman: Washington State University Department of Anthropology; Seattle: National Park Service, Pacific Northwest Regional Office, 1994.

Jewitt, John R. *A Journal Kept at Nootka Sound.* Fairfield, WA: Ye Galleon Press, 1988.

Jochelson, Waldemar. *History, Ethnology and Anthropology of the Aleut.* Salt Lake City: University of Utah Press, 1933.

Kopperl, Robert E. "Cultural Complexity and Resource Intensification on Kodiak Island, Alaska." PhD diss., University of Washington, 2003.

de Laguna, Frederica. *Under Mount Saint Elias: The History and Culture of the Yakutat Tlingit.* Washington, DC: Smithsonian Institution Press, 1972.

Lyman, R. Lee. "Bone Density and Differential Survivorship of Fossil Classes." *Journal of Anthropological Archaeology* 3 (1984): 259–299.

———. *Vertebrate Taphonomy.* Cambridge: Cambridge University Press, 1994.

Matson, R. G. "The Evolution of Northwest Coast Subsistence." In *Long-Term Subsistence Change in Prehistoric North America,* Dale R. Croes, Rebecca A. Hawkins, and Barry L. Isaac, eds., pp. 367–428. Research in Economic Anthropology, Supplement 6. Greenwich, CT: JAI Press, 1992.

Matson, R. G., and Gary Coupland. *Prehistory of the Northwest Coast.* New York: Academic Press, 1995.

McKechnie, Iain. "Five Thousand Years of Fishing at a Shell Midden in the Broken Group Islands, Barkley Sound, British Columbia." Master's thesis, Simon Fraser University, 2005.

Morales, Arturo. "A Study of the Representativity and Taxonomy of the Fish Faunas from Two Mousterian Sites in Northern Spain with Special Reference to Trout (*Salmo trutta* L., 1758)." In *2nd Fish Osteo-archaeology Meeting,* Nathalie Desse-Berset, ed., pp. 41–59. Notes et Monographies Techniques 16. Paris: Centre de Recherches Archeologiques, 1984.

Moss, Madonna L. "Archaeology and Cultural Ecology of the Prehistoric Angoon Tlingit." PhD diss., University of California, Santa Barbara. Ann Arbor, MI: University Microfilms, 1989.

———. *Archaeological Investigation of Cape Addington Rockshelter: Human Occupation of the Rugged Seacoast on the Outer Prince of Wales Archipelago, Alaska.* University of Oregon Anthropological Paper No. 63. Eugene: Museum of Natural History, 2004.

Newton, Richard G., and Madonna L. Moss. *Haa Atxaayi Haa Kusteeyix Sitee, Our Food Is Our Tlingit Way of Life, Excerpts from Oral Interviews.* Juneau, AK: USDA Forest Service, Alaska Region, 2005.

Oberg, Kalervo. *The Social Economy of the Tlingit Indians.* Seattle: University of Washington Press, 1973.

Orchard, Trevor J. "The Role of Selected Fish Species in Aleut Paleodiet." Master's thesis, University of Victoria, 2001.

Partlow, Megan A. "Salmon Intensification and Changing Household Organization in the Kodiak Archipelago." PhD diss., University of Wisconsin–Madison, 2000.

Partlow, Megan A., and Robert E. Kopperl. "Exploring the Skeletal Evidence for Dried Cod in the North Pacific." Poster presented at the 34th annual meeting of the Alaska Anthropological Association, Fairbanks, 2007.

Perdikaris, Sophia, and Thomas H. McGovern. "Viking Age Economics and the Origins of Commercial Cod Fisheries in the North Atlantic." In *Beyond the Catch,* Louis Sicking and Darlene Abreu Ferreira, eds., pp. 61–90. Boston: Brill, 2009.

Sauvage, H. E. "On Fishing in the Reindeer Period." In *Reliquiae Aquitanicae; Being Contributions to the Archaeology of Perigord and the Adjoining Provinces of Southern France,* E. Lartet and H. Christy, eds., pp. 219–225. London: Williams and Norgate, 1875.

Smith, Ross E. "Structural Bone Density of Pacific Cod (*Gadus macrocephalus*) and Halibut (*Hippoglossus stenolepis*): Taphonomic and Archaeological Implications." Master's thesis, Portland State University, 2008.

Swan, James G. *The Indians of Cape Flattery at the Entrance to the Strait of Fuca, Washington Territory.* Smithsonian Contributions to Knowledge. Philadelphia: Collins Printer, 1870.

Thomas, David Hurst, and Deborah Mayer. "Behavioral Faunal Analysis of Selected Horizons." In *The Archaeology of Monitor Valley 2: Gatecliff Shelter,* David Hurst Thomas, ed., pp. 353–391. Anthropological Papers No. 59. New York: American Museum of Natural History, 1983.

Wigen, Rebecca J. "Fish Remains from Sites 45SJ165 and 45SJ169." In *Archaeological Investigations at Sites 45SJ165 and 45SJ169, Decatur Island, San Juan County, Washington,* Sara L. Walker, ed., pp. 227–308. Eastern Washington University Reports in Archaeology and History 100–118. Cheney, WA: Archaeological and Historical Services, Eastern Washington University, 2003.

———. "Vertebrate Fauna." In *The Hoko River Archaeological Site Complex: The Rockshelter (45CA21), 1,000–100 BP, Olympic Peninsula, Washington,* Dale R. Croes, ed., pp.71–105. Pullman: Washington State University, 2005.

CHAPTER 5

Site-Specific Salmon Fisheries
on the Central Coast of British Columbia

Aubrey Cannon, *Department of Anthropology, McMaster University*
Dongya Yang, *Department of Archaeology, Simon Fraser University*
Camilla Speller, *Department of Archaeology, Simon Fraser University*

Until recently, archaeological debate concerning Northwest Coast salmon fishing was generally cast in relation to presumptions of long-term evolutionary developments toward a more intensive, salmon-based storage economy, which was considered the basis for surplus production, social ranking, and cultural elaboration (Matson 1992). Increasingly, detailed analyses of fish remains from a variety of archaeological contexts are showing a wider range of variability in fisheries at particular sites. As papers in this volume demonstrate, fisheries other than salmon were also economically important on the Northwest Coast, and we are now beginning to see widespread evidence of local and regional variability in the timing and relative intensity of salmon fishing (Cannon 1996; Monks, Moss, Orchard, Trost et al., this volume).

Although development of an intensive salmon-fishing, storage-based economy is no longer considered the essential starting point for evolutionary developments leading to the complex social organizations recorded ethnographically on the Northwest Coast (Cannon and Yang 2006; Moss and Erlandson 1995), debate over this issue did provide an important impetus for research that has led to more refined understanding of the various ways that fishing was organized. One track in this research has focused on patterns of fishing and processing evident at particular sites, while another has documented local and regional variability in the history and intensity of salmon fishing. We present research below that incorporates ancient DNA-based salmon species identification supplemented by observations of skeletal element representation to characterize particular salmon-fishing strategies on the central coast of British Columbia. On the basis of our results we argue for a strong orientation toward salmon fisheries available in the immediate vicinity of village sites.

Concern with identifying the inception of large-scale storage was responsible for a number of methodological advances in the analysis of fish bone assemblage composition and taphonomy, driven largely by the assumption that salmon processed for storage would result in assemblages dominated by vertebrae over cranial elements. Although the premises for this distinction as a basis for identifying storage-based economies were largely without foundation (Cannon 2001), detailed ethnoarchaeological and archaeological studies have shown variability in the elemental composition of assemblages from different types of sites. Preservation may account for the abundance of more delicate facial bones relative to vertebrae in some cases (Butler and Chatters 1994), but variations in processing are also likely to affect the representation of skeletal elements (Hoffman, Czederpiltz, and Partlow 2000). This approach provides one level of information regarding where salmon were caught, processed, and consumed, but species identification and better understanding of relationships between the distribution and locations of harvest facilities, salmon runs, and village sites (Brewster and Martindale, this volume; Cannon and Yang 2006; Lepofsky, Trost, and Morin 2007; White 2006, this volume) provide insight into a broader scale of variation in Northwest Coast fishing strategies.

Although regional, multisite investigations have yet to incorporate fully the emerging understanding of how processing and preservation affect fish bone assemblages, there is now the beginning of an empirical, conceptual, and methodological basis for understanding the nature and extent of variation in how salmon fisheries were organized. Although likely not exhaustive of the possibilities, four basic patterns can be identified on the basis of our present knowledge and understanding. One consists of the large-scale regional aggregations that occurred at the locations of major salmon fisheries, including the mouths and lower reaches of the Skeena, Fraser, and Columbia rivers (Butler 2000; Coupland, Martindale, and Marsden 2001; Coupland, Stewart, and Patton 2010; Grier 2003:173–174; Lepofsky et al. 2005), which could draw populations from a wide region and result in the transport of large quantities of processed and dried salmon to distant residential sites for later consumption. The archaeological signature of this strategy may be partially evident in the size and number of sites at major fishing locations, though ethnographic documentation remains the primary evidence for the form and importance of these fisheries. The archaeological evidence for transport to remote locations for consumption is much less clear (see Brewster and Martindale, this volume). The extent of this type of population movement for resource acquisition prior to European contact has also been questioned for not taking into account the effects of depopulation and other changes on ethnographically recorded patterns (Ford 1989), but from ethnographic accounts and available archaeological evidence it seems reasonable to assume some level of regional population aggregation occurred to take advantage of the large salmon runs available at the mouths and lower reaches of major river systems.

A second type of organization for which there is some further archaeological evidence is one involving more limited regional access to common fisheries. An abundance of salmon remains and a common species profile at sites that lack fisheries in their immediate vicinities has been interpreted in one example on the south coast of British Columbia as an indication of shared access to regional resources achieved and maintained through social negotiations between settlements (Lepofsky, Trost, and Morin 2007). A third and contrasting form of fisheries organization is one focused largely or even exclusively on salmon streams in the immediate vicinity of residential sites (e.g., Maschner 1997; Moss, this volume). This pattern might be suggested by site location but requires further evidence for verification. Finally, a fourth type of salmon fishing strategy is one involving the use of numerous, relatively small-scale salmon fishing facilities regionally distributed at locations away from major residential sites. Although fish traps are found in the immediate vicinity of village sites, as, for example, at Namu and Kisameet, on the central British Columbia coast, this regionally focused strategy entails greater if not primary emphasis on harvest locations not associated with evidence of major or long-term residential occupation. These would include fish trap sites where small groups camped while they caught and processed salmon for storage and later consumption at residential sites (Langdon 2006, 2007; White 2006, this volume). Langdon (2006:41) has also suggested that some smaller traps could have been regularly accessed by canoe. Trapped fish could be transferred to the canoe, allowing fishermen to visit a number of trap locations in a regular circuit. A further variation on this type of strategy is the apparently unique and highly specialized reef netting practiced in the Gulf of Georgia area to harvest salmon runs returning to the Fraser River system to spawn. This particular practice may have emerged relatively late in the pre–European contact history of the region (Easton 1990).

In different parts of the coast, at particular residential sites and at various times, salmon could have been acquired through any number of organizational strategies. Some may be difficult to identify, but the major patterns outlined above can be differentiated archaeologically to varying degrees. One implication of a local-stream-based strategy, for example, is an emphasis on the salmon available in the immediate vicinity of sites. Ancient DNA identification of salmon species from village sites on the central coast of British Columbia points very clearly to this type of strategy.

Namu Salmon Fisheries

The archaeological history of salmon fishing on the central British Columbia coast has developed out of the initial analysis and interpretation of fish remains recovered from excavations at the site of Namu, located within the traditional territory of the Heiltsuk First Nation (Figure 5.1; Cannon 1988, 1991, 1995, 2000a, 2001; Cannon and Densmore 2008; Cannon and Yang 2006; Carlson 1979, 1996). These analyses had raised questions concerning temporal variability in the abundance of salmon, which, though overwhelmingly dominant among larger fish remains in all periods, peaked in abundance during the period 6000–4000 cal BP at 97 percent of the identified fish. This was an increase from already high numbers of salmon (89 percent of fish) in earlier deposits (7000–6000 cal BP), which represent the initial period of well-preserved faunal remains at the site. The more interesting observation was the lesser abundance of salmon in periods after 4000 cal BP. Revised figures based on recent analysis of fish remains from the last 2,000 years show

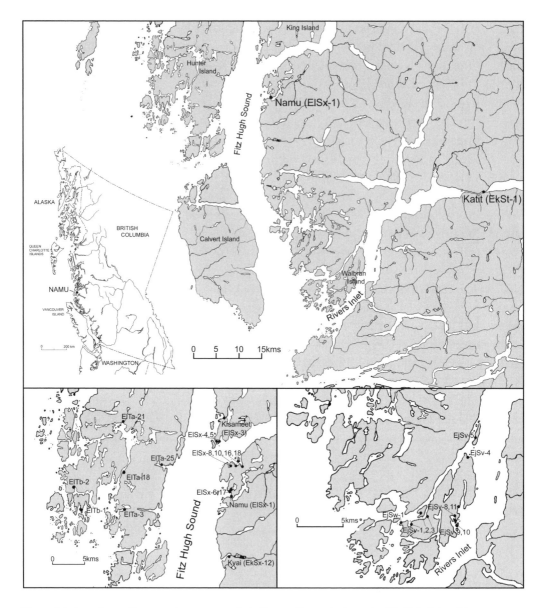

Figure 5.1. Locations of core and auger-sampled sites on the central coast of British Columbia.

that salmon made up 82 to 85 percent of identified remains during the periods between 4000 cal BP and the time of European contact. The percentage of salmon fell to its lowest level (ca. 55–75 percent) in deposits dating to 2000–1000 cal BP but recovered to near peak levels of over 95 percent of larger fish remains by about 1,000 years ago (Cannon and Densmore 2008).

Over the years since the initial observation of this apparent later decline in the Namu salmon fishery, a wide variety of ancillary studies has confirmed that the decline was real and not just a function of deposition patterns, taphonomy, or an increase in the relative proportions of other fish taxa. The interpretation of a reduction in salmon abundance and consequent shortfalls in the staple food supply is supported by evidence indicative of responses to food shortage. These include increased consumption of resources that might be considered less nutritious, abundant, or reliable than salmon, including ratfish and deer (Cannon 1995). Periods that exhibit a lower relative abundance of salmon remains also have a greater proportion of deer phalanges that exhibit transverse breaks, which are consistent with human processing for marrow extraction (Binford 1978:148; Zita 1997). Isotopic analysis of dog bones from these periods also shows more reliance on terrestrial sources of protein and increased consumption of shellfish (Cannon, Schwarcz, and Knyf 1999). The considerable variability in isotope values of dogs in the period after 4000 cal BP further suggests that some dogs consumed marine-based foods in quantities comparable to the periods of peak salmon fishing, while others ate considerably more terrestrial-based protein. This result is consistent with a fishery that was subject to periodic failures rather than an overall decline.

Recent aDNA analysis of dogs from Namu has indicated a dramatic decrease in the diversity of haplotype groups after 4000 cal BP (Barta 2006). This genetic bottleneck suggests potential culling of the dog population or a die-off due to periodic starvation resulting from failures in the salmon fishery. This interpretation of the genetic evidence is supported by ethnographies from the Kamchatka Peninsula in Siberia, which record the large numbers of salmon needed to support dogs and the culling of dog populations or starvation that typically occurs when the fisheries fail (Shnirelman 1994).

The array of independent evidence for responses to food shortage that coincides with periods of reduced salmon abundance strongly suggests the residential population at Namu periodically experienced some level of privation as the result of failures in the mainstay salmon fishery. Given the coarse temporal resolution available through radiocarbon dating of shell midden deposits, it is impossible to determine whether these failures were restricted to the occasional bad year or involved longer periods of lesser productivity. The wide array of factors that can affect salmon productivity on a range of temporal scales may also make it impossible to determine the precise nature and cause of periodic failures (Cannon 1996), but whatever the resolution of specific events, the archaeological evidence from Namu is consistent with views based on Northwest Coast ethnography that highlight the frequent incidence of starvation, even in the context of a generally rich natural environment (Suttles 1968).

Based solely on data from Namu it remains an open question whether failures in the salmon fishery were due to fluctuations in local stream and lake conditions or broader environmental conditions that affected productivity throughout the region. One possible reason to think that reduced productivity might have been a broader regional phenomenon is the apparent failure to take advantage of alternative fishing locations to compensate for the Namu shortages. However, investigation of variability in salmon and herring fisheries at other shell midden sites in the area shows a different pattern (Cannon 2000a, 2000b). Auger samples from nearby villages at the mouth of the Kisameet River, 12 km to the north of Namu, and the Kvai (Koeye) River, 9 km to the south, show densities of salmon remains comparable to those from the peak period of salmon fishing at Namu (Table 5.1). The dates of these villages to within the last 3,000 years suggest periodic failures in the Namu fishery were local rather than regional phenomena. The further implication is that fisheries on the central coast were primarily local in orientation and in the case of these particular villages focused on the stream in the immediate site vicinity. The evidence of food shortages

Table 5.1. Number and density of fish remains in auger samples from central coast sites

Site	Site Type[1]	Sample Volume (l)		Number and Density[2] of Fish Remains						# of Taxa
				Salmon		Herring		Other		
		total	<2 mm	n	d	n	d	n	d	
Namu (ElSx-1)										
2000 cal BP–Contact	WV	19.2	16.3	795	48.7	769	47.1	74	4.5	11
4000–2000 cal BP	WV	12.2	8.8	388	44.1	325	36.9	97	11.0	8
5000–4000 cal BP	WV	15.0	11.8	927	78.5	636	53.9	41	3.5	9
6000–5000 cal BP	WV	7.9	6.6	572	86.3	328	49.5	22	3.3	8
7000–6000 cal BP	WV	8.4	7.6	540	71.3	340	44.9	16	2.1	9
Namu (Combined)	WV	62.7	51.1	3222	63.1	2398	46.9	250	4.9	15
Namu Region										
Kvai (EkSx-12)	WV	26.2	23.0	1599	69.4	198	8.6	195	8.5	18
Kisameet (ElSx-3)	WV	30.7	22.9	1555	67.9	1609	70.3	192	8.4	14
ElSx-4	C	6.0	4.5	27	6.0	17	3.8	2	0.4	3
ElSx-5	BC	21.4	14.4	363	25.2	608	42.1	18	1.2	8
ElSx-6	C	0.5	0.4	9	23.7	16	42.1	1	2.6	3
ElSx-8	C	4.3	2.8	0	0.0	5	1.8	2	0.5	2
ElSx-10	BC	27.6	20.0	481	24.0	895	44.7	57	2.8	9
ElSx-16	C	4.1	2.5	16	6.3	152	59.7	2	0.8	3
ElSx-17	C	0.8	0.7	1	1.5	0	0.0	0	0.0	1
ElSx-18	BC	13.2	8.5	274	32.3	270	31.9	23	2.7	9
ElTa-3	C	2.4	2.0	86	43.8	16	8.1	35	17.8	5
ElTa-18	C	12.1	10.7	27	2.5	1	0.1	3	0.3	3
ElTa-21	C	6.2	5.3	14	2.7	4	0.8	1	0.2	3
ElTa-25	C	16.6	6.0	50	8.3	37	6.1	18	3.0	6
ElTb-1	S/SV	29.8	24.3	423	17.4	4142	170.6	237	9.8	13
ElTb-2	C	3.0	2.5	5	2.0	37	14.9	55	22.1	8
Rivers Inlet										
Katit (EkSt-1)	WV	18.1	7.9	1503	190.3	738	93.4	294	37.2	5
Cockmi (EjSw-1)	BC	56.0	36.4	1374	37.7	1465	40.2	274	7.5	7
EjSv-1	C	4.1	2.2	39	17.7	43	19.5	40	18.2	3
EjSv-2	C	7.2	5.8	123	21.3	211	36.5	40	6.9	6
EjSv-3	C	3.8	2.7	41	15.0	56	20.6	43	15.7	3
EjSv-4	C	3.8	3.0	16	5.3	36	11.9	12	4.0	5
EjSv-5	C	7.4	5.3	105	19.9	218	41.3	4	0.8	4
EjSv-8	C	9.2	5.8	91	15.8	106	18.4	25	4.3	4
EjSv-9	BC	7.4	5.1	202	39.3	232	45.1	117	22.8	5
EjSv-10	C	32.7	26.1	345	13.2	472	18.1	88	3.4	7
EjSv-11	C	4.2	3.7	61	16.5	53	14.3	13	3.5	5

1. Site Types: WV = Winter Village, S/SV = Spring/Summer Village, BC = Base Camp, C = Camp.
2. Density calculated as the NISP/litre of fine (<2mm) matrix to avoid distortions introduced through the exaggerated volumes that may result from large quantities of coarsely broken or whole shell or rock.

predicated on shortfalls in stored salmon production, and evidence that productivity was higher at other nearby streams, suggests the organization of salmon fishing at Namu did not have the capacity to make up for local shortage through the use of a broader regional fishing strategy. This localized orientation contrasts with a pattern of regional production based on dispersed fish trap locations, which operated in other parts of traditional Heilt-suk territory (White 2006, this volume).

Ancient DNA identification of salmon species provides another line of evidence that stresses the critical importance of salmon in the Namu village subsistence economy (Cannon and Yang 2006). The results presented here from the analysis of salmon remains recovered at a broader range of sites also add weight to the interpretation of a strong local orientation in village fisheries. The Namu results showed use of a full range of Pacific salmon species, with a clear emphasis on pink, followed by sockeye and chum (Table 5.2). The significance of pink salmon is its low-fat quality, which makes it ideal for processing for long-term storage (Romanoff 1985). Chum salmon also possesses this quality but was either not as readily available at Namu or was not selected as often, possibly because the smaller size of pink salmon made it easier to process and dry for long-term storage. The abundance of sockeye is easy to understand because its high oil content and rich flavor

Table 5.2. Salmon species by site

Site	Site type[1]	Recovery method	Sockeye	Pink	Chum	Chinook	Coho
Namu (ElSx-1)							
2000 cal BP–Contact	WV	Excavation/auger/wet	7	11	5	1	2
4000–2000 cal BP	WV	Excavation/wet	15	5	11	0	0
5000–4000 cal BP	WV	Excavation/wet	4	9	6	0	1
6000–5000 cal BP	WV	Excavation/wet	2	13	2	0	0
7000–6000 cal BP	WV	Excavation/wet	4	11	5	0	2
Namu Total			32	49	29	1	5
McNaughton (ElTb-10)	WV	Excavation/dry	0	3	17	0	1
FbSx-6	?	Excavation/dry	0	5	0	1	0
Kisameet (ElSx-3)	WV	Excavation/dry	0	4	21	0	0
Kisameet (ElSx-3)	WV	Auger/wet 2mm	0	10	8	0	0
Kvai (EkSx-12)	WV	Auger/wet 2mm	1	20	0	0	0
ElTb-1	S/SV	Auger/wet 2mm	0	13	1	0	0
ElSx-5	BC	Auger/wet 2mm	1	5	1	0	0
ElSx-10	BC	Auger/wet 2mm	0	3	0	0	0
ElSx-18	BC	Auger/wet 2mm	0	1	0	0	0
ElSx-4	C	Auger/wet 2mm	0	2	0	0	0
ElTa-25	C	Auger/wet 2mm	0	0	4	0	0
Katit (EkSt-1)	WV	Grab/riverbank	7	0	2	5	0
Katit (EkSt-1)	WV	Auger/wet 2mm	22	2	0	1	0

1. Site types: WV = winter village; S/SV = spring/summer village; BC = base camp; C = camp.

make it well suited for fresh consumption in the summer and fall. Although sockeye can also be smoked for storage, its high oil content makes it more susceptible to spoilage if kept too long (Romanoff 1985; White, this volume).

The aDNA analysis also indicated that periods of lesser salmon abundance could be attributed almost entirely to reduction in pink salmon, which is proportionately much less abundant in period assemblages that contain fewer salmon remains overall. If pink salmon was typically a critical part of winter food stores, as its superior storage qualities and abundance during periods of peak production suggest, then its loss due to failures in the annual spawning run would have had a severe impact on local residents. Period fish assemblages that have fewer salmon also contain proportionately more sockeye. Given the earlier timing of sockeye runs and their lesser suitability for long-term storage, it seems unlikely sockeye would have had the same potential for sustaining the village population through the winter, and this is consistent with evidence of food shortages in these periods.

Contemporary pink salmon populations exhibit pronounced two-year cycle dominance in many areas of the coast, though recent commercial fishing may have exaggerated the degree and persistence of these cycles (Heard 1991:204–207). The pattern discernible in the Namu evidence is of at least an annual failure that recurred at unknown intervals over an extended period of time. The total duration of this period of occasional pink fishery failures was approximately three millennia. A major vulnerability of local pink runs to periodic failure is the fact that pink salmon almost invariably spawn at two years of age. Therefore, any environmental circumstance that affected spawning success in a given year, such as a short-term excess in stream outflow that would wash out the eggs in nests located in the lower reaches of streams, would result in a much smaller return of spawning salmon two years later. In contrast, spawning runs of chum salmon, which also nest in the lower reaches of streams, typically consist of fish of multiple ages, thereby reducing the impact of adverse conditions in any given year on the size of spawning runs in subsequent years.

The inverse relationship between the relative abundance of pink and sockeye salmon over time is the focus of an ongoing aDNA study of a much larger and more systematically selected sample, which will make it possible to refine the chronology of disruption in the pink fishery. The initial results of those analyses confirm the existence of the inverse relationship and indicate that decline and recovery in the overall abundance of salmon is closely tied to the relative abundance of pink salmon in particular. Since the spawning times of these two species overlap in late summer, it seems likely that recognized failure in pink salmon runs resulted in more effort to harvest and process sockeye salmon as an alternative. A combination of lesser quantities of available sockeye salmon and a greater propensity for sockeye stores not to last as well throughout the winter would account for periodic food shortages at Namu.

Local Orientation in Central Coast Fisheries

The variable abundance and species composition of the Namu salmon fishery along with multiple indicators of responses consistent with food shortage suggest the local population was tied conceptually and physically to the productivity of its own immediately available salmon resource. This interpretation of the evidence does not diminish the importance of the wide range of other locally and regionally available food resources or the capacity to supplement salmon caught locally with supplies caught elsewhere, but it does suggest the storage-based subsistence economy was largely focused on the salmon available in the immediate site vicinity. Shell and vertebrate faunal remains show the use of a wide range of food resources that might have been obtained locally or from other localities throughout the area, and though plant remains are all but absent in the Namu midden matrix, a variety of berries and other plant foods were no doubt also important parts of the diet. Nevertheless, the ubiquity of salmon bones in the excavations and in auger samples, and the sometimes even greater abundance of herring bones in the finely screened auger samples, are enough to identify these particular resources as the staple bases of sedentary village settlement. Further, despite whatever capacity existed to harvest these particular resources regionally, the orientation of the staple resource economy appears to have been very local, largely centered on the fish available within the immediate site vicinity.

An interpretation of local orientation in the Namu salmon fishery derives its strength from the consilience of multiple lines of evidence, but its power is limited by a focus on just one site. This recognized need to address the history of fisheries at Namu within a broader regional context was the motivation for undertaking broader programs of regional, multisite investigation based on auger sampling of shell midden sites, first at sixteen sites in 1996–1997 in the Namu vicinity, within the traditional territory of the Heiltsuk First Nation, and then at another eleven sites in 2005–2006 in the area of Rivers Inlet to the south of Namu, within the traditional territory of the Wuikinuxv (Oweekeno) First Nation (Figure 5.1). These studies have provided a basis for characterizing the regional nature and local variability of fisheries. Identification of fish remains recovered by washing auger samples in the lab through 2 mm mesh screen shows a common focus at almost all sites in the region on salmon and herring as the staple fish resources. Analysis also shows supplementary reliance on a wide range of other fish species and differences in the abundance and variety of fish remains according to site size, location, and type (Table 5.1; Cannon 2000a, 2002).

The availability of a small number of whole vertebrae among the much more abundant salmon vertebra fragments in the auger sample matrix allowed for the additional application of aDNA analysis to identify the species focus of local fisheries. The results show variability in salmon species depending on their local availability, which indicates a broader regional pattern of village dependence on the streams in their immediate vicinity. This multisite program of ancient DNA analysis for the central coast was conducted on the basis of the whole vertebrae derived from bucket-auger sampling and others selected from curated collections of faunal remains from past excavations.

Existing faunal collections had been considered to have great potential for aDNA analysis, though the results of our study show that the recovery methods used in past excavations can place severe limitations on the utility of recovered remains for identifying the species profiles of salmon fisheries. Randomly selected samples of salmon vertebrae from three sites excavated in the early 1970s—McNaughton Island (ElTb-10) (Pomeroy 1980), Kisameet (ElSx-3), and FbSx-6 on Roscoe Inlet in the northern part of Heiltsuk territory (Luebbers 1978)—show the presence of a range of species (Table 5.2), but the majority are pink and chum salmon. These results are generally consistent with the results of the Namu study, but the predominance of larger chum salmon vertebrae at the McNaughton Island site, excavated in 1972, and at Kisameet, excavated in 1970, suggests a bias from the effects of larger screen size and a lack of water screening in the recovery of salmon bone. This is confirmed by a comparison of the vertebrae recovered in auger samples at Kisameet in 1996 with those from the excavation in 1970. Both sets of samples show greater emphasis on chum than was evident at Namu, but the smaller pink remains are much more common in the auger samples, which were water-screened in the lab through 2 mm mesh (Table 5.2). The emphasis on chum at Kisameet indicates a subtle but potentially important difference between this site and Namu. Kisameet is located only 12 km north of Namu, and both sites are similarly situated at the mouths of small rivers draining major lakes. Both sites have stone fish traps near the river mouths that would have been used to trap salmon as they schooled in the bay before spawning. The slightly greater emphasis on chum salmon at Kisameet is likely a function of local species availability. Fisheries and Oceans Canada (2001) escapement data show larger numbers of chum in the Kisameet River than in the Namu River. Although both chum and pink are low in oil content, making them well suited for winter preservation, chum are larger and more variable in size, which could potentially make them more work to process in large numbers for winter storage. If this favored the selection of pink over chum, the slightly greater emphasis on chum at Kisameet could suggest a less optimal resource base at that location, though the difference is admittedly very slight.

The Kisameet village was established around 2800 cal BP, during the period when Namu was experiencing periodic shortfall in its pink salmon fishery. The village on the Kvai River, 9 km south of Namu, was founded ca. 2200–1900 cal BP. Vertebrae recovered in the auger samples from the Kvai village site are almost exclusively pink salmon. Only a single sockeye specimen is present among the twenty-one identified specimens. These results are also consistent with contemporary species availability; the Kvai River is home to the largest pink fishery on this part of the central coast (Fisheries and Oceans Canada 2001).

Given the apparent regional preference for pink salmon, the Kvai River could reasonably have been a preferred village location earlier than 2000 cal BP. It is possible the river has eroded evidence of earlier village settlement, but the virtual absence of herring among the fish remains at this site could also point to a major disadvantage to winter village settlement at this location (Table 5.1). Contemporary fishermen confirm the lack of herring in this immediate area. The sandy beach deposits at the mouth of the river, which would have buried any permanent constructions, and the lack of exposed rocks also likely precluded the type of stone-walled fish trap favored at Namu and Kisameet, though alternative means for harvesting salmon were obviously available.

Auger-collected samples of salmon remains were available from one other major residential site located in the outer coast area just west of Namu. The site of ElTb-1, on Hurricane Island, which has been characterized as a possible spring/summer aggregate village (Cannon 2002), is not sited on or in the vicinity of a salmon stream. The extraordinary abundance of herring remains at this site suggests it was occupied in late winter/early spring at the peak of that fishery. Ancient DNA analysis of a sample of fourteen vertebrae recovered from auger samples at this site shows a predominance of pink salmon remains, typical for other sites in the region. These may represent the last of winter stores, brought to the site to sustain populations waiting for the herring fishery to begin.

Additional auger-collected samples from a variety of major and minor campsites throughout the Namu vicinity yielded smaller numbers of one to seven whole salmon vertebrae per site. Although these samples are too small to interpret very far, the combined sample of eighteen vertebrae shows a range of species. The majority is pink, but chum and sockeye are also present. Sockeye is present at one major site (ElSx-5) thought to have been a residential base camp. A specialized camp for shellfish harvest (ElTa-25) contained only chum vertebrae. This variety could represent opportunistic harvest or a combination of fresh-caught and preserved fish, but the sample sizes are too small to say much more.

Recent core- and auger-sampling investigation of shell midden sites in Rivers Inlet shows the same focus on salmon and herring fisheries evident in the Namu vicinity and on the outer coast (Table 5.1). The only major winter village site included in the investigation in Rivers Inlet was the site of Katit on the Wannock River, which drains Oweekeno Lake. According to Olson (1954), all of the Wuikinuxv First Nation was concentrated in a village of perhaps a dozen houses on this tiny islet in the river from about 1900 to 1935. These were the remnants of a decimated population that formerly had been living in permanent villages ranging from the Kvai River to the head of Oweekeno Lake stretching far into the interior. The village was destroyed by fire in 1935, and the community resettled in its present location on the north shore of the river beginning in 1940. Olson could provide no indication of the date of initial settlement at the Katit site, but radiocarbon dates from the base of the cultural deposits at two locations on the eastern end of the island, where core and auger samples were obtained, returned dates from the last few centuries before initial European contact.

The cultural deposits at the Katit site consist of two meters of organic- and charcoal-rich deposits above a pure clay substrate that defines the original island. The island itself is rapidly eroding and is probably only a fraction of the size it was when occupied. The eroding deposits from the river banks on the south side of the island show an abundance of fire-broken rock and layers of almost pure salmon bone. Ancient DNA analysis of a grab sample of vertebrae from the eroded riverbank showed the presence of sockeye, chinook, and chum. Analysis of a more systematically collected sample, consisting of all the whole vertebrae contained in auger samples from one location on the eastern end of the island, shows a clear predominance of sockeye among the salmon from this site (Table 5.2). This is not surprising since Rivers Inlet was home to the largest sockeye fishery on the central coast, up until its recent collapse (McKinnell et al. 2001). In ethnographic accounts, Rivers Inlet is described as a major summer destination for Kwakwaka'wakw populations to the south, who traveled there to gain access to the sockeye runs (Rohner 1967:24).

All indications from the density of fish bones, the size of the site, and the records from the period of European contact are that Katit was a permanent winter village throughout its occupation. Given the clear preference for pink salmon over sockeye in the Namu vicinity, it is interesting that a major winter village settlement could be sustained at Katit primarily on the basis of sockeye. If the preference for pink salmon in the Namu area was in part due to its better preservation qualities, then one explanation for how sockeye could sustain the Katit village may be the presence of eulachon in the same river. Eulachon grease was used as a food preservative, and its oil, together with other foods, such as plant

Figure 5.2. Examples of salmon gill rakers from the site of Katit.

foods, might have been enough in combination with the rich sockeye fishery to sustain the population throughout the winter. It is also possible that pink salmon were caught and processed in large numbers at the nearby Kilbella River, which supports a large pink fishery, but if so then they were evidently brought back to Katit without their vertebrae.

The match between the aDNA evidence and the productivity of the local salmon fishery is enough to make the case for a subsistence economy focused primarily on fish caught and processed at the residential site location. Additional evidence for on-site salmon processing comes in the form of numerous gill rakers found in the auger samples from Katit (Figure 5.2). These were also found in samples from sites in the Namu vicinity, but in nothing like the quantities found at Katit. While there may be some ambiguity in using head bones as indicators of fresh consumption versus processing for storage, since salmon heads could also be processed for storage, gills and the attached gill rakers would have been discarded with the initial gutting of fish. Their abundance at Katit is attributable to the processing of fish caught in the immediate site vicinity and the absence of a beach processing area in front of the site. The relative rarity of this element at other sites is likely due to initial processing and discard of fish offal taking place on the beach in front of shell midden deposits.

It is not clear that village sites were established at particular localities primarily to take advantage of locally available salmon runs, nor is it certain that winter villages at Namu, Kisameet, Kvai, and Katit were mainly dependent on stores of preserved salmon to sustain populations through the winter, though the predominance of salmon remains and their seasonal availability in late summer/early autumn ahead of the winter season supports this interpretation. From the auger samples, the only other resource that matches the ubiquity of salmon is herring. Although it is impossible to compare their relative food contributions directly on the basis of preserved and recovered remains, herring bones occur in numbers that typically equal or exceed the abundance of identified salmon remains, with the notable exception of the Kvai River village. Herring and herring roe can also be preserved in a variety of ways for later consumption (McKechnie 2005:102–103), but the timing of the peak fishery in late winter/early spring makes it an unlikely basis for winter storage-based subsistence. Clearly, central coast villages were largely reliant on preserved salmon stores as a winter staple, and individual villages subsisted on a species profile matched to runs in their immediate vicinity.

Discussion

The results of aDNA-based identification of salmon species from auger-recovered vertebrae from four winter villages and one possible spring/summer aggregate village show clear differences in salmon species emphasis between locales. The immediate implication is that village settlements were sustainable on the basis of a wide range of specific fisheries. The further implication is that site locations were not chosen to optimize access to any one particular species or range of species, though pink salmon and herring may have provided an optimal bridge between late autumn and early spring, when other resources were unavailable or difficult to access.

Salmon species identifications further indicate an emphasis on local resource acquisition, despite ample knowledge of and capacity for travel throughout the region as a whole. A similar pattern is also apparent for shellfish gathering (Cannon and Burchell 2009; Cannon, Burchell, and Bathurst 2008) and has been suggested by MacDonald et al. (2008) to apply to the acquisition of red ochre. Specific locales were important for a variety of reasons apart from salmon fisheries, and villagers seem to have adjusted themselves largely to what was available locally, sometimes, as at Namu ca. 4000–1000 cal BP, even in the context of periodically acute food shortages. The picture that is emerging on the basis of multisite investigations on the central coast is one of hyperlocal orientation in staple subsistence and other resource activities. Residential groups were oriented toward the use of their own particular resources rather than being engaged in regional systems of resource acquisition involving shared access or exchange.

It is impossible, of course, to rule out the harvest and processing of salmon at other localities and transport to residential sites for consumption in a way that did not involve return of large numbers of skeletal remains. Salmon could have been caught at other locations and returned to villages as processed and even preserved fillets. Fish traps are located at small streams throughout the region and most were specifically used for salmon (White 2006, this volume). The absence of large residential areas in association with many fish traps indicates that salmon caught at these sites were consumed elsewhere. Dates associated with three such sites in the outer coast west of Namu (ElTa-3, ElTa-21, and ElTb-2) suggest harvest at these locations increased relatively late in the archaeological history of the region, within the past 2,000 years (Cannon 2002). These remote camp and fish trap sites are not present in the area of Fitz Hugh Sound, where the Kisameet, Namu, and Kvai villages are located.

The use of short-term camps at fish trap sites may have been a more important strategy in parts of Heiltsuk territory other than the Namu vicinity, but the same local orientation to a specific suite of salmon species is also evident within this fishing strategy. Based on knowledge of fish trap use from living memory and oral tradition, Elroy White (2006:95–98, this volume) documents regional patterns in the targeting of particular species, with traps on the outer coast focused on sockeye and coho; those in the inner waterways and inlets focused on a variety of species, but especially pink and chum; and those in an area on the north side of King Island focused on chum. Clearly, in addition to village fisheries oriented toward the runs in immediately adjacent rivers, there was also a system of dispersed regional fisheries focused on fish traps at the mouths of small streams. The extent to which these systems overlapped or operated as parallel but independent regional fishing strategies is impossible to determine on the basis of presently available archaeological evidence. It is possible that some winter villages subsisted on fish acquired both from the immediate site vicinity and from camps dispersed over a wider area, or it may be that one system served primarily one set of winter residential sites while the other system served another set. An argument can be made that if salmon from other locales could have been readily acquired and was a regular part of their overall fishing strategy, the Namu residents should not have needed to resort to marginal alternative foods when local pink salmon was in short supply. The profile of the salmon fishery also would not have shifted so dramatically from an emphasis on pink to a focus on sockeye if local variability could have been offset through a broader regional fishing strategy. The salmon fishery at Kisameet village might also be expected to have focused less on chum and to match more closely the emphasis on pink salmon evident at other sites if it was drawing from the same regional pool of fisheries. We would also expect a greater proportion of pink remains among the fish bones from the Katit village if these were being acquired at the nearby Kilbella River and transported to Katit. In place of a hypothetical homogenized regional pattern or an optimal species profile based on processing and storage qualities, Katit, Kvai, Namu, and Kisameet exhibit heterogeneity matched to the species available in adjacent rivers.

For the central British Columbia coast, any generalization concerning the nature and focus of salmon fisheries will mask potentially important local variation. In place of a regional pattern, the picture emerging from detailed aDNA analysis of central coast fisheries is one of a mosaic of productive locales, each somewhat unique, autonomous, and apparently largely self-sustaining. In combination with intensive production of seasonally available resources, such as shellfish, plant foods, and fish, even possibly including some salmon, at other locations, some village locales were sufficiently productive to maintain and constrain sustainable scales of production. The settlement of villages, such as Namu, for millennia without archaeologically measurable interruption, is also testament to the capacity of local fisheries to sustain viable residential populations, even in the face of periodic shortages, which in the case of Namu evidently occurred over an extended period of time. The prevailing organization of central coast salmon fisheries throughout the past 7,000 years was one oriented to local streams, though this may have been augmented or complemented in parallel by more recent development of a system based on harvest at numerous remote fish-trap locations.

This interpretation of the organization of salmon fishing on the central coast is very different from the pattern of large-scale production, regional exchange, and larger-scale village aggregations seen in association with the major fisheries of the Skeena, Fraser, and Columbia rivers. It also differs from the pattern of shared regional access to a common resource base recently inferred for Indian Arm and Burrard Inlet on the south coast (Lepofsky, Trost, and Morin 2007). The pattern on the central coast is one of sustained and sustainable small village settlements seemingly well matched to the resources of particular locales. The success of this subsistence-settlement pattern in the Namu area is evident in the sustained occupations of village sites, which were never permanently abandoned up to the time of European contact (Cannon 2003). This is in clear contrast to the volatility of settlement and permanent village abandonment in areas such as Prince Rupert Harbour (Archer 2001), which likely relied in part on the regional fisheries of the Skeena and Nass rivers. It even contrasts with the more common abandonment of smaller sites in nearby Rivers Inlet. Within the Namu vicinity, village dependence on locally available resources, especially salmon, was not only sustainable but potentially helped to minimize the type of intense regional conflict that may have led to the abandonment of villages in Prince Rupert Harbour.

Ancient DNA species identification adds another level of particular detail to what is already emerging as an increasingly complex picture of the archaeological history of salmon fishing on the Northwest Coast. There is evidence now for considerable temporal and regional variability in the initial timing, intensity, methods, regional organization, and species focus of salmon fisheries. Archaeological histories of fisheries in different regions now regularly show varied patterns, combinations, and shifts over time in the organization of salmon fishing (McMillan et al. 2008; Mitchell 1988; Monks, this volume; Orchard, this volume). As generations of fisheries biologists have learned in the course of decades of intensive research, the study of Pacific Northwest salmon ecology is enormously complex (Groot and Margolis 1991). It is not surprising that the study becomes even more complex when millennia of human histories are also taken into account. As research continues and methods are developed and refined, archaeology will continue to acquire the data and the tools necessary to resolve some of the broader trends from the increasingly complex picture that is now just beginning to emerge. We see aDNA identification of species and regional multisite investigation of salmon fisheries as just part of this overall trend in Northwest Coast archaeology. Further application of these and other methods of investigation will undoubtedly continue to reveal new dimensions in our understanding of the long-term histories of salmon and other fisheries.

Acknowledgments

Site investigations in the Namu vicinity and at Rivers Inlet were supported by SSHRC grants to Aubrey Cannon. Ancient DNA analysis of salmon vertebrae was supported by SSHRC and SSHRC-RDI grants to Dongya Yang.

References Cited

Archer, David J. W. "Village Patterns and the Emergence of Ranked Society in the Prince Rupert Area." In *Perspectives on Northern Northwest Coast Prehistory*, Jerome S. Cybulski, ed., pp. 203–219. Archaeological Survey of Canada, Mercury Series 160. Hull: Canadian Museum of Civilization, 2001.

Barta, Jodi Lynn. "Addressing Issues of Domestication and Cultural Continuity on the Northwest Coast Using Ancient DNA and Dogs." PhD diss., McMaster University, 2006.

Binford, Lewis R. *Nunamiut Ethnoarchaeology*. New York: Academic Press, 1978.

Butler, Virginia L. "Resource Depression on the Northwest Coast of North America." *Antiquity* 74 (2000): 649–661.

Butler, Virginia L., and James C. Chatters. "The Role of Bone Density in Structuring Prehistoric Salmon Bone Assemblages." *Journal of Archaeological Science* 21 (1994): 413–424.

Cannon, Aubrey. "Radiographic Age Determination of Pacific Salmon: Species and Seasonal Inferences." *Journal of Field Archaeology* 15 (1988): 103–108.

———. *The Economic Prehistory of Namu: Patterns in Vertebrate Fauna*. Burnaby, BC: Department of Archaeology, Simon Fraser University, Publication 19, 1991.

———. "The Ratfish and Marine Resource Deficiencies on the Northwest Coast." *Canadian Journal of Archaeology* 19 (1995): 49–60.

———. "Scales of Variability in Northwest Salmon Fishing." In *Prehistoric Hunter-Gatherer Fishing Strategies*, Mark G. Plew, ed., pp. 25-40. Boise, ID: Boise State University Monograph Series, 1996.

———. "Assessing Variability in Northwest Coast Salmon and Herring Fisheries: Bucket-Auger Sampling of Shell Midden Sites on the Central Coast of British Columbia." *Journal of Archaeological Science* 27 (2000a): 725–737.

———. "Settlement and Sea-Levels on the Central Coast of British Columbia: Evidence from Shell Midden Cores." *American Antiquity* 65 (2000b): 67–77.

———. "Was Salmon Important in Northwest Coast Prehistory?" In *People and Wildlife in Northern North America: Essays in Honor of R. Dale Guthrie*, S. Craig Gerlach and Maribeth S. Murray, eds., pp. 178–187. Oxford: BAR International Series S944, 2001.

———. "Sacred Power and Seasonal Settlement on the Central Northwest Coast." In *Beyond Foraging and Collecting: Evolutionary Change in Hunter-Gatherer Settlement Systems*, Ben Fitzhugh and Junko Habu, eds., pp. 311–328. New York: Kluwer-Plenum, 2002.

———. "Long-term Continuity in Central Northwest Coast Settlement Patterns." In *Archaeology of Coastal British Columbia: Essays in Honour of Professor Philip Hobler*, Roy L. Carlson, ed., pp. 1–15. Burnaby, BC: Archaeology Press, Simon Fraser University, 2003.

Cannon, Aubrey, and Meghan Burchell. "Clam Growth-Stage Profiles as a Measure of Harvest Intensity and Resource Management on the Central Coast of British Columbia." *Journal of Archaeological Science* 36 (2009): 1050–1060.

Cannon, Aubrey, Meghan Burchell, and Rhonda Bathurst. "Trends and Strategies in Shellfish Gathering on the Pacific Northwest Coast of North America." In *Early Human Impact on Megamolluscs*, A. Antczak and R. Ciprian, eds., pp. 7–22. BAR International Series. Oxford: Archaeopress, 2008.

Cannon, Aubrey, and Nadia Densmore. "A Revised Assessment of Late Period (AD 1–European Contact) Fisheries at Namu, British Columbia." *Canadian Zooarchaeology* 25 (2008): 3–13.

Cannon, Aubrey, Henry P. Schwarz, and Martin Knyf. "Marine-based Subsistence Trends and the Stable Isotope Analysis of Dog Bones from Namu, British Columbia." *Journal of Archaeological Science* 26 (1999): 399–407.

Cannon, Aubrey, and Dongya Y. Yang. "Early Storage and Sedentism on the Pacific Northwest Coast: Ancient DNA Analysis of Salmon Remains from Namu, British Columbia." *American Antiquity* 71 (2006): 123–140.

Carlson, Roy L. "The Early Period on the Central Coast of British Columbia." *Canadian Journal of Archaeology* 3 (1979): 211-228.

———. "Early Namu." In *Early Human Occupation in British Columbia*, Roy L. Carlson and Luke Dalla Bona, eds., pp. 83–102. Vancouver: UBC Press, 1996.

Coupland, Gary, Andrew R. C. Martindale, and Susan Marsden. "Does Resource Abundance Explain Local Group Rank among the Coast Tsimshian?" In *Perspectives on Northern Northwest Coast Prehistory*, Jerome S. Cybulski, ed., pp. 223–248. Archaeological Survey of Canada, Mercury Series Paper 160. Hull: Canadian Museum of Civilization, 2001.

Coupland, Gary, Kathlyn Stewart, and Katherine Patton. "Do You Never Get Tired of Salmon? Evidence for Extreme Salmon Specialization at Prince Rupert Harbour, British Columbia." *Journal of Anthropological Archaeology* 29 (2010): 189–207.

Easton, N. Alexander. "The Archaeology of Straits Salish Reef Netting: Past and Future Research Strategies." *Northwest Anthropological Research Notes* 24 (1990): 161–177.

Fisheries and Oceans Canada. Statistical Areas—North Coast. http://www.pac.dfo-mpo.gc.ca/northcoast/Areas/Area8/, 2001.

Ford, Pamela J. "Archaeological and Ethnographic Correlates of Seasonality: Problems and Solutions on the Northwest Coast." *Canadian Journal of Archaeology* 13 (1989): 133–150.

Grier, Colin. "Dimensions of Regional Interaction in the Prehistoric Gulf of Georgia." In *Emerging from the Mist: Studies in Northwest Coast Culture History*, R. G. Matson, Gary Coupland, and Quentin Mackie, eds., pp. 170–187. Vancouver: UBC Press, 2003.

Groot, C., and L. Margolis, eds. *Pacific Salmon Life Histories*. Vancouver: UBC Press, 1991.

Heard, William R. "Life History of Pink Salmon (*Oncorhynchus gorbuscha*)." In *Pacific Salmon Life Histories*, C. Groot and L. Margolis, eds., pp. 121–230. Vancouver: UBC Press, 1991.

Hoffman, Brian W., Jessica M. Z. Czederpiltz, and Megan A. Partlow. "Heads or Tails: The Zooarchaeology of Aleut Salmon Storage on Unimak Island, Alaska." *Journal of Archaeological Science* 27 (2000): 699–708.

Langdon, Stephen J. "Tidal Pulse Fishing: Selective Traditional Tlingit Fishing Techniques on the West Coast of the Prince of Wales Archipelago." In *Traditional Ecological Knowledge and Natural Resource Management*, Charles R. Menzies, ed., pp. 21–46. Lincoln: University of Nebraska Press, 2006.

———. "Sustaining a Relationship: Inquiry into the Emergence of a Logic of Engagement with Salmon among the Southern Tlingit." In *Native Americans and the Environment: Perspectives on the Ecological Indian*, Michael E. Harkin and David Rich Lewis, eds., pp. 233–273. Lincoln: University of Nebraska Press, 2007.

Lepofsky, Dana, Ken Lertzman, Douglas Hallett, and Rolf Mathewes. "Climate Change and Culture Change on the Southern Coast of British Columbia 2400–1200 Cal BP: An Hypothesis." *American Antiquity* 70 (2005): 267–293.

Lepofsky, Dana, Teresa Trost, and Jesse Morin. "Coast Salish Interaction: A View from the Inlets." *Canadian Journal of Archaeology* 31 (2007): 190–223.

Luebbers, Roger. "Excavations: Stratigraphy and Artifacts." In *Studies in Bella Bella Prehistory*, J. J. Hester and S. M. Nelson, eds., pp. 11–66. Burnaby, BC: Simon Fraser University, Publications in Archaeology No. 5, 1978.

MacDonald, Brandi Lee, R. G. V. Hancock, Alice Pidruczny, and Aubrey Cannon. "Neutron Activation Analysis of Archaeological Ochres from Coastal British Columbia." *Antiquity Online* 82 (2008): 316, http://antiquity.ac.uk/ProjGall/macdonald/index.html.

Maschner, Herbert D. G. "The Evolution of Northwest Coast Warfare." In *Troubled Times: Violence and Warfare in the Past*, D. L. Martin and D. W. Frayer, eds., pp. 267–302. Amsterdam: Gordon and Breach, 1997.

Matson, R. G. "The Evolution of Northwest Coast Subsistence." In *Long-Term Subsistence Change in Prehistoric North America*, Dale R. Croes, Rebecca A. Hawkins, and Barry L. Isaac, eds., pp. 367–428. Research in Economic Anthropology, Supplement 6. Greenwich, CT: JAI Press, 1992.

McKechnie, Iain. "Five Thousand Years of Fishing at a Shell Midden in the Broken Group Islands, Barkley Sound, British Columbia." Master's thesis, Simon Fraser University, 2005.

McKinnell, S. M., C. C. Wood, D. T. Rutherford, K. D. Hyatt, and D. W. Welch. "The Demise of Owikeno Lake Sockeye Salmon." *North American Journal of Fisheries Management* 21 (2001): 774–791.

McMillan, Alan D., Iain McKechnie, Denis E. St. Claire, and S. Gay Frederick. "Exploring Variability in Maritime Resource Use on the Northwest Coast: A Case Study from Barkley Sound, Western Vancouver Island." *Canadian Journal of Archaeology* 32 (2008): 214–238.

Mitchell, Donald. "Changing Patterns of Resource Use in the Prehistory of Queen Charlotte Strait, British Columbia." In *Prehistoric Economies of the Pacific Northwest Coast*, Barry L. Isaac, ed., pp. 245–290. Research in Economic Anthropology, Supplement 3. Greenwich, CT: JAI Press, 1988.

Moss, Madonna L., and Jon M. Erlandson. "Reflections on North American Pacific Coast Prehistory." *Journal of World Prehistory* 9 (1995): 1–45.

Olson, Ronald L. *Social Life of the Owikeno Kwakiutl*. Berkeley: University of California Anthropological Records 14(3): 213–259, 1954.

Pomeroy, John Anthony. "Bella Bella Settlement and Subsistence." PhD diss., Simon Fraser University, 1980.

Rohner, Ronald P. *The People of Gilford: A Contemporary Kwakiutl Village*. Ottawa: National Museum of Canada Bulletin No. 225, Anthropological Series No. 83, 1967.

Romanoff, Steven. "Fraser Lillooet Salmon Fishing." *Northwest Anthropological Research Notes* 19 (1985): 119–160.

Shnirelman, Victor A. "Cherchez le Chien: Perspectives on the Economy of the Traditional Fishing-Oriented People of Kamchatka." In *Key Issues in Hunter-Gatherer Research*, Ernest S. Burch and Linda J. Ellanna, eds., pp. 169–188. Oxford: Berg, 1994.

Suttles, Wayne. "Coping with Abundance: Subsistence on the Northwest Coast." In *Man the Hunter*, Richard B. Lee and Irven Devore, eds., pp. 56–68. Chicago: Aldine, 1968.

White, Elroy A. F. "Heiltsuk Stone Fish Traps: Products of My Ancestors' Labour." Master's thesis, Simon Fraser University, 2006.

Zita, Paul. "Hard Times on the Northwest Coast: Deer Phalanx Marrow Extraction at Namu, British Columbia." In *Drawing Our Own Conclusions, Proceedings of the 1997 McMaster Anthropology Society Students Research Forum*, Alexis Dolphin and Dara Strauss, eds., pp. 62–71. Hamilton, ON: McMaster Anthropology Society, 1997.

CHAPTER 6

Heiltsuk Stone Fish Traps
on the Central Coast of British Columbia

Elroy White, *Archaeology Heritage Consulting, Bella Bella, British Columbia*

This chapter documents a small part of my master's research (White 2006), which I conducted on Heiltsuk stone fish traps on the central coast of British Columbia. This ancient fishing technology is poorly represented in the archaeological literature, and the central coast had not received any archaeological attention since Pomeroy's (1980) research. Stone fish traps possess more than antiquarian interest for the Heiltsuk community. In the course of my research, I visited forty-two of the more than 250 fish traps recorded on the central coast, capturing them on video to aid in my interviews with members of the Heiltsuk community who have direct experience and memories of the use of such traps. Many Heiltsuk have been aware of these ancient fishing technologies, which their ancestors built to implement a sustainable fishery for subsistence purposes, but they lacked a forum in which to express their knowledge.

The traps vary in shape, size, and location. For this chapter, I focus on one type, the beach stone fish trap. They are the products of my ancestors' labor and represent a legacy we have inherited. During the interviews, I noticed that my consultants used *fish* in a generic sense to designate salmon species—pink, chum, sockeye, coho, and spring—but they used specific names for other varieties of fish. Using the generic term *fish* could lead to many speculations about the actual function of traps found at rivers, streams, and creek sites. In addition to salmon, fish includes ooligans, herring, cod, halibut, shiners, and perch. Therefore, the traps could theoretically have been used to capture all types of fish. In order to resolve this issue, in my interviews I included all the fish types and found that the traps in question were not multifunctional to capture all fish, and that *fish* actually meant salmon. All interview participants used the common term *fish* in reference to salmon and I was careful to ensure that they actually specified salmon.

My research was a collaborative effort conducted with the aid of twelve Heiltsuk oral historians to present a Heiltsuk perspective on their ancient and recent history through oral accounts. This is the first attempt to bring archaeology into the scope of Heiltsuk thought with the intent of promoting that perspective in our dialogue with Canadian archaeology. As an archaeologist of Heiltsuk descent, I examined local knowledge and ethnographic narratives to enhance my understanding of stone fish traps from an Internalist perspective, as discussed by Yellowhorn (2002). Although stone fish traps are classified as material culture, my research objective is unique in that I deemphasized data such as length, width, and height in favor of the view that these stone fish traps are products of my ancestors' labor. I intend to link oral history to beach stone fish trap technology in relation to the Smokehouse Days, the time in the year when Heiltsuk families relocated to their traditional campsites for at least two months to catch, process, and smoke-dry salmon for storage and redistribution.

The Role of Oral Narratives

First Nations people are interested in connecting with their ancient past. According to their oral narratives, they firmly believe their ancestors were responsible for creating the archaeological record, but they lack the opportunity and means to present their own perspectives on the past under their own terms (Trigger 2003:65). In academia, archaeology is the career choice for a growing core of First Nations students who are confident in appropriating the methods of archaeology and feel fully able to construct or contest archaeological theories. They will become researchers whose Internalist perspectives will balance the interpretations created in the mainstream (Linklater 1994; Ouellet 2005; Yellowhorn 2002). Their strength emanates from their ability to combine professional training with community collaboration to create an approach that encourages dialogue within their communities on the nature of antiquity (Yellowhorn 2002).

I collected oral accounts by appropriating the methods of ethnology to infuse my collaborative research with greater meaning. The interviews with twelve Heiltsuk elderly historians grew from my interest in contemporary oral history as a valid source of knowledge about stone fish trap location, function, operation, seasonal use, age, and ownership. The perspective presented here elaborates and expands upon previous research, especially on the beach stone fish traps used during the Smokehouse Days. Lifelong familiarity with Heiltsuk traditional territory, with oral history, and with practical wisdom and life on the coast informs this vernacular discourse. It grows from lived experience in seafood gathering, camping excursions, work in commercial enterprises, and in viewing these stone trap walls firsthand.

The Heiltsuk

Since time immemorial, Heiltsuk ancestors have occupied the deep mainland inlets, protected inner waterways, and outer islands in permanent winter villages and in small seasonal campsites. Ancestral stories narrate supernatural actions associated with the formation of villages, dwellings, people, slaves, material possessions, and intangible rights to songs, dances, and resource use areas (Hogan et al. 2005; Waterfall, Hopwood, and Gill 2001). They inform researchers about material culture such as shell middens, stone fish traps, rock art (including paintings and carvings), burials, culturally modified trees, and canoe skids.

The modern Heiltsuk population descended from a larger group of Hailhzaqv-speaking tribes. Linguistically, the Heiltsuk speak Hailhzaqvala and are part of the Northern Wakashan language family along with the Xaixais, Haisla, Oweekeno, and Kwakwaka'wakw (Black 1997). Heiltsuk is an Anglicized version of the name Hailhzaqv, which literally translates as "to speak and act correctly" (Black 1997:9). Presently, only five surviving groups are recognized and remembered: the Wuyalitxv, Yisdaitxv, Wuithitxv, Qvuqva'aitxv, and Xaixais (Hogan et al. 2005:1; Figure 6.1).

Recent generations of Heiltsuk people do as their ancestors once did in utilizing the land, salmon rivers, and the sea. In the process, they have developed a long-term familiarity with the territory that they have proudly inherited. Of particular interest for this paper are the beach stone fish traps that allude to a fishery management strategy that began in antiquity.

The Heiltsuk are descendants of observant marine-based people who acquired their primary subsistence from the sea and numerous salmon rivers and streams. Highly skilled at subsistence procurement, processing, and storage, they gathered seafood for immediate consumption and for long-term storage. Along intertidal zones and at the mouth of streambeds, Heiltsuk ancestors left evidence of their labor dedicated to the salmon fishery. Salmon was the main food resource because of the fish's predictable instinct to return to its natal spawning streams. The Heiltsuk have maintained a strong reciprocal relationship with salmon and constructed this elegant fishing technology to capture them. In the deepest layers of the Heiltsuk village at Namu, there is ample evidence that salmon fishing has a time depth of at least 7,000 years (Cannon 1991; Cannon and Yang 2006; Cannon, Yang, and Speller, this volume).

Trapping devices to catch and harvest large quantities of seafood at one time were designed to intercept seasonally abundant migrating species. In the archaeological record and ethnographic accounts, two common types are known: stone fish trap structures and wooden fish weirs. Stone wall traps and wooden weirs are widely distributed in varying quantities along the Northwest Coast. Some Northwest Coast groups built their trapping devices mainly out of wood, or built stone wall structures augmented by wood, while other trapping devices were built exclusively of beach cobbles from local sources placed in streambeds or on tidal flats. Weirs typically consist of hemlock and cedar stakes driven into the muddy sediment. Conical or funnel-shaped baskets usually form part of a weir structure or were used individually as a separate fishing implement. Stone and wooden materials were often combined (Boas 1975/1909; Moss and Erlandson 1998; Stewart 1977). My master's thesis focused on the mapping and video documentation of forty-two stone fish traps.

Four of the six Pacific salmon regularly migrate to Heiltsuk rivers, streams, and creeks. The species are chum (*Oncorhynchus keta*), pink (*O. gorbuscha*), coho (*O. kisutch*), and sockeye (*O. nerka*). Chinook (*O. tshawytscha*) no longer spawn in Heiltsuk watersheds (Jones 2000; Pomeroy 1980). Each salmon species has a set of distinctive traits such as life

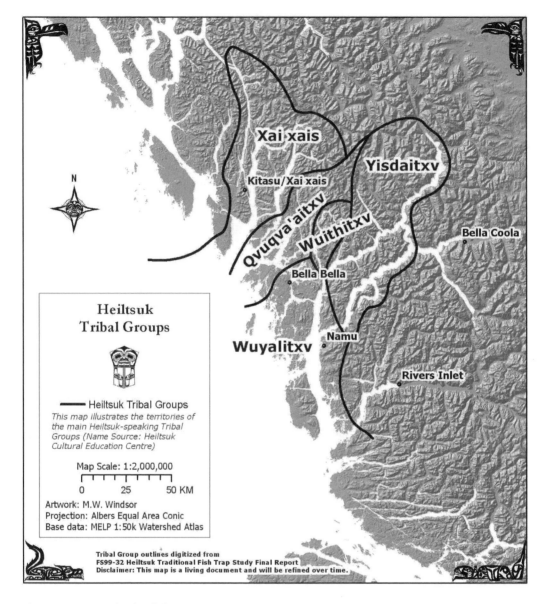

Figure 6.1. Five Heiltsuk tribal groups.

span, spawning age, season and duration of migrations, and morphological changes during spawning seasons. For this chapter, pink and chum will mainly be elaborated upon as they were the target species during the Smokehouse Days.

Heiltsuk Oral Histories of Fish Trap Use

I include below a short biography of each consultant, who is listed according to tribal group in respect to his or her family lineage origins. Each participant is the successful product of his or her ancestors' knowledge. They regularly cited their sources and often correlated knowledge to other archaeological sites that make up the material culture history of the Heiltsuk. All have participated in both the commercial and food fisheries in various capacities. Nine of the participants live in Bella Bella and three call Vancouver their home. All ages given are those at the time of the interviews in 2004.

Wuyalitxv—Outside People

Reginald Moody Humchitt is the current hereditary head chief of the Heiltsuk Nation. His family ancestry originates from Huyat (FaTa-10; R. Humchitt 2004).* His father was Charlie Moody, who shared his knowledge with James Hester's Bella Bella Prehistory project in 1968. Like his late father, Reginald shared memories about the function, use, and operation of the Gullchuk Bay trap site (FaTa-2).

    Edward Martin, Sr., was seventy-three when we sat down for our interview (Martin 2004). Unfortunately, he joined his ancestors in the spirit world on November 27, 2005. He attributed his subsistence and linguistic education to his grandmother, Emma Starr, who shared her knowledge with ethnographer Ronald Olson in 1935 and 1949. Edward was from the village of Huyat (FaTa-10), where he lived from 1935 to 1940. At Huyat, he observed his immediate and extended family members catch, process (clean), and smoke-dry thousands of lean chum salmon for their winter food supply.

    Peggy Housty is a high-ranking woman whose ancestral name comes from the story of Tsumclaqs (Cumqlaqs; Housty and Housty 2004). Peggy and her family members met with me to discuss their knowledge about this ancestral narrative and granted me permission to use her story in my research. Peggy shared her memories about the operation of the salmon trap (FaTa-2) at Gullchuk Bay.

Qvuqva'aitxv—Calm Water People

Carmen Humchitt is the hereditary chief of the Qvuqva'aitxv (C. Humchitt 2004). He was seventy-five years old at the time of the interview. He traced his ancestry to villages at Qaba (FbTb-4), Gale Creek (FbTb-13), and Gayaxti (FbTb-3). As a child in the early 1930s, Carmen observed his family using a stone trap to catch chum and pink salmon at Kunsoot River at the head of Gullchuk Bay (FaTa-2).

    Don Vickers's ancestry is from the Qvuqva'aitxv (Vickers 2004). He grew up under the mentorship of his wise granny, Magaga (Lucy Windsor), while living in a small smokehouse camp at Huyat (FaTa-10) in the early 1930s. Lucy and her husband, Charley Windsor, shared their knowledge with ethnographers Boas (1973:41) and Olson (1955). Don witnessed a non–First Nations fishery officer dismantle a stone fish trap at Huyat (FaTa-10) and shared his painful experience about the imbalance of authority and statutory appropriation of streams and salmon care.

    Emma Reid grew up at Huyat (FaTa-10) during the Smokehouse Days of the 1950s (E. Reid 2004). She lived in a sectioned smokehouse with an extended family network. She recalls Don Vickers's family living in the same house and other families living in close proximity to them. Her main recollections were about salmon processing and storage

---

* Taped interviews will be cited when the cultural expert is first introduced and for direct quotes.

after modern conveniences were introduced. Her family gathered other food items such as berries, fruits, and Indian medicines along the riverbanks and she fondly recalled cleaning salmon eggs in the river. During our conversation, Emma shared information about smaller stone fish traps. These traps are no longer visible along the shorelines.

Yisdaitxv—People of Isda

Fred Reid is from the Yisdaitxv (F. Reid 2004). His attention to detail of names, dates, places, and relationships to land, water, and people are his strong points. Fred identified many Heiltsuk elders who were his sources of knowledge. Fred correlated salmon species to Heiltsuk rivers, streams, and creeks. Stone fish traps are found at many of these locations. He shared insights about marine subsistence locations, fishing methods, food storage, and resource use and ownership.

George Housty held the chiefly name Naci, which means "wisdom place" (Housty and Housty 2004). George now carries the family name Dukaiaisla, "looking either onto or out from shore." George shared knowledge about the operation of a Gullchuk Bay beach stone fish trap (FaTa-2) at Kunsoot River.

Edward White, Sr., my father, was sixty-five years old at the time of the interview (White 2004). He holds the chief name Yaimadzalas, from his Martin family. Edward is very well informed about the salmon smoking process and has an acute relationship with his environment as a result of his education as a child in subsistence seafood gathering. Later in his life, he participated in the commercial fishing industry. He attributes his familiarity with the central coast to his uncles, grandfathers, and maternal grandmother, Annie Larsen.

Wuithitxv—People of the Inlet

Bobby Jackson was seventy-five years old at the time of the interview (Jackson 2004). His family smoke-dried both pink and chum salmon in October and November at Kadjusdis River (FaTa-46) and at Neekas (FcTa-6), where a stone fish trap can be found. He is also familiar with the salmon traps (FcSx-3) at Clatse Bay in Roscoe Inlet. His grandfather, Samuel Jackson, worked with Franz Boas, who recorded the Eagle man from La'latsa (Clatsa) story (Boas 1973:64).

Xaixais—Downriver People

Bill Wilson (Mnigalis) traces his ancestry to the Xaixais (Wilson 2004). When he was six or seven years old, around 1951–1952, his family made frequent day trips or weekend trips to Huyat (FaTa-10) in September and October. Like Emma Reid's family, Bill's family did not use the long stone fish trap (FaTa-11). His family members built smaller salmon traps near the grassy flats. These traps are no longer visible and thus have not been recorded. Bill recalled fond memories of assisting his grandmother, Magaga (Lucy Windsor), who directed all smokehouse activities, including salmon trap construction and dismantling, smoke-drying, and jarring large quantities of salmon. Lucy Windsor narrated stories to the ethnographers Franz Boas and Ronald Olson.

Ooweekeno—Rivers Inlet Nation

Evelyn Windsor (Nuakawa, or "wise one") is a high-ranking woman with the title of an Umaqs, from Rivers Inlet of the Ooweekeno Nation (Windsor 2004). She is a certified linguist and Native language instructor with considerable knowledge of culturally related topics such as potlatch ceremonies, oral history, family genealogies, medicinal plants, place-names, salmon harvesting and processing, and seafood subsistence gathering. She has worked with ethnobotanist Nancy Turner and other researchers such as anthropologist Michael Harkin and linguist John Rath. Evelyn is familiar with wooden fish weirs and stone fish traps in Rivers Inlet Lake and shared memories about her late husband, Marshall Windsor, who spent his youth at Huyat (FaTa-10).

Prey Species: Salmon

Initially, I wanted to learn about stonefish trap function, whether the target was salmon, herring, seal, cod, shiner/perch, porpoise, or clams, and to focus on technical details about stone fish traps. After reviewing the Heiltsuk interview data, salmon stood out as the main prey, followed by herring; seals, cod, shiner/perch, porpoise, and clams were not considered primary targets and may have been trapped incidentally. Edward Martin, Sr., Fred Reid, Reginald Moody Humchitt, and Carmen Humchitt are aware of trap locations for herring and shared insightful information about their operation and wall morphology. Unfortunately, I was unable to visit these traps to capture footage of them or to include them in my site mapping.

Table 6.1 lists the seasonal salmon runs to Heiltsuk streams and the preferred method of preparation, either for long-term winter use or for immediate consumption. The oilier salmon such as sockeye and coho could be smoke-dried but would have required more work. If the smoking process was not fully completed, there was the risk of the food product becoming moldy and rotten and unfit for consumption. This is based upon a consensus from the oral historians on practices prior to the introduction of modern conveniences such as deep freezers, gas boats, drag seines, and jarring and canning methods. As for the salmon migrations, some runs began earlier or occurred later than others in specific streams.

Tony Pomeroy (1976:166) boldly asserted that "almost every stream, small or large, has some type of stone fish trap associated with it." Considering more than 1,800 freshwater

Table 6.1. Heiltsuk salmon seasonal migrations, salmon smokehouse preparation, and storage techniques prior to the introduction of modern conveniences

Salmon	Migration season	Technique	Consumption
Pink	Late August, September, and into October	Smoke dry	Winter supply
Chum	September, October, and into November	Smoke dry	Winter supply
Coho	Late fall–October	Half smoke	Immediate
Sockeye	Spring–June	Half smoke	Immediate

sources empty into Heiltsuk waters, his assertion was of course an exaggeration. Only 250 traps have been documented. I can only surmise that when he said fish, he actually referred to salmon, as did the participants in my interviews. All other researchers had noted this pattern of targeting salmon. For example, Ronald Olson (1955:320–321) concluded from his research that all people of Hunter Island secured "most of their salmon in Kildidt lagoon." According to Heiltsuk consultants, there are coho and sockeye streams on the northwest side and chum on the northeast side of Kildidt lagoon.

Table 6.2 lists identifications of target species for thirty of the forty-two sites that were the focus of my videography and mapping efforts. The participants provided the information for each site and the target species. I used the following abbreviations: P = pink, Ch = chum, Co = coho, Sp = spring, and So = sockeye. The absence of data in some areas is a function of the interview process. The consultants only provided salmon and stream associations when asked about them. The information in Table 6.2 does not represent the full extent of my consultants' knowledge of the use of fish traps in the area.

My main objective was to ask questions about stone fish trap operation and their technical aspects. Carmen Humchitt, Reginald Moody Humchitt, Peggy Housty, and Bobby Jackson contributed data about the Gullchuk Bay traps. Don Vickers, Bill Wilson, George Housty, Edward Martin, Sr., Evelyn Windsor, and Emma Reid related their knowledge about the Huyat Bay traps. Fred Reid and Edward White, Sr., answered my questions about which streams provided which salmon species. Heiltsuk consultants identified the outer islands of the Wuyalitxv as having more sockeye and coho creeks and the inner waterways and inlets as having all species, especially pink and chum (Figure 6.2).

In Heiltsuk traditional territory, sockeye migrate to their natal streams in June and coho in October. These species are found in other locations in Heiltsuk traditional territory. In the inner waterways and inlets, all species are present, with pink and chum being more common, especially at Gullchuk Bay (FaTa-2, FaTa-46), Huyat Bay (FaTa-11), Clatse Bay (FcSx-3), and Port John (FaSx-1). In Evans Arm, chum is most common in streams at sites FaSw-1, FaSw-2, FaSw-3, and FaSx-8. Finally, Namu (ElSx-1) is known for its productive sockeye runs in June, followed by pink in August and September and coho in October. Rarely is there mention of chum at Namu.

The Smokehouse Days

> The most important food for the First Generation was smoked and dried salmon.... There were three salmon traps in the creek at Mauwash. (Angus Campbell [1898–1948] in Storie and Gould 1973:53)

In his own words, Heiltsuk historian Angus Campbell, whose rights come from the ancient Heiltsuk village site of Namu, informed an audience about the significance of a place-name that once was commonly used for Namu. He called it Mauwash. Campbell defined it as "you just have to go and ask me" (Storie and Gould 1973:53). According to Campbell, salmon was plentiful during the days of the First Generation of ancestors. The small river always had a continuous supply of salmon compared to the bigger rivers such as the Bella

Table 6.2. Identifications of target species at fish trap locations

Borden number	Edward Martin	Carmen Humchitt	Don Vickers	Fred Reid	Reginald Moody Humchitt	Bill Wilson	Bobby Jackson	Edward White, Sr.
FaTa-2 Kunsoot R.		P, Ch			P, Ch, Co			P, Ch
FaTa-1 Huyat	P, Ch, Co		So	So	P, Ch, So, Co	P, Ch, So, Co		P, Ch, So, Co
FaTa-34	Co							Co
FaTa-46		So, Co			So		Ch, Co, So	P, Ch, So, Co
FaTa-69								So
FaTa-70	Co				Co			Co
FaTa-71								Co
FbTa-3								P, Ch, So, Co
FbTa-8								P, Ch
FbTa-9								P, Ch
FbSx-2								Ch
FcSx-3					P, Ch			P, Ch,
FdSx-1								P, Ch,
FaSx-1				Ch	P, Ch, So, Co			P, Ch, So, Co
FaSw-1				Ch				Ch
FaSw-2				Ch				Ch
FaSw-3				Ch				Ch
ElTb-2				So, Co				So, Co
ElTb-10								So
ElTb-15								So, Co
ElTb-18								So
ElTb-20								So, Co
ElTb-21								So, Co
ElTb-32	So, Co		Co	So, Co				So, Co
ElTb-33	So		Co	So, Co				So, Co
ElTa-3								Co
ElTa-6	Ch							Ch
ElTa-8	So, Co		Co	Co				So, Co
ElTa-19	Ch		Ch					Ch
ElSx-1 Namu					P, So			P, So, Co

P = pink, Ch = chum, Co = coho, Sp = spring, and So = sockeye

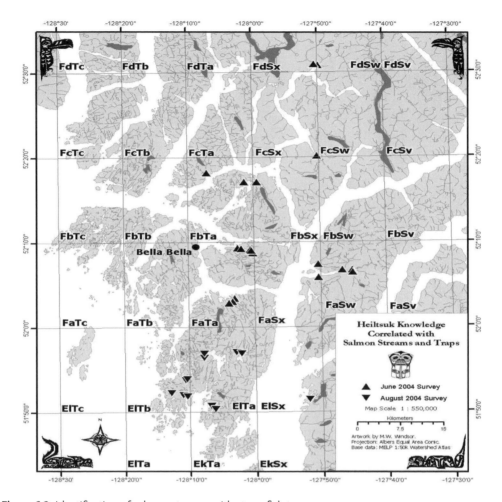

Figure 6.2. Identification of salmon streams with stone fish traps.

Coola, Rivers Inlet, Kitimaat, Nass, and Skeena. The people who relied on their rivers for their salmon approached Campbell's ancestors, asking for permission to obtain their salmon supplies, use their stone fish traps, and, apparently, smoke-dry their salmon. This one Heiltsuk narrative links salmon, salmon traps, and the smoke-drying process as part of an elaborate time-honored tradition of the Smokehouse Days.

Many elderly Heiltsuk fondly recall the Smokehouse Days as an important time in their young lives. Semicircular beach stone fish traps were still used to capture salmon, before the advent of modern conveniences such as deep freezers, gas boats, drag seines, and jarring and canning methods. The capture process relied on the daily tidal rhythms as salmon tended to school together over the tidal flats. In August, September, and October, when families used these traps, there was always a pool of water inside the enclosures that produced a holding pen for live salmon.

Prior to 1935, the Heiltsuk consultants recalled, the stone walls of fish traps reached heights of 90 to 120 cm. Subsequently, due to siltation build-up or to sinking action or both, these stone walls are now much lower, with measured heights of 46 to 51 cm. Therefore, within seventy years, there has been a loss of at least 60 to 90 cm of height. At the time of operation, these walls must have been visually impressive, with wide flat bases tapering to the narrow tops. On a recent field trip to Gullchuk Bay, I noticed the disappointment on Edward White's face when he saw how low the walls were now. He said, "The walls were higher, at least four feet. They have been washed down or someone moved them" (White 2004).

Typically, for eight months of the year, these traps lay unused as Heiltsuk families traveled in spring and early summer to their different campsites. They would gather a variety of important seafood provisions of herring, seaweeds, shellfish, and cod. In August, Heiltsuk families slowly made their way back to these smokehouse campsites in the inner waterways and into the deep meandering inlets to intercept the pink and chum migrations. Some families stayed from August until the first week of November. Carmen Humchitt (2004) recalled, "One time we stayed for a whole month; one time we stayed there until after Halloween."

As salmon mill about the tidal flats, their silvery bodies begin to physically change. The males of each species form large humps on their backs and their jaws take on a hook-billed appearance reminiscent of dogs. Females do not undergo as drastic a physical change. Streamlined bodies slowly become thinner as their oil content decreases. Color changes are notable as the shiny, bright texture fades to dull dark colors of black, red, green, and purple. These desirable traits reveal the clues for salmon in perfect condition for smoke-drying and hence selection from the schools in the traps.

Huyat Bay and Gullchuk Bay were particularly important locales during the Smoke-house Days. At Huyat Bay (FaTa-10) at least five to six small smokehouses that measured on average 7.3 by 9.1 m once stood along these shorelines on the west side of the river. Family members were interested in catching, processing, and smoke-drying thousands of lean pink and chum. Each summer the families of Albert Humchitt, Johnny White, Charley and Lucy Windsor, Paul Brown, William Dixon, and Albert Cuyler called Huyat their home.

These extended families lived together for at least five to six months in small-sectioned houses that served both as living quarters and salmon-smoking facility. George Housty recalled that "they smoked fish in a big smokehouse with rooms in each corner for different families. The rooms were not sectioned off. They were large enough to sleep in" (Housty and Housty 2004). Housty described triangular (gabled) roofs with ventilation openings that allowed the smoke to filter out. Some smokehouses had three fires whereas others only had one. The fires did not produce open flames; their main objective was to create continuous smoke, an important factor for smoke-drying. Edward Martin (2004) emphasized that "the fires never went out and I seen fourteen hundred salmon smoking at one time."

Edward Martin and Edward White both mentioned that there was so much salmon that it was easy to capture them. Edward Martin, Sr. (2004), noted that "most of the time there were little drag seines or simple little gill nets. You didn't need…a massive net because there were so much fish [salmon]." He also recalled that they "never had to worry about working hard to trap them [salmon] because they were so much of them." Carmen Humchitt's (2004) observations attested to the efficiency of the trap walls when salmon were plentiful: "It used to be just plugged with fish [salmon]! Fish [salmon] were so easy to trap!"

At Gullchuk Bay, numerous smokehouses once lined the entire inlet from the entrance where the Kadjusdis River empties. Families built their smokehouses on the ancient village that takes the name of the river. Carmen Humchitt, Adam Dixon, Louie Hall, Willie West, Nathan Wilson, Charlie Moody, and Susan Campbell lived here for

Figure 6.3. Location of opening on FaTa-2 on the curved part of the stone wall observed at a rising tide. Inside these walls, salmon were trapped alive and lean ones were selected for smoke drying.

months at a time. Five beach stone fish traps are located here, and Carmen Humchitt, Reginald Moody Humchitt, Peggy Housty, and Bobby Jackson shared their recollections about the operation of these traps.

Operating a Beach Stone Fish Trap

[T]hey are pretty simple rig [beach stone fish trap] to operate given the date [early 1930s] that they were used when there were thousands of salmon when I was there. (Edward Martin, Sr., 2004)

Typically a stone trap wall should have at least one or two openings along its length at specific intervals that were intentionally placed there at the end of the previous summer. The openings consist of 1 m long gaps from which all stones were pushed away to the outside (Figure 6.3). This intentional act was part of a time-honored Heiltsuk conservation process to provide an escape route for any salmon when the traps were not in use. When the season ended in the first week of November, one of the last tasks for the men to perform was to provide an escape route.

Bobby Jackson and Billy Wilson used the term "gate" for these gap openings. Initially, this confused me, since gate implied a wooden fence being placed in these open-

Figure 6.4. Semicircular beach stone fish trap at Huyat Bay (FaTa-11).

ings. According to Bobby Jackson, a gate is a pile of rocks that were removed quickly and temporarily. This responsibility fell onto his uncles and older men, who simply kicked about ten to fifteen stones outwards from the wall. Bobby Jackson (2004) described it thus: "[T]ide brings fish in, and when tide rises, fish [salmon] swim out. Open a gate and let rest of fish [salmon] out again otherwise they get rotten."

At Huyat, I was surprised to learn that this impressive semicircular beach stone fish trap (FaTa-11) was not used at all during the Smokehouse Days (Figure 6.4). Edward Martin told me that they were always aware of its presence. Emma Reid and Bill Wilson recalled the adults making smaller stone fish trap versions near the grass flats and on the beachfront that parallels the smokehouse campsite. The families used these shorter semi-circular stone walls more as a containment area to deposit their daily catch of salmon. They became pools of salmon inside an enclosed area near the smokehouse buildings. Bill Wilson (2004) called it a "trap inside the trap" as he noted:

> I remember the big one (FaTa-11). It was part of the main one but they just blocked it off, make small portion. There were many formations there, smaller ones were within the big one, not always left up, took rocks from the main one. Traps were little offshoots. Walls not very high.

Emma Reid informed me about the different stone wall formations she was familiar with and how the men created openings in the wall. She stated:

The one [trap] we had close to the house here…have to be right close to the grass line. They just rolled them, rolled them away from this one. And there is some they didn't touch. And the littler rocks just moved or tossed out into the river and guess they picked them up again. (E. Reid 2004)

All consultants attributed their subsistence education about the Smokehouse Days to their grandparents. For example, Bill Wilson (2004) recalled the following about his grandmother, Lucy Windsor (Magaga), who related several stories and oral history to Boas (1973) and Olson (1955):

Great granny [Magaga] observed salmon, could estimate if going to catch lots or hardly any fish [salmon]. When tide went out, used just go inside the walls and pick the fish [salmon] out. Put fish [salmon] into wooden baskets and brought to the campsite.

The stone fish traps represent expansive holding areas or pens for live salmon, which became trapped inside the walls at lowering tides. Since the traps never dried up, the enclosed area became a virtual pool of live salmon in all stages of spawning maturation. Peggy Housty stated that "the inside never dried up at all, fish released at high tide. Salmon never dried up. Little bit of water inside the walls" (Housty and Housty 2004). The traps caught live salmon, though the occasional dead one would turn up. Reginald Moody Humchitt (2004) shared this information with me:

Trap never dried. Never used the whole area inside the walls. The further down you go, the fish [salmon] school tighter, didn't have to run around too much, fish [salmon] were tight in a ball.

Some were healthy, others were near spawning time, and others were near death. Since the traps did not discriminate, the families chose or selected only the lean chum salmon for smoke-drying. Desirable characteristics were that the fish were thin (lean) and dark-colored. The objective was to search for the salmon with these characteristics to remove from the trap. The lean salmon smoked easily in five to seven days, providing the smoldering fires continued to burn. All oil and moisture was completely removed, leaving the salmon dry and easily storable in any condition without worrying about waste or spoilage. All historians emphasized important factors for successful smoking procedures. Evelyn Windsor (2004) said:

Everything had to be dried when the fish were lean, no more fat and because if they still had fat, or got it too early, then, it would probably mould and go rancid. And it was put away after it was dried, smoked-dried, and it was put away um, ah, cedar boxes for storages for the winter.

Conclusion

Rather than focusing on gathering and analyzing the metric data on the stone fish traps from a purely scientific approach, my priority as a Heiltsuk member myself, familiar with many of the stone fish traps on central coast, was to record the oral history from family members who once used these traps at the smokehouse campsites. My ancestors inherited a simple physical legacy for the capture of migrating salmon. Their ancestors designed the stone wall structures to minimize maintenance and to maximize efficiency in the capture of thousands of migrating salmon. Rather than reinventing the wheel, the modern Heiltsuk continued a time-honored tradition of utilizing the semicircular beach stone traps that impressively occupy vast tidal flats next to the salmons' natal streams.

The accounts from twelve oral historians provide a relatively unknown perspective in the academic realm, but it is common knowledge among the modern Heiltsuk and their First Nations neighbors. The smoke-drying process was laborious and time intensive, requiring cooperation from all family members. The main objective was to smoke-dry thousands of salmon during the salmon migration season at family-owned smokehouse campsites. In order to sustainably capture so many salmon at one time, the modern Heiltsuk inherited a simple selective fishery system that began in antiquity, the semicircular beach stone fish trap.

References Cited

Black, Martha. *Bella Bella: A Season of Heiltsuk Art*. Toronto: Royal Ontario Museum, 1997.

Boas, Franz. *The Kwakiutl of Vancouver Island*. The Jesup North Pacific Expedition 5(2): 301–522. New York: AMS Reprint, 1975 [1909].

———. *Bella Bella Tales*. The American Folk-lore Society. Millwood, NY: Kraus Reprint, 1973 [1932].

Cannon, Aubrey. *The Economic Prehistory of Namu: Patterns in Vertebrate Fauna*. Burnaby, BC: Department of Archaeology, Simon Fraser University, Publication 19, 1991.

Cannon, Aubrey, and Dongya Y. Yang. "Early Storage and Sedentism on the Pacific Northwest Coast: Ancient DNA Analysis of Salmon Remains from Namu, British Columbia." *American Antiquity* 71 (2006): 123–140.

Hogan, Philip, Kelly Brown, Eric Eno Tamm, and Karen Peachy, eds. "For Our Children's Tomorrows." Executive Summary of the Heiltsuk Land Use Plan, MS on file, Heiltsuk Tribal Council, Waglisla (Bella Bella), BC, 2005.

Housty, George, and Peggy Housty. Taped interview with Elroy White, September 2004. Audiotape and transcript on file, Heiltsuk Cultural Education Centre, Waglisla (Bella Bella), BC, 2004.

Humchitt, Carmen. Taped interview with Elroy White, September 2004. Audiotape and transcript on file, Heiltsuk Cultural Education Centre, Waglisla (Bella Bella), BC, 2004.

Humchitt, Reginald Moody (Toby). Taped interview with Elroy White, September 2004, Vancouver. Audiotape and transcript on file, Heiltsuk Cultural Education Centre, Waglisla (Bella Bella), BC, 2004.

Jackson, Robert (Bobby). Taped interview with Elroy White, September 2004. Audiotape and transcript on file, Heiltsuk Cultural Education Centre, Waglisla (Bella Bella), BC, 2004.

Jones, Jim. "'We Looked After All the Salmon Streams,' Traditional Heiltsuk Cultural Stewardship of Salmon and Salmon Streams: A Preliminary Analysis." Master's thesis, University of Victoria, 2000.

Linklater, Eva Mary Mina. "The Footprints of Wasahkacahk: The Churchill River Diversion Project and Destruction of the Nelson House Cree Historical Landscape." Master's thesis, Simon Fraser University, 1994.

Martin, Edward, Sr. Taped interview with Elroy White, September 2004. Audiotape and transcript on file, Heiltsuk Cultural Education Centre, Waglisla (Bella Bella), BC, 2004.

Moss, Madonna L., and Jon M. Erlandson. "A Comparative Chronology of Northwest Coast Fishing Features." In *Hidden Dimensions: The Cultural Significance of Wetland Archaeology*, Kathryn Bernick, ed., pp. 180–198. Vancouver: UBC Press, 1998.

Olson, Ronald L. *Notes on the Bella Bella Kwakiutl.* Berkeley: University of California Anthropological Records 14(5), 1955.

Ouellet, Richard Andre. "Tales of Empowerment: Cultural Continuity within an Evolving Identity in the Upper Athabasca Valley." Master's thesis, Simon Fraser University, 2005.

Pomeroy, John Anthony. "Stone Fish Traps of the Bella Bella Region." In *Current Research Reports,* Roy L. Carlson, ed., pp. 165–173. Burnaby, BC: Department of Archaeology, Simon Fraser University, Publication 3, 1976.

———. "Bella Bella Settlement and Subsistence." PhD diss., Simon Fraser University, 1980.

Reid, Emma. Taped interview with Elroy White, September 2004. Audiotape and transcript on file, Heiltsuk Cultural Education Centre, Waglisla (Bella Bella), BC, 2004.

Reid, Fred. Taped interview with Elroy White, September 2004. Audiotape and transcript on file, Heiltsuk Cultural Education Centre, Waglisla (Bella Bella), BC, 2004.

Stewart, Hilary. *Indian Fishing: Early Methods on the Northwest Coast.* Seattle: University of Washington Press, 1977.

Storie, Susanne, and Jennifer Gould, eds. "Bella Bella Stories, Told by the People of Bella Bella." MS on file, British Columbia Indian Advisory Committee, Victoria, BC, 1973.

Trigger, Bruce. *Artifacts and Ideas: Essays in Archaeology.* New Brunswick, NJ: Transactions Publishers, 2003.

Vickers, Don (Jasper). Taped interview with Elroy White, September 2004. Audiotape and transcript on file, Heiltsuk Cultural Education Centre, Waglisla (Bella Bella), BC, 2004.

Waterfall, Pauline (Hilistis), Doug Hopwood, and Ian Gill. "People and the Land." In *North of Caution: A Journey through the Conservation Economy on the Northwest Coast of British Columbia*, by Ian Gill, Terry Glavin, Doug Hopwood, Richard Manning, Ben Parfitt, Alex Rose, and Pauline Waterfall, pp. 101–127. Vancouver: Ecotrust Canada, 2001.

White, Edward, Sr. Taped interview with Elroy White, September 2004, Ladner, BC. Audiotape and transcript on file, Heiltsuk Cultural Education Centre, Waglisla (Bella Bella), BC, 2004.

White, Elroy. "Heiltsuk Stone Fish Traps: Products of My Ancestors' Labour." Master's thesis, Simon Fraser University, 2006.

Wilson, Bill. Taped interview with Elroy White, September 2004, Tsawwassen, BC. Audiotape and transcript on file, Heiltsuk Cultural Education Centre, Waglisla (Bella Bella), BC, 2004.

Windsor, Evelyn. Taped interview with Elroy White, September 2004. Audiotape and transcript on file, Heiltsuk Cultural Education Centre, Waglisla (Bella Bella), BC, 2004.

Yellowhorn, Eldon. "Awakening Internalist Archaeology in the Aboriginal World." PhD diss., McGill University, 2002.

CHAPTER 7

Riverine Salmon Harvesting and Processing Technology in Northern British Columbia

Paul Prince, *Grant MacEwan University*

For people living inland along the major Pacific rivers of northern British Columbia, such as the Nass and Skeena, salmon were and are the crucial staple (Figure 7.1). Nothing else is as abundant, predictable, and storable, as archaeologists formerly assumed was the case for salmon throughout the Northwest Coast (Ames 1994; Matson 1992). Ethnographically, the Tsimshian, Nisga'a, Gitksan, and Witsuwit'en people situated their lives to best advantage around this resource. From an archaeological perspective, however, the importance of salmon in much of this riverine environment is difficult to detect. I will argue that while faunal evidence for a salmon-based economy is inadequate, indirect evidence of harvesting, and especially processing technologies, is abundant throughout the region's prehistory. The chronological scale at which I am able to discuss these variables is crude, but the evidence suggests that salmon fishing was always a focal point. This result informs both regional and pan-Pacific discussions of when intensive salmon fishing may have begun, and what the various kinds of evidence for it are.

Importance of Salmon in Riverine Economies

The peoples of the Skeena and Nass valleys used a wide range of plant and animal foods from terrestrial and aquatic environments (Daly 2005; Halpin and Seguin 1990; 'Ksan 1980; Millennia Research 1995). Salmon, however, was clearly the focal point that provided food

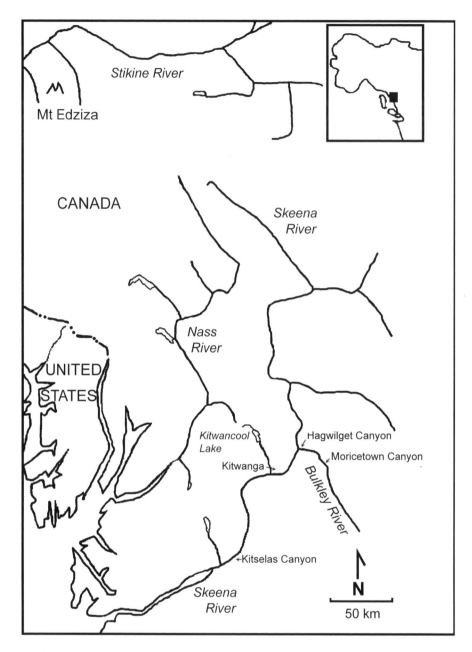

Figure 7.1. Skeena and Nass River locations discussed in text.

surpluses and supported the characteristic large villages occupied on a nearly permanent basis during the historic period.

I have argued elsewhere (Prince 2005) that the importance of salmon is indicated by the contingencies people developed to deal with the threat of a failed salmon run. The best-known example of such failure is a landslide in the 1820s that blocked salmon from reaching Moricetown Canyon on the Bulkley River (Daly 2005:134). The solution for the Witsuwit'en was for some of their house groups to negotiate rights with their Gitksan allies to establish a village downstream at Hagwilget Canyon. On a broader scale, Adams (1973) explained the entire network of relationships between Gitksan lineages and local groups and their associated kinship, territory, and feasting systems as satisfying a need to move people and resources from areas of shortfall to plenty. The resulting fluidity of house groups is a common theme of the oral traditions of the area, which are also a code for moral conduct toward animals, including salmon.

The human ecology of the Skeena and Nass watersheds is presented as though it were in a stable equilibrium (Ames 1979a; Daly 2005). But stories of shortfalls and the morality of proper respect for salmon remind us of the serious risks involved in anticipating the return of resources that must travel a long distance and are vulnerable to stream blockages, water temperature, and nutrient levels affecting their spawning and rearing habitats. The cultural response was to find ways of maintaining an intensive focus upon salmon, supplemented by many other resources. This was accomplished by situating settlements near a diversity of spawning runs, establishing social networks that provided physical access to different watersheds or the salmon harvested from them, and developing technologies of mass capture, processing, and storage. From an archaeological perspective, an important task has been to identify such economic, social, and technological adaptations and explain their development (Ames 1998; Coupland 1988).

Intensive Salmon Fishing

Any archaeological consideration of salmon-focused economies provokes more general questions about intensive resource use and acceptable evidence for it. Virginia Butler (2008) raised these questions in her discussion of the session that was the genesis of this volume, and their relevance to my chapter is briefly considered here. Binford (2001:210) defines subsistence intensification as "tactical or strategic practices that tend to increase the exploitation of resources in circumscribed or shrinking food patches." Butler herself sees intensification as a matter of increasing foraging productivity through human agency, including technologies of mass capture, processing, and storage, and the logistical organization of labor and habitat use (Butler and Campbell 2004:328). In either case, the net effect is a concentration upon a particular resource, by virtue of the characteristics of the resource and people's adaptation to it. In the case of riverine salmon fishing, Binford's circumscribing factors include the salmon's drive to return to specific natal waters at predictable times, and their concentration at major constrictions in their migratory routes, such as canyons, waterfalls, and rapids, where they can be caught in large numbers (Kew 1992).

Specific archaeological evidence used to argue for intensive salmon fishing in the greater North Pacific includes fishing weirs (Moss, Erlandson, and Stuckenrath 1990) and net sinkers (Fitzhugh 2002:272), indicating mass capture; specialized tools such as slate knives and ulus (Fitzhugh 2002:271; Graesch 2007) and microblades or microliths (Matson 1992), indicating efficient large-scale processing and organization of labor; and storage facilities, including bentwood boxes, cache pits, and planked houses (Ames and Maschner 1999:140–141), which are taken to indicate sedentism and a residential focus on salmon fishing areas. Archaeologists routinely refer to such characteristics to distinguish intensive salmon fishing economies from the mere use of salmon, although these characteristics are often difficult to discern archaeologically. Typically, intensive salmon fishing is considered relatively recent and linked to resource ownership, control of labor, and other aspects of "cultural complexity," while the use of salmon has a much longer history (Ames 1994; Ames and Maschner 1999; Matson and Coupland 1995).

Butler and Campbell (2004:329) do not question the validity of steplike evolutionary models for the development of technologies and organizational principles requisite to intensive salmon-based economies, only the circularity of arguing from the basis of limited evidence. Instead, Butler and Campbell argue for basing interpretations upon rigorous faunal analysis. Cannon and Yang (2006), however, explicitly questions the logic of assuming that mass harvesting and storage of salmon (intensive exploitation) lagged significantly behind the development of salmon use, preceded sedentism, or is linked in a causal way to social inequality. Cannon has repeatedly argued that such variables should be examined as a matter of historical contingency in local and regional instances, rather than at a culture area level (Cannon 1996, 1998). Like Butler and Campbell, Cannon cautions against using indirect evidence to distinguish intensive salmon fishing from more casual use, and favors using multiple lines of direct faunal evidence (Cannon and Yang 2006). In the case at hand, though, faunal evidence for fishing is poor (as explained below), and multiple lines of indirect evidence must be used instead. I argue below that doing so reveals the difficulty of characterizing intensive salmon fishing and casts doubt on expectations that it was evolutionarily dependent upon storage, sedentism, and, in this case, coastal influences.

Faunal Evidence

Very few sites have been excavated in the middle to upper parts of the Skeena watershed, and none in the Nass. Faunal preservation at such sites is very poor, and even at major fishing locations there is sparse mammal bone and virtually no fish bone. This is likely due to a combination of acidic soil and salmon bone disposal practices, which historically included burning and discard in the rivers ('Ksan 1980:30). Even where eighth-inch mesh screens were used, hardly any salmon bone was recovered (Prince 2007; Traces 2005). The most robust salmon assemblage is from eighteenth-century house floor hearths at the Kitwanga Hill Fort, GgTa-1 (Figure 7.2). Here, salmon was the predominant taxon by NISP. This is especially significant at this site because (1) the remains of large mammals were rare, (2) small mammals are not significantly meatier than fish, and (3) small mammals may have been more important for their pelts in the fur trade than for food.

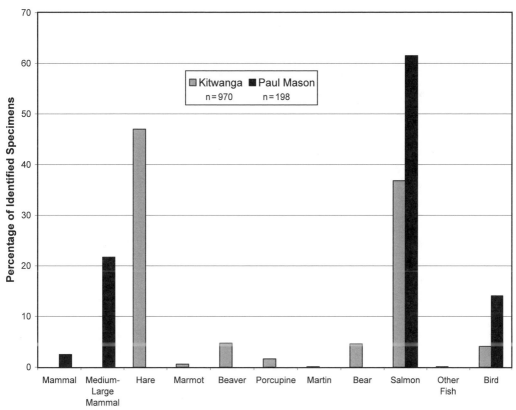

Figure 7.2. Major faunal categories identified from the Kitwanga Hill Fort and Paul Mason site house floor assemblages.

A comparable collection is from house floors at the Paul Mason Site in Kitselas Canyon with radiocarbon dates ranging from 3130 ± 100 BP to 2750 ± 90 BP (Coupland 1988:146). There, 94.6 percent of the specimens are unidentified due to fragmentation, leaving only 198 identified specimens (Coupland 1988:381–382). Of these, fish, mostly salmon, comprise the majority. Considering only the hearths of the houses, 80 percent of the identified specimens are salmon (Coupland 1993:56). In both the Kitwanga Fort and Paul Mason site cases, although 2,500 years apart, salmon remains are dominant. Otherwise, faunal evidence has not been very informative of subsistence throughout the region, and some other means of gaining an appreciation of the salmon economy is required. I will thus introduce some other indications of salmon harvesting and processing.

Harvesting Facilities and Site Locations

Various mass harvesting strategies were employed in this area in historic times, including wooden weirs, boulder impoundments and basket traps in shallow water, and dip-netting, gaffing, or lowering baskets into eddies at canyon rapids and falls (Daly 2005; 'Ksan 1980). Archaeologically, several weirs are known from the outlet of Kitwancool Lake dated between 770 ± 40 BP and the early twentieth century (Prince 2005:71). These provide relatively recent evidence of intensive fishing.

Site locations provide less direct evidence for mass harvesting (Fitzhugh 2002:277). The surveyed portions of major canyons such as Kitselas, Kitsumkalum, and Hagwilget have very high densities of archaeological sites (Archer 1987; Carlson and Bussey 1990; Coupland 1988), which undoubtedly reflect salmon harvesting, processing, and storage activities, although most are undated.

Canyons are also notably important as locations for villages, from the roughly 3,000-year-old Paul Mason village at Kitselas to the numerous Gitksan and Witsuwit'en villages of the historic period (Coupland 1988; Prince 1998, 2000). These villages reflect the abundance and ease of harvesting salmon at canyons. Other historic villages are also located in areas of salmon abundance, and, in the case of most Gitksan settlements, tend towards stream confluences where residents could fish both the main stem of the river and a tributary, providing some insurance against shortfalls in the returns of a local stream (Ames 1979a). By contrast, the historic period Witsuwit'en and their Babine Carrier neighbors were dependent on a single stream, subject to occasional shortfalls. In response, they maintained smaller villages and exercised seasonal mobility (Daly 2005:132–139).

Exceptions to the Gitksan pattern are the Gitanyow settlements, located up the Kitwanga branch of the Skeena. I have argued that the Kitwanga River was a suitable focal point for village settlement for at least 1,500 years owing to the diversity of species and individual populations of salmon that return to its waters (Prince 2005).

Salmon Processing Technology

While settlement locations may reflect the importance of intensive salmon harvesting, few of the recovered artifacts directly relate to fishing. Bone leister prongs are known from Moricetown Canyon (Traces 2005), and a single harpoon was recovered from Kitwanga Hill Fort (Prince 2001), but most fishing gear was likely made of wood, bone, and fiber that did not survive. Rolls of birch bark have been excavated at Moricetown, Hagwilget, and Kitselas canyons in contexts dating as early as 4100 ± 310 BP (Allaire 1979:31). These have been interpreted as torch lights to attract fish at night (Albright 1986:33; Matson and Coupland 1995:129). Birch bark was also used to wrap dried or smoked fish for storage ('Ksan 1980:22). Either way, birch bark may be related to the fishing economy.

While artifacts used in fishing are rare, tools suited to and employed in salmon processing may be quite common (Morin 2004; Rousseau 2004). I follow Rousseau's (2004) suggestions, for the Canadian Plateau, that a number of common tool types—including utilized cobble spalls, large bifaces, small convex scrapers, and microblades—may have been frequently used to decapitate, clean, eviscerate, and fillet salmon; these can thus serve as important indirect evidence for fish processing.

In the following analyses I compare excavated or intensively surface-collected assemblages with these tool types in the Skeena watershed from Kitselas Canyon upstream. All of the sites are from either the main stems of the Skeena and Bulkley at major canyon constrictions or from the Kitwanga River system, which is a particularly rich and diverse salmon habitat. As a baseline, assemblages excavated from the Mount Edziza region of the Stikine watershed, in nonsalmon-bearing habitats, are included.

Sites Employed in the Study and Their Chronological Placement

Various chronologies have been utilized to organize archaeological information from this study area, including broad developmental periods that fit either the entire North Coast of British Columbia or the Interior Plateau and Intermountain areas (Clark 1991; Fladmark 1985, 1986; Fladmark, Ames, and Sutherland 1990; MacDonald and Inglis 1981). A fine-grained chronology of phases has been developed for stratified sites at Kitselas Canyon (Allaire 1978, 1979; Coupland 1988, 1993). The latter scheme has been most compelling for researchers organizing materials far upstream in the Skeena watershed (Albright 1986; Ames 1979b; Ives 1987). Yet neither the chronology of absolute dates from upstream sites nor the diagnostic artifact types convincingly fit the Kitselas Canyon phases. Instead, I employ a more general chronology developed for the vast intermountain interior regions of central and northern British Columbia by Magne and Matson (2008) that fits the sites examined here into two broad periods: Middle Prehistoric (7000–3500 BP), and Late Prehistoric (3500–100 BP). With this chronology I can more confidently draw comparisons between loosely dated assemblages from the Skeena watershed and beyond to the Mount Edziza area. The Skeena watershed sites are described in geographic order, starting farthest upstream.

Skeena Watershed Sites

Site GgSt-2, at Moricetown Canyon on the Bulkley River, is approximately 30 km upstream from its confluence with the Skeena. This is one of the most important sites in northern British Columbia, representing perhaps up to 6,000 years of occupation, yet it has been repeatedly disturbed and remains poorly understood. The site was first investigated by Turnbull in 1966, who uncovered over 1 m of stratified cultural deposits (MacDonald and Inglis 1981:38). In 1985, Albright (1986) excavated a controlled trench perpendicular to a campground service trench that had exposed cultural deposits and collected 114 artifacts (aside from debitage), which are included in my analyses. She identified three main stratigraphic layers of occupation, with a range of radiocarbon dates (Albright 1986:36). Layer A produced a single date of 1650 ± 90 BP. Layer B has three dates ranging from 1700 ± 80 BP to 1960 ± 80 BP, and a fourth date of 3680 ± 170 BP, which she considered anomalous (Albright 1986:36). Layer C was divided into two strata: Layer C1 was above Layer C Orange (Albright 1986:32). She obtained a date of 2860 ±110 BP for Layer C1 and dates of 4700 ±130 BP and 5660 ± 90 BP from post molds that extended into C Orange (Albright 1986:36). The date for the transition between Layers B and C is thus confusing. Further, most artifacts are said to be from Layer C1 (Albright 1986:32), and it is not clear if any are associated with the earliest dates, although the posts are certainly signs of occupation. Based on my understanding, all the artifacts could belong to chronological contexts from roughly 3850 BP to 1560 BP. I thus group them together in the Late Prehistoric Period.

A much larger artifact assemblage was collected from construction backfill in 2005 (Traces 2005). A salvage crew screened three of eight dump truck loads of matrix removed from an area 10 m x 18 m x 1 m deep during construction (Traces 2005:10). They sampled dump truck loads from the crude stratigraphic contexts "top" and "bottom," but found these proveniences too general and unreliable to pursue during analysis (Traces 2005:10).

It is likely that they collected material from all occupational periods. In my analyses, I lump Traces' collection together with Albright's in the Late Prehistoric period, although some of it may belong to the Middle period.

Site GhSv-2 is a multicomponent site at Hagwilget Canyon first tested by MacDonald (1967). The materials included in the present analyses were excavated by Ames (1979b) from the same area as MacDonald. Ames recorded three occupational zones. The earliest, Zone A, is a multipurpose occupation layer with a diverse lithic assemblage, storage pits, and large hearths. A single radiocarbon date of 3430 ± 200 BP was derived from material midway through the stratum in MacDonald's excavation, postdating the inception of occupation by an unknown time interval (Ames 1979a:208). Zone B lacks the features and material diversity of earlier deposits and was interpreted as representing a long span of limited use as a fishing site. Zone C represents a historic period Witsuwit'en village. Because the precise stratigraphic position of the radiocarbon date is unknown, and no divisions are reported in Zone A, I consider the entire site assemblage as belonging to the Late Prehistoric period.

Ames's artifact tallies are difficult to compare to other assemblages because he often presented these as percentages of artifact types, without clear reference to the whole from which they were derived. He also did not report cobble spall tools, and possibly did not consider them a diagnostic tool type. Therefore, the figures reported here should be considered conservative estimates of key artifact types.

Site GhSv-A was intensively surface collected in 1985 by Albright. The site is located at Hagwilget Canyon, approximately 1.25 km upstream from GhSv-2 (Albright 1986:53). Artifacts were collected from a cultivated field, which is a portion of a larger site (Albright 1986:53). Based on comparisons to material from Kitselas Canyon, Albright dates the site to between 3500 and 2000 BP and suggests it may have been a village contemporaneous with the hypothesized hiatus in village occupation at nearby site GhSv-2 (Albright 1986:60–62). It would therefore represent a residential shift upstream.

The Kitwanga Hill Fort, GgTa-1, is a small fortified village near the confluence of the Skeena and Kitwanga rivers. MacDonald excavated the site in 1979 and proposed two components: one fixed at AD 1550–1650, and the other from the late eighteenth to nineteenth centuries (MacDonald 1989:74). Based on analysis of the materials and excavation records, I assigned both components to the eighteenth–nineteenth centuries (Prince 2001). For the purposes of the comparisons made here, the entire assemblage is treated as Late Prehistoric, and European trade goods are not included.

Site GiTa-19 is a multicomponent planked house village at the outlet of Kitwancool Lake into the Kitwanga River. My excavations at this site yielded dates of 2240 ± 40 BP and 1500 ± 50 BP on midden deposits and 320 ± 30 BP on terminal occupation architecture (Prince 2005:71). The materials analyzed here were excavated mainly from a house floor corresponding to the latter date, and are grouped with the Late Prehistoric Period assemblages.

The Paul Mason Site (GdTc-16) is located at Kitselas Canyon on the middle Skeena River. The material was excavated and reported in detail by Coupland (1988), who arranged it to build upon the chronology of the canyon developed earlier by Allaire (1978, 1979). Coupland (1988:242) identified occupations belonging to the Bornite Phase (5000–4300 BP), Gitaus Phase (4300–3600 BP), and Paul Mason Phase (3200–2700 BP).

The Paul Mason Phase occupation, as noted earlier, has planked house platforms comprising a village reminiscent of later periods on the Northwest Coast.

Gitaus (GdTc-2) is a rich, multicomponent site downstream from Paul Mason, at Kitselas Canyon, excavated by Louis Allaire (1978), who defined the Gitaus Phase and identified other components named the Skeena Phase (3600–3200 BP) and Kleanza Phase (2500–1500 BP). Taken together, the Gitaus and Paul Mason sites present a nearly continuous culture-historical record of 3,500 years. Drawing comparisons to Allaire's material from Gitaus, however, poses a number of difficulties, as he excavated in arbitrary levels and reports stratigraphically mixed layers. Further, the primary report on the analysis includes typographical errors in the tabulation of Zones V and VI, which are switched in places (Allaire 1978:27). In the following comparisons I thus omitted Zones V and VI and the stratigraphically mixed layers (Zones VI + VII and III + IV) from the Skeena Phase totals, using only the unmixed Zone III. I also omitted his Gitaus Phase material entirely (Zones IV, V, and VII), relying instead upon Coupland's clearer description of this phase from the Paul Mason Site. As described earlier, I chose to lump phases together to facilitate comparisons to the other sites in this study. Therefore, the Bornite and Gitaus phase artifacts of the Paul Mason Site are grouped together in the Middle Prehistoric period, and the Skeena and Kleanza phase materials of the Gitaus site are grouped together with the Paul Mason Phase as Late Prehistoric.

The Mount Edziza Sites

The Mount Edziza region lies on the Stikine-Yukon Plateau and is above the limits of modern salmon runs in the Stikine River system. Historically, there was a great deal of communication between peoples of the Stikine, Nass, and Skeena watersheds, as their headwaters are close together (MacDonald 1989:20). The artifacts included in these analyses were excavated from two sites at Mount Edziza by Fladmark (1985) and are included here as a point of comparison to the salmon fishing areas of the Skeena watershed, where the subsistence economy was different but contact was frequent.

The Grizzly Run Site (HiTp-63) is on a terrace along a route of travel to a high mountain pass. Fladmark (1985:155–156) identified two components: Component 1, dating to 3910 ± 120 BP, and Component 2, dating to 4870 ± 120 BP. Component 2 was interpreted as an obsidian knapping station, producing 19,318 pieces of detritus and numerous early stage bifaces but few finished artifacts (Fladmark 1985:156). Component 1 produced a diverse collection of tools, including bifaces with use wear and retouched flakes, and was interpreted as a small camp or cache (Fladmark 1985:157). For my purposes, both components are lumped together as Middle Prehistoric based on their ages.

The Wet Creek Site (HiTp-1) is located at the confluence of two small streams that flow to the Iskut River and thence to the Stikine (Fladmark 1985:109–111). A focus of excavation was a small natural depression that Fladmark believed was used repeatedly as a shelter. He obtained dates of 260 ± 80 BP and 2850 ± 160 BP from the depression (Fladmark 1985:132–134) and interpreted the bulk of the excavated collection as corresponding to the earlier date and reflecting obsidian knapping geared toward biface preform manufacture.

Excavations closer to the creek, in "Area 2," produced a range of finished tools from a multipurpose campsite (Fladmark 1985:138). Two radiocarbon dates were derived, 1430 ± 60 BP and 1140 ± 80 BP, but Fladmark (1985:142) preferred to consider the bulk of the material as dating closer to 4000 BP based on the presence of microblades. As discussed below, though, microblades are not necessarily temporally discrete. Based on the material and radiocarbon dates, I thus consider all of the Wet Creek artifacts to be Late Prehistoric in age.

Comparisons of Potential Salmon Processing Assemblages: Bone, Ground Stone, and Chipped Stone Technologies

Previous studies of material culture in the Skeena watershed have focused on constructing a culture-history, emphasizing changes in the presence or absence of a few tool types and noting an overall trend toward increasing use of ground stone tools (Allaire 1978, 1979; Coupland 1988, 1993; Matson and Coupland 1995). The general argument is that over time inland people participated in the dominant Northwest Coast cultural sphere to a greater extent, such that by 2500 BP there was a more or less homogeneous culture type up the Skeena at least as far as Hagwilget Canyon (Fladmark, Ames, and Sutherland 1990:239; Matson and Coupland 1995:238; Sutherland 2001). Allaire (1979), Coupland (1988), and Ives (1987) debated the extent to which this may have occurred as a result of acculturation of interior peoples by coastal cultures, episodes of migration inland, or in situ evolution. Whatever the mechanisms, the development of intensive salmon harvesting and storage are generally considered interior variants of Northwest Coast culture and to have occurred through the gradual incorporation of specialized technologies, culminating in increased ground stone and bone tools and eventually prestige items (Coupland 1993:68–69; Fladmark, Ames, and Sutherland 1990:239; Matson and Coupland 1995:187). When the proportions of bone, ground stone, and chipped stone artifacts (excluding unused debitage) are compared, however, chipped stone technology is uniformly predominant in the Skeena sites in both the Middle Prehistoric and Late Prehistoric periods (Table 7.1). My subsequent comparisons are thus of potential salmon processing tools as a percentage of all chipped stone in each component.

Table 7.1. Bone, ground stone, and chipped stone tools, excluding debitage and hammerstones, at sites in the study

	Middle Prehistoric		Late Prehistoric							
	Paul Mason GdTc-16	Grizzly Run HiTp-63	Morice town GgSt-2	Hagwilget Canyon GhSv-2	GhSv-A	Kitwanga	GiTa-19	Gitaus	Paul Mason	Wet Creek
Bone tools	–	–	36	24	–	6	1	–	–	–
Ground stone	31	–	3	22	6	17	11	90	113	1
Chipped stone	358	82	790	59	50	67	47	1328	503	165
Total no. of tools	389	82	829	105	56	90	59	1418	616	166

Cobble Spalls and Cobble Flake Tools

One ubiquitous tool type throughout the Pacific Northwest is the utilized cobble spall. This was almost certainly a multipurpose tool, but Rousseau (2004) emphasized its utility in decapitating salmon. Rousseau (2004:27) reasons these tools were preferred over finer-edged knives because they could withstand the force needed to cut bone and are easily replaced when dull. These tools are so common and crude (Figure 7.3) that they may go unreported or unrecognized in the field, and variability in their form is not often considered. These factors may account for some of the variability in their frequencies evident from Tables 7.2 and 7.3. They are not a recorded type at GhSv-2 in Hagwilget Canyon but are present at all other Skeena watershed sites in both periods. Significantly, they do not occur at either of the Edziza sites, where salmon fishing did not occur. At the least, cobble spall tools appear where we know people fished salmon, and in the earliest occupations. Coupland (1996:162) cautiously attributes the earliest occupation at the Paul Mason Site to salmon fishing but does not argue for large-scale harvesting until village settlement is evident, approximately 3200–2700 BP. This argument can be revised, however, if cobble spalls are considered a sign of salmon processing.

Table 7.2. Potential fish processing tools as portion of all chipped stone at middle period sites

	Cobble spall & flake tools		Bifacial cutting tools		Small scrapers		Microblades & cores		All chipped stone
	Freq	%	Freq	%	Freq	%	Freq	%	Freq
Paul Mason (GdTc-16)	51	14.2	2	0.5	17	4.7	15	4.2	358
Grizzly Run (HiTp-63)	–	–	16	19.5	2	2.4	55	67.1	82

Table 7.3. Potential fish processing tools as portion of all chipped stone at late period sites

	Cobble spall & flake tools		Bifacial cutting tools		Small scrapers		Microblades & cores		All chipped stone
	Freq	%	Freq	%	Freq	%	Freq	%	Freq
Moricetown (GgSt-2)	75	9.5	149	18.9	38	4.8	20	2.5	790
Hagwilget Canyon (GhSv-2)	–	–	21	35.6	15	25.4	–	–	59
GhSv-A	7	14.0	10	20.0	8	16.0	–	–	50
Kitwanga Fort (GgTa-1)	44	65.7	3	4.5	1	1.5	–	–	67
GiTa-19	1	2.1	2	4.2	2	4.2	1	2.1	47
Gitaus (GdTc-2)	879	66.2	42	3.2	47	3.5	–	–	1328
Paul Mason (GdTc-16)	114	22.7	25	4.9	41	8.2	25	4.9	503
Wet Creek (HiTp-1)	–	–	139	84.2	6	3.6	10	6.1	165

Figure 7.3. Cobble spall tool from Site GiTa-19.

Bifaces as Cutting Tools

Rousseau (2004) suggests many biface types (aside from those typically interpreted as knives) may be fish cleaning and filleting tools, as does Fitzhugh (2002:282). This includes many artifacts commonly referred to as projectile points. Rousseau (2004:18–20) posits that bifaces with blunt tips and excurvate edges would be particularly effective at slitting open salmon and in collaring the head. In contrast to the southern British Columbia plateau, where biface typologies are well defined and agreed upon, in the Skeena watershed a wide range of idiosyncratic terminology has been applied to biface forms, many of which are likely similar. Here I list as potential bifacial cutting tools those that are not strictly projectile points—including those described as knives, large, stemmed, asymmetrical, excurvate, leaf, lanceolate, and bipoint shaped—and eliminate those with small triangular blades and notches for hafting to projectiles. This creates a conservative estimate of what may be bifacial cutting tools, although they likely were not limited to processing fish. Particularly provocative in this regard are bipoints, which could slit fish bellies, and large lanceolate blades with convex bases (Figures 7.4 and 7.5). The latter is a common variation on what are sometimes referred to as "Skeena Points" (Magne and Matson 2008), but this type is not easily separated from the broader category "lanceolates" in most reports. Following Rousseau's (2004) observation of a large biface with a convex, unifacial base and his experiments with a hypothesized composite knife/scraper, I suggest these large bifaces could also perform a dual purpose, with the blade used for butchering and the convex base used to scrape salmon guts.

Bifacial cutting tools comprise a small percentage of the chipped stone at the Paul Mason Site in the Middle Prehistoric Period, and are more common at the Grizzly Run Site (Table 7.2). The difference here may reflect the function of the Grizzly Run site as an obsidian knapping station geared toward biface production, many of which may be crude

Figure 7.4. Bipointed biface fragment from Site GiTa-19.

Figure 7.5. Examples of large bifaces from Moricetown Canyon, GgSt-2. Reproduced with the permission of Frank Craig.

preforms rather than finished tools, such that hafting elements and their final basal shapes were not rendered. Indeed, Fladmark (1985:155) remarks on the lack of use wear on the early component artifacts, which comprise the majority of the assemblage.

In the Late Prehistoric period, bifacial cutting tools are more common and show a general increase in frequency among the Skeena sites as distance inland increases, although they peak at Hagwilget Canyon rather than Moricetown Canyon (Table 7.3). The importance of bifaces inland has traditionally been explained as reflecting a greater emphasis on land mammal hunting (Ames and Maschner 1999:126; Coupland 1996:162), but given the locations of sites, these bifaces could just as easily represent fish processing. Mount Edziza's Wet Creek Site has an extremely large percentage of nonprojectile-point bifaces, but as with the earlier Grizzly Run assemblage, these numbers likely reflect preform production.

Small Scrapers

Rousseau (2004) also suggests that small unifacial scrapers with convex working edges less than 4 cm long, typically interpreted as hide scrapers, are actually effective salmon cleaning tools, especially in removing the kidney and connective membranes. He goes so far as to hypothesize that small scrapers were used in conjunction with bifaces as part of a composite cutting and scraping tool. I found no particular association, however, in the ratio of bifacial cutting tools to scrapers. There is no reason to expect they formed one half of a hypothesized composite tool, though both could have been used in processing fish.

As with bifaces, the nomenclature used to describe scrapers in the Skeena assemblages is highly variable and the shape of the cutting edges is often not specified. I attempted a conservative tabulation of scrapers, regardless of edge shape, in the size range specified by Rousseau. These are present in small quantities in the earliest components at both the Paul Mason and Grizzly Run sites, and are therefore hard to attribute to salmon processing. In the Late Prehistoric Skeena assemblages, they are most frequent at the Hagwilget Canyon sites, but they do not significantly differ in their proportions at the other Skeena sites, including the most interior location (Moricetown Canyon), or at Wet Creek on Mount Edziza. Therefore, they are more closely associated with at least one salmon fishing locale than with interior hunting areas.

Microblades

The final tools I consider here are microblades and microblade cores. Microblade technology has been considered variously to be an early tool type (Coupland 1996; Fladmark 1985) and ethnically specific to Athapaskan cultures (Carlson 1996:217). It now seems microblades are too broadly distributed in time and space to support either argument unequivocally (Moss 2004:184–185). The function of microblades is also debated, but one suggestion is that they were used in fish processing (Coupland 1996:165–166), especially in the defleshing (Rousseau 2004:27) and initial incision stages (Morin 2004:313). In the Middle Period, microblades are a minor occurrence at Kitselas Canyon but comprise a major portion of the Grizzly Run assemblage, which seems to be geared in part to their manufacture, perhaps for broader distribution down the Stikine, Nass, and Skeena rivers to salmon fishing areas. Microblades and microblade cores are present in several but not all Late Prehistoric Skeena assemblages in roughly the same proportions as the earlier Paul Mason Site and the Edziza region's Wet Creek Site. Therefore, while they consistently occur at fishing locations, they do not seem to change in relative abundance over time, nor are they limited to fishing sites.

Discussion

The inland portions of the Skeena and Nass river valleys have some very productive spawning and rearing habitats and excellent places for the mass capture of salmon. In the historic period, this was clearly the focus of Aboriginal economies. The prehistory of the Nass River is virtually unknown, but the Skeena watershed has seen large-scale surveys and several excavations over the past forty years at major salmon fishing areas. While the faunal record is very poor, the Paul Mason and Kitwanga Hill Fort assemblages indicate little variation from salmon-focused subsistence over a span of 2,500 years. Although most sites identified by survey are undated, notably high densities of sites are located where salmon are readily captured, such as stream confluences and especially canyons (Prince 2000), the latter of which have the oldest sites. Of the undated sites, the most common types are cache pits, indicative of the widespread importance of the "delayed return" nature of the salmon economy (Prince 2004).

Most importantly for this discussion, technologies do not appear to change much over the nearly 6,000 years examined here. It thus becomes difficult to support arguments for either a sudden or steplike evolutionary change in subsistence toward a salmon-based economy with special preparatory techniques for storage. Instead, I argue, the requisite processing technology for intensive use of salmon was always present. This technology does not exactly fit expectations of what intensive processing should look like, as I have proposed that generalized, rather than highly specialized, tools were used. Specifically, lithic tools of the Middle and Late Prehistoric periods were predominantly chipped stone, which is most often intuitively equated with terrestrial mammal hunting and processing, while ground stone tools such as slate knives are taken as indications of intensive fish processing (Graesch 2007). In the Skeena watershed, however, ground slate knives are absent, and all types of ground stone are rare, despite the historical focus upon salmon fishing and the long-term continuity in use of productive fishing sites. If we entertain Rousseau's suggestions from the southern Plateau that chipped stone such as large bifaces, small convex-edged scrapers, microblades, and especially cobble spalls could be effective in salmon processing, then these activities become visible throughout the archaeological record of the Skeena, and there seems to have been little change. Further, when drawing comparisons to the Mount Edziza region, the contrast between salmon harvesting and nonsalmon-harvesting locales in the abundance of cobble spalls, and perhaps large formed bifaces, takes on significance.

Conclusion

Direct faunal evidence for the development of intensive salmon fishing in the interior of northern British Columbia is still lacking. In the absence of such evidence, the economic prehistory of the entire region has been viewed through the lens of the evolutionary sequence developed for Kitselas Canyon. Coupland (1996:192) argued that salmon fishing occurred from the outset of occupation there but was not a focus until the Paul Mason phase, 3200–2700 BP. Coupland reasoned that sedentary settlement (i.e., the planked house village) would not have been possible without intensive salmon harvesting and storage. In the conventional argument, the other characteristics of Northwest Coast culture, including social inequality, developed relatively quickly out of this intensive economic base and spread inland, either through exchange or migration (Coupland 1988; Matson and Coupland 1995). This scenario is consistent with and supports general evolutionary models that view intensive salmon fishing as emerging only after a long stage of salmon use as part of a broader foraging economy, laying the foundation for the development of cultural complexity (Ames and Maschner 1999; Matson 1992; Matson and Coupland 1995).

I have argued that the abundance of sites located at good fishing locations, the presence of salmon harvesting facilities, and the abundance of potential processing tools demonstrate long-term continuity in the maintenance of a salmon-based economy inland in the Skeena watershed. I acknowledge Rousseau's caution that experimentation and tool edge use wear studies are required to definitively associate chipped stone tools with salmon processing activities. The suitability of cobble spalls, large bifaces, small convex-edged scrapers, and microblades to such tasks is evident, however, and their predominance at prime salmon fishing locales is provocative.

At a broader level, these results inform more general discussions of what may be useful evidence for salmon-based economies, and what may be a link to other practices such as social differentiation and sedentary settlement. Cannon has maintained that because the evidence for intensive salmon harvesting in some cases is ancient, general arguments for an evolutionarily requisite relationship to sedentism and cultural complexity are weak (Cannon 1996, 1998; Cannon and Yang 2006). Other archaeologists have also questioned the relationship between salmon harvesting, storage, and other traits of Northwest Coast culture (Monks 1987; Moss and Erlandson 1995:34). In the case at hand, I do not think the economic importance of salmon harvesting and storage has been exaggerated. Instead, their antiquity has been underestimated by the dearth of faunal evidence and expectations of a steplike sequence of economic and technological development. The long record of indirect evidence for an intensive salmon-based economy suggests that peoples inland on the Skeena did not have to wait for the technologies to generate large stores of salmon to either evolve or be introduced from the coast.

Acknowledgments

Many thanks to Frank Craig for sharing the results of his salvage operations at Moricetown Canyon and for permission to use his photographs. The quality of this chapter has been greatly improved by the comments of Aubrey Cannon and Madonna Moss, whom I also thank for the invitation to contribute to the original symposium and this volume. I also appreciate the comments of Virginia Butler and Anthony Graesch on my original draft. This research has been supported by the Grant MacEwan University Arts and Science Research Fund, and the Faculty Research and Scholarly Activity Fund.

References Cited

Adams, John. *The Gitksan Potlatch: Population Flux, Resource Ownership and Reciprocity.* Toronto: Holt, Rinehart and Winston, 1973.

Albright, Sylvia. Report on 1985 Archaeological Investigations of Gitksan-Wet'suwet'en Villages. MS on file, Gitksan-Wet'suwet'en Tribal Council, Smithers, BC, 1986.

Allaire, Louis. *L'Archéologie des Kitselas d'après le Site Stratifié de Gitaus (GdTc-2) sur la Rivière Skeena en Colombie Britannique.* Ottawa: National Museum of Man Mercury Series 72, 1978.

———. "The Cultural Sequence at Gitaus: A Case of Prehistoric Acculturation." In *Skeena River Prehistory*, R. Inglis and G. F. MacDonald, eds., pp. 18–52. Ottawa: National Museum of Man Mercury Series 87, 1979.

Ames, Kenneth M. "Stable and Resilient Systems along the Skeena River: The Gitksan-Carrier Boundary." In *Skeena River Prehistory*, R. Inglis and G. F. MacDonald, eds., pp. 219–243. Ottawa: National Museum of Man Mercury Series 87, 1979a.

———. "Report on Excavations at GhSv-2, Hagwilget Canyon." In *Skeena River Prehistory*, R. Inglis and G. F. MacDonald, eds., pp. 181–218. Ottawa: National Museum of Man Mercury Series 87, 1979b.

———. "The Northwest Coast: Complex Hunter-Gatherers, Ecology, and Social Evolution." *Annual Review of Anthropology* 12 (1994): 1–40.

———. "Prehistory of the Northern British Columbia Coast." *Arctic Anthropology* 35 (1998): 68–87.

Ames, Kenneth M., and Herbert D. G. Maschner. *Peoples of the Northwest Coast: Their Archaeology and Prehistory.* London: Thames and Hudson, 1999.

Archer, David J. W. The Kitsumkalum Heritage Survey Project: A Report on the 1986 Field Season, Permit 1986–17. MS on file, BC Archaeology Branch, Victoria, 1987.

Binford, Lewis R. *Constructing Frames of Reference: An Analytical Method for Archaeological Theory Building Using Hunter-Gatherer and Environmental Data Sets.* Berkeley: University of California Press, 2001.

Butler, Virginia L. "Discussion of Red Fish, White Fish, Big Fish, Small Fish: The Archaeology of North Pacific Fisheries Symposium." Paper presented at the 73rd annual meeting of the Society for American Archaeology, Vancouver, 2008.

Butler, Virginia L., and Sarah K. Campbell. "Resource Intensification and Resource Depression in the Pacific Northwest of North America: A Zooarchaeological Review." *Journal of World Prehistory* 18 (2004): 327–405.

Cannon, Aubrey. "Scales of Variability in Northwest Salmon Fishing." In *Prehistoric Hunter-Gatherer Fishing Strategies*, M. G. Plew, ed., pp. 25–40. Boise, ID: Boise State University, 1996.

———. "Contingency and Agency in the Growth of Northwest Coast Maritime Economies." *Arctic Anthropology* 35 (1998): 57–67.

Cannon, Aubrey, and Dongya Y. Yang. "Early Storage and Sedentism on the Pacific Northwest Coast: Ancient DNA Analysis of Salmon Remains from Namu, British Columbia." *American Antiquity* 71 (2006): 123–140.

Carlson, Arne, and Jean Bussey. Archaeological Inventory and Impact Assessment, Bulkley River, Lower Hagwilget Canyon, Permit 1989–102. MS on file, BC Archaeology Branch, Victoria, 1990.

Carlson, Roy L. "The Later Prehistory of British Columbia." In *Early Human Occupation in British Columbia,* R. L. Carlson and L. Dalla Bona, eds., pp. 215–226. Vancouver: UBC Press, 1996.

Clark, Donald W. *Western Subarctic Prehistory*. Ottawa: Canadian Museum of Civilization, 1991.

Coupland, Gary. *Prehistoric Cultural Change at Kitselas Canyon*. Ottawa: Canadian Museum of Civilization Mercury Series 138, 1988.

———. "Recent Research on the Northern Coast." *BC Studies* 99 (1993): 53–76.

———. "The Early Prehistoric Occupation of Kitselas Canyon." In *Early Human Occupation in British Columbia,* R. L. Carlson and L. Dalla Bona, eds., pp. 159–166. Vancouver: UBC Press, 1996.

Daly, Richard. *Our Box Was Full: An Ethnography of the Delgamuukw Plaintiffs*. Vancouver: UBC Press, 2005.

Fitzhugh, Ben. "Residential and Logistical Strategies in the Evolution of Complex Hunter-Gatherers on the Kodiak Archipelago." In *Beyond Foraging and Collecting: Evolutionary Change in Hunter-Gatherer Settlement Systems*, B. Fitzhugh and J. Habu, eds., pp. 257–306. New York: Kluwer Academic, 2002.

Fladmark, Knut R. *Glass and Ice: The Archaeology of Mt. Edziza*. Burnaby, BC: Simon Fraser University, Department of Archaeology Publication 14, 1985.

———. *British Columbia Prehistory*. Ottawa: National Museums of Canada, 1986.

Fladmark, Knut R., Kenneth M. Ames, and Patricia D. Sutherland. "Prehistory of the Northern Coast of British Columbia." In *Northwest Coast*, Handbook of North American Indians, vol. 7, W. Suttles, ed., pp. 229–239. Washington, DC: Smithsonian Institution, 1990.

Graesch, Anthony P. "Modelling Ground Slate Knife Production and Implications for the Study of Household Labor Contributions to Salmon Fishing on the Pacific Northwest." *Journal of Anthropological Archaeology* 26 (2007): 576–606.

Halpin, Marjorie, and Margaret Seguin. "Tsimshian Peoples: Southern Tsimshian, Coast Tsimshian, Nishga and Gitksan." In *Northwest Coast*, Handbook of North American Indians, vol. 7, W. Suttles, ed., pp. 267–284. Washington, DC: Smithsonian Institution, 1990.

Ives, John W. "The Tsimshian Are Carrier." In *Ethnicity and Culture*, R. Auger, M. F. Glass, S. MacEachern, and P. McCartney, eds., pp. 209–225. Calgary: University of Calgary Archaeological Association, 1987.

Kew, Michael. "Salmon Availability, Technology and Cultural Adaptation in the Fraser River Watershed." In *A Complex Culture of the British Columbia Plateau: Traditional Stl'atl'imx Resource Use*, B. Hayden, ed., pp. 177–221. Vancouver: UBC Press, 1992.

'Ksan, People of. *Gathering What the Great Nature Provided: Food Traditions of the Gitksan*. Vancouver: Douglas and McIntyre, 1980.

MacDonald, George F. Archaeological Reconnaissance in the Tsimshian Area. MS on file, BC Archaeology Branch, Victoria, 1967.

———. *Kitwanga Fort Report*. Ottawa: Canadian Museum of Civilization, 1989.

MacDonald, George F., and Richard Inglis. "An Overview of the North Coast Prehistory Project (1966–1980)." *BC Studies* 48 (1981): 37–63.

Magne, Martin, and R. G. Matson. "Projectile Points of Central and Northern Interior British Columbia." In *Projectile Point Sequences in Northwestern North America*, R. L. Carlson and M. P. R. Magne, eds., pp. 273–292. Burnaby, BC: Simon Fraser University Archaeology Press, 2008.

Matson, R. G. "The Evolution of Northwest Coast Subsistence." In *Long-Term Subsistence Change in Prehistoric North America,* B. L. Isaac, ed., pp. 367–428. Research in Economic Anthropology Supplement 6. Greenwich, CT: JAI Press, 1992.

Matson, R. G., and Gary Coupland. *The Prehistory of the Northwest Coast.* San Diego: Academic Press, 1995.

Millennia Research. Overview Mapping of Archaeological Resource Potential in the Bulkley and Kispiox LRMP Areas. MS on file, BC Archaeology Branch, Victoria, 1995.

Monks, Gregory G. "Prey as Bait: The Deep Bay Example." *Canadian Journal of Archaeology* 11 (1987): 119–142.

Morin, Jesse. "Cutting Edges and Salmon Skin: Variations in Salmon Processing Technology on the Northwest Coast." *Canadian Journal of Archaeology* 28 (2004): 281-318.

Moss, Madonna L. "The Status of Archaeology and Archaeological Practice in Southeast Alaska in Relation to the Larger Northwest Coast." *Arctic Anthropology* 41 (2004): 177–196.

Moss, Madonna L., and Jon M. Erlandson. "Reflections on North American Pacific Coast Prehistory." *Journal of World Prehistory* 9 (1995): 1–45.

Moss, Madonna L., Jon M. Erlandson, and Robert Stuckenrath. "Wood Stake Weirs and Salmon Fishing on the Northwest Coast: Evidence from Southeast Alaska." *Canadian Journal of Archaeology* 14 (1990): 143–158.

Prince, Paul. "Settlement, Trade and Social Ranking at Kitwanga." PhD diss., McMaster University, 1998.

———. "Protohistoric Settlement and Interaction on the Upper Skeena in Long-Term Perspective." In *The Entangled Past: Integrating History and Archaeology*, M. Boyd, J. Erwin, and M. Hendrickson, eds., pp. 83–89. Calgary: Archaeological Association of the University of Calgary, 2000.

———. "Artifact Distributions at the Kitwanga Hillfort: Protohistoric Competition and Trade on the Upper Skeena." In *Perspectives on Northern Northwest Coast Prehistory*, J. S. Cybulski, ed., pp. 249–268. Hull: Canadian Museum of Civilization Mercury Series 160, 2001.

———. "Ridge-top Storage and Defensive Sites: New Evidence of Conflict in Northern British Columbia." *North American Archaeologist* 25 (2004): 35–56.

———. "Fish Weirs, Salmon Productivity and Village Settlement on an Upper Skeena Tributary." *Canadian Journal of Archaeology* 29 (2005): 68–87.

———. "Determinants and Implications of Bone Grease Rendering: a Pacific Northwest Example." *North American Archaeologist* 28 (2007): 1–28.

Rousseau, Mike K. "Old Cuts and Scrapes: Composite Chipped Stone Knives on the Canadian Plateau." *Canadian Journal of Archaeology* 28 (2004): 1–31.

Sutherland, Patricia D. "Revisiting an Old Concept: The North Coast Interaction Sphere." In *Perspectives on Northern Northwest Coast Prehistory*, J. S. Cybulski, ed., pp. 49–60. Hull: Canadian Museum of Civilization Mercury Series 160, 2001.

Traces Archaeological Research and Consulting. Archaeological Materials and Recovery Project for Kya Wiget, Site GgSt-2 in Moricetown Canyon. MS on file, Moricetown Band, Moricetown, BC, 2005.

CHAPTER 8

Late Holocene Fisheries in Gwaii Haanas
Species Composition, Trends in Abundance, and Environmental or Cultural Explanations

Trevor J. Orchard, *Department of Anthropology, University of Toronto*

As discussed elsewhere in this volume, fish were universally important in the subsistence practices of Northwest Coast First Nations both in precontact (Butler and Campbell 2004; Hanson 1991; Orchard and Clark 2005) and ethnographic (Donald 2003; Suttles 1990) times, and they are important resources for contemporary First Nations cultures in the region. Salmon are notable as a significant resource throughout the culture area, though recent research has highlighted variability in the timing and nature of a near-universal intensification of salmon harvesting (Cannon 2000a; Coupland, Stewart, and Patton 2010; McMillan et al. 2008; Monks 2006). There has also recently been increasing acknowledgment of the variable importance of numerous other fish taxa (Butler and Campbell 2004; Hanson 1991; McMillan et al. 2008; Monks 2006). Southern Haida Gwaii, off the northern British Columbia coast, provides a useful case for examining these patterns (Orchard 2007, 2009; Orchard and Clark 2005).

Our knowledge of Late Holocene (ca. 1500–200 BP) fisheries in Haida Gwaii derives primarily from the region of Gwaii Haanas National Park Reserve and Haida Heritage Site (hereafter Gwaii Haanas) at the southern end of the archipelago (Figure 8.1). Two multiyear regional projects have provided the bulk of the archaeological record for this time period. The Kunghit Haida Culture History Project involved excavations at eighteen sites in southernmost Gwaii Haanas, all of which date more recently than 2000 BP (Acheson 1998). Subsequently, the Gwaii Haanas Environmental Archaeology Project (GHEAP) involved

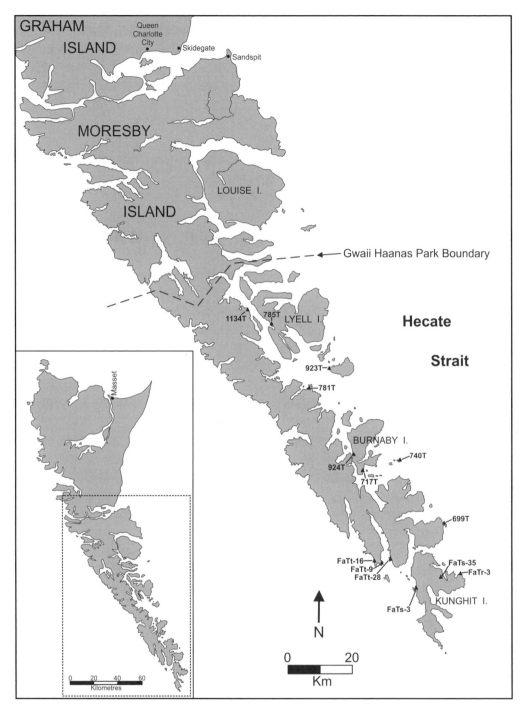

Figure 8.1. Map of study area showing locations of sites mentioned in the text.

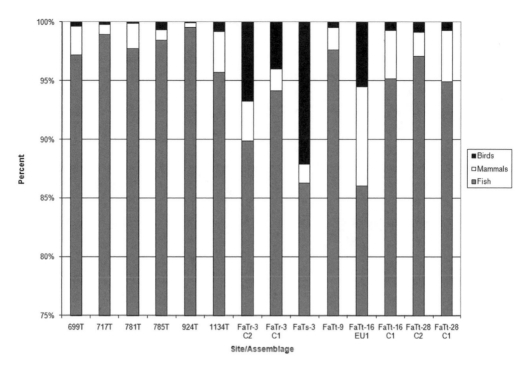

Figure 8.2. Relative proportions of birds, mammals, and fish in Gwaii Haanas village assemblages, as percent of total vertebrate remains identified to at least the class level.

excavations at eight sites throughout eastern Gwaii Haanas, all dating to roughly 1500 BP or later (Orchard 2007, 2009).

A strong maritime focus in Late Holocene Haida Gwaii is recorded in the presence of numerous remains of marine mammals and birds, and particularly in the dominance of marine fishes in faunal assemblages from Gwaii Haanas (Figure 8.2). In most village sites fish comprise greater than 80 percent of the total sample of bones identified to at least the class level, and in many cases contribute greater than 90 percent of the vertebrate assemblage. Two sites from the GHEAP project (740T and 923T*) contained proportionately fewer fish and more birds, but these have been excluded from the present analysis because of relatively small samples at these more limited-activity locations (Orchard 2009). Among the larger village sites, however, marine fishes are consistently the dominant component of the vertebrate assemblages.

Among the fish remains, salmon (*Oncorhynchus* spp.), rockfish (*Sebastes* spp.), and Pacific herring (*Clupea pallasii*) together contribute greater than 80 percent of the total fish NISP in all cases (Table 8.1). Gadids and halibut (*Hippoglossus stenolepis*) are also relatively consistent contributors to fish assemblages, with gadids present in ten and halibut present in thirteen of the fourteen assemblages. Gadids, however, typically comprise only a fraction of a percent of the total fish assemblages, showing somewhat higher abundance (2 percent) in site 785T. Halibut are slightly more common, contributing 2–3 percent of

---

* Sites recorded in Gwaii Haanas subsequent to the designation of the area as a National Park Reserve are assigned a site code in the Parks Canada system, including a site number and a "T" designation (e.g. 740T). As requested by the Haida, these sites have not been assigned Borden Numbers and are not included in the provincial site registry.

Table 8.1. Relative proportion of fish taxa in Gwaii Haanas village assemblages, as percent of total fish remains (NISP) identified to family, genus, or species

Assemblage	Salmon	Rockfish	Herring	Gadid	Halibut	Other fish	Total NISP
699T	59.9	2.1	27.7	0.0	1.2	9.1	11651
717T	54.7	0.1	43.0	0.0	0.2	2.0	6877
781T	5.1	3.5	80.3	0.1	0.1	10.8	3495
785T	76.8	7.9	3.3	2.1		9.8	6317
924T	34.7	0.1	63.1	0.1	0.0	2.0	16721
1134T	85.0	14.5			0.1	0.4	904
FaTr-3 C2	78.3	13.0	0.2	0.1	2.2	6.2	8777
FaTr-3 C1	25.7	62.0	0.8	0.0	2.6	9.0	11240
FaTs-3	10.0	76.6	0.3		0.9	12.3	1608
FaTt-9	83.0	4.1	8.5	0.1	2.4	2.0	45100
FaTt-16 EU1	85.0	8.1	1.4		2.7	2.8	567
FaTt-16 C1	10.0	74.1	0.4		6.1	9.5	529
FaTt-28 C2	57.3	22.5	1.2	0.2	8.7	10.1	6416
FaTt-28 C1	87.6	8.5		0.0	2.1	1.7	2636

Note: First six assemblages from the GHEAP project (Orchard 2007, 2009), bottom eight assemblages from the Kunghit project (Acheson 1998; Wigen 1990).

several of the fish assemblages, and reaching levels of roughly 6 percent in component one of site FaTt-16 and almost 9 percent in component two of site FaTt-28 (Wigen 1990). These halibut remains almost certainly underrepresent Haida use of halibut due to a combination of cultural and natural taphonomic factors (Orchard 2009; Orchard and Wigen 2008; Wigen 1990; Wigen and Stucki 1988).

The highly variable relative proportions of herring are in large part a factor of sampling. Acheson's (1998) Kunghit project employed minimal, unsystematic use of eighth-inch screens during excavations, with the majority of the excavated matrices only screened through quarter-inch mesh. This certainly accounts for the relative lack of herring in the Kunghit assemblages (Table 8.1). In contrast, eighth-inch screens were systematically used throughout the GHEAP project (Orchard 2009), accounting for the relatively higher proportions of herring in these assemblages. Notably, even eighth-inch screens have been shown to miss significant numbers of herring bones while the variable use of wet screening and the degree of experience possessed by field analysts all contribute to greater or lesser recovery rates for herring and other small taxa (Butler 1993; Cannon 2000b; Casteel 1976; Kopperl 2001; Moss, Butler, and Elder, this volume). Regardless, the GHEAP assemblages reveal a wide range of herring abundance (0–80 percent of fish NISP), likely resulting from differential access to herring. Herring are the dominant fish taxon, for example, in sites 781T (80 percent) and 924T (63 percent), suggesting that these sites were located near prominent herring aggregations and were occupied during the late winter and spring, when herring aggregate in nearshore waters to spawn (Hart 1973:97). Lack of comparable sampling across the assemblages examined here, and insufficient data to conclusively interpret the variable abundances of herring

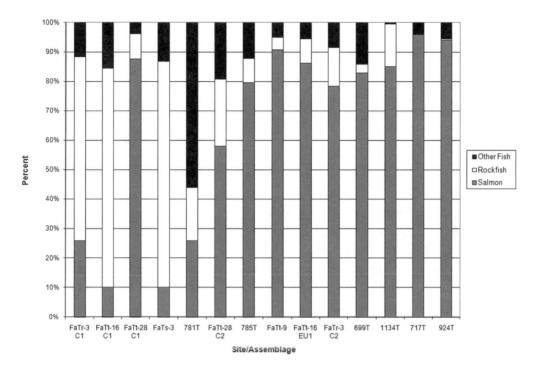

Figure 8.3. Relative proportion of fish taxa in Gwaii Haanas village assemblages, as percent of total fish remains (NISP) identified to family, genus, or species. Herring have been removed from consideration to eliminate inconsistencies arising from inconsistent sampling. Assemblages are ordered in rough chronological order based on available dates (Acheson 1998; Orchard 2007, 2009), with older assemblages on the left and younger assemblages on the right.

across these assemblages, mean that a more detailed consideration of herring use in Gwaii Haanas will require further research, and is beyond the scope of this paper.

With herring excluded, salmon and/or rockfish dominate all fish assemblages from excavated village sites in Gwaii Haanas, except for site 781T (Figure 8.3). The relative proportions of salmon and rockfish, however, are highly variable. The major division between salmon-dominated assemblages and those with a focus on rockfish has been identified as the primary factor affecting the variability in faunal assemblages from southern Haida Gwaii (Orchard and Clark 2005; Orchard 2009). This has been linked to a temporal shift in Haida economies towards a greater focus on intensified use of salmon after 1200 BP (Orchard 2009; Orchard and Clark 2005; cf. Acheson 1998; Wigen 1990). The nature of this economic change remains unclear, and colleagues have suggested to me that salmon populations in Haida Gwaii were insufficient to support intensive harvesting. Alternately, salmon in Haida Gwaii archaeological assemblages may have been harvested through trolling in open waters, targeting migrating populations, or obtained from larger mainland sources via trade, raiding, or direct harvesting.

In the remainder of this chapter, I use metric, radiographic, and isotopic approaches to interpret the species composition of salmon assemblages from Gwaii Haanas (Orchard and Szpak, this volume). Knowledge of salmon species in these assemblages, in combination with historic escapement data for the region, provide a basis for inferring that Gwaii Haanas salmon resulted from local harvest. While a complete explanation of the factors underlying a

Late Holocene intensification of salmon fishing in Gwaii Haanas will require more research, I conclude the chapter by exploring some possible causes for this local shift in fishing.

Species Composition of Gwaii Haanas Salmon Assemblages

For this project, 1,908 salmon vertebrae from five sites in Gwaii Haanas were analyzed using the metric and radiographic approach outlined in Chapter 2 (Orchard and Szpak, this volume). A total of 1,741 vertebrae, or 91 percent, showed two radiographic annuli, with relatively small numbers of vertebrae showing one (n = 97, 5 percent) or three (n = 70, 4 percent) annuli (Table 8.2). Sizes of vertebrae ranged from 5.1 mm to 12.5 mm, with a mean width of 9.5 mm and a standard deviation of 1.2 mm. In the context of the species characteristics of Pacific salmon (Orchard and Szpak, this volume), the lack of any vertebrae in the less than 8 mm size category exhibiting three or four growth annuli suggests that sockeye are completely absent from the samples. While the vertebrae less than 10.5 mm in width that exhibit two or three annuli could include some sockeye, the presence of sockeye in any numbers would almost certainly also yield four- or five-year-old fish (three or four annuli respectively) in the less than 8 mm category. Similarly, chinook salmon should yield vertebrae with both three and four annuli in the greater than 10.5 mm size category, and may also result in small numbers of vertebrae exhibiting five, six, or seven annuli. Additionally, the generally very large size of chinook salmon (Orchard and Szpak, this volume, Table 2.1) suggests that their presence should result in a size range greatly exceeding the maximum size of 12.5 mm present in the Gwaii Haanas assemblages. While small numbers of either sockeye or chinook may be present in these assemblages, the most parsimonious explanation for the observed results is that the Gwaii Haanas assemblages are comprised of chum, pink, and coho salmon. If sockeye and chinook are thus removed from consideration, 97 vertebrae (5 percent) can be attributed to pink salmon while 446 vertebrae (23 percent) can be attributed to chum. The remaining 1,365 vertebrae (72 percent) may represent pink, chum, or coho salmon.

Table 8.2. Results of radiographic and metric analysis of salmon vertebrae from Gwaii Haanas

Transverse diameter	Growth annuli	Vertebrae	Possible species
≤ 8 mm	1	91	Pink
	2	139	Pink; Coho; (sockeye)
8–10.5 mm	1	6	Pink; (chum)
	2	1226	Pink; Coho; Chum; (chinook); (sockeye)
	3	31	Chum; Sockeye; Chinook (coho)
≥ 10.5 mm	2	376	Chum; (chinook)
	3	39	Chum; Chinook

Note: The most likely taxa are capitalized, while less likely taxa are in parentheses (after Orchard and Szpak, this volume, Table 2.3).

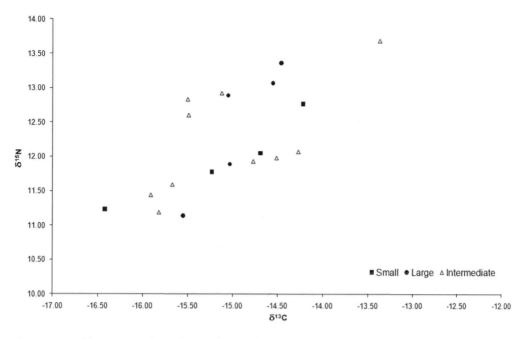

Figure 8.4. Stable isotope values (δ^{13}C and δ^{15}N) for salmon vertebrae from Gwaii Haanas sites. Results are distinguished by size categories of vertebral transverse diameters: <8.0 mm (Small), 8.0–10.5 mm (Intermediate), >10.5 mm (Large).

Stable isotope analysis (Figure 8.4) provides additional information on the species composition of these assemblages (Orchard and Szpak, this volume). Stable isotope values were analyzed by Paul Szpak, using the McMaster Stable Isotope Research Laboratory, as part of a project examining food-web dynamics of taxa present in Gwaii Haanas assemblages (Szpak, Orchard, and Gröcke 2009). Bone collagen was extracted from twenty-one salmon vertebrae and measured for carbon (δ^{13}C) and nitrogen (δ^{15}N) isotopes, with results reported relative to VPDB and atmospheric nitrogen respectively. Two vertebrae produced C:N ratios outside of the accepted range (DeNiro 1985) and were excluded from further consideration. Among the remaining nineteen vertebrae, values for δ^{13}C range from –16.42‰ to –13.37‰. Values for δ^{15}N range from 11.14‰ to 13.68‰, falling into two separate groupings and pointing to a relatively limited range of trophic levels for these salmon. These values cannot be directly compared to those from previous salmon analyses (Satterfield and Finney 2002), as the current data represent isotope values from bone collagen and previous values are from muscle tissue. Given the lack of distinction between isotope values for small and large vertebrae in the Gwaii Haanas sample, however, these vertebrae fit best within the pink, chum, and sockeye group of Satterfield and Finney (2002), with a few specimens possibly falling within the coho, chum, and sockeye group. The propensity for coho and sockeye to spend a significant portion of their life in freshwater suggests that these species should demonstrate significantly depleted δ^{13}C values, which may distinguish them from pink and chum salmon (Orchard and Szpak, this volume). These results support the argument that the Gwaii Haanas assemblages are comprised primarily of chum, pink, and possibly coho salmon.

Origin of Gwaii Haanas Salmon Assemblages

While metric, radiographic, and isotopic data point to the dominance of chum and pink salmon and the possible presence of coho in the Gwaii Haanas salmon fisheries during the Late Holocene, it remains to be determined whether these assemblages resulted from local or mainland sources. Comparing the reconstructed assemblages with historic records of salmon populations in Gwaii Haanas provides a basis for assessing this question. The Department of Fisheries and Oceans Canada (DFO) has systematically collected salmon escapement data for southern Haida Gwaii since 1947 (DFO 2007; Krishka 1997). A close fit between these historic records and the reconstructed precontact salmon assemblages can reasonably be argued to indicate a strong likelihood that salmon harvesting targeted locally spawning populations.

Importantly, escapement surveys only record estimates of the numbers of salmon *returning to streams and rivers to spawn*. Since commercial salmon fishing began around 1900 in the Gwaii Haanas region (Krishka 1997; cf. Carrothers 1941), an unknown portion of the total salmon population is missing from escapement estimates. Additionally, commercial activities, including logging and mining, were common in Gwaii Haanas prior to the earliest escapement surveys (Dalzell 1968), and such activities undoubtedly negatively impacted salmon populations (Hewes 1973; Krishka 1997; Selbie et al. 2007). The twenty-four Gwaii Haanas streams monitored after 1947 reveal a dramatic decline in chum salmon from a peak of more than 400,000 fish in 1947, and an average exceeding 150,000 fish prior to 1952, to an average of under 20,000 fish in the early 1990s (Krishka 1997:56–59). These data suggest the possibility of much larger salmon populations prior to the period of active monitoring (Northcote and Atagi 1997).

Historic escapement estimates, given these limitations, provide a *minimal* estimate of the numbers of salmon spawning in the region. Escapement data for Gwaii Haanas throughout the second half of the twentieth century reveal large numbers of chum salmon, with reasonably high numbers of pink salmon, low numbers of coho salmon, and essentially no sockeye or chinook (Table 8.3). This pattern of abundance provides a remarkably good fit with the composition of archaeological assemblages from the region, suggesting that salmon harvesting likely targeted these local populations.

The question of scale also needs to be considered. Were the Gwaii Haanas salmon populations large enough to support an intensive fishery by Haida living in the region? Although we do not know precisely how many people lived on Haida Gwaii during different time periods, twentieth-century escapement figures suggest that well over a million salmon were available annually (Table 8.4). This must be considered a minimum estimate, considering historic impacts to fisheries. Therefore, local waters would appear to be the most likely source of salmon used by Haida Gwaii residents.

Other Lines of Evidence

Although most Late Holocene archaeological sites excavated in Gwaii Haanas have been villages and large shell middens, other site types provide support for the interpretation

Table 8.3. Department of Fisheries and Oceans escapement data for Gwaii Haanas for the years 1950 to 1999

Species	Mean annual escapement					
	1950s	1960s	1970s	1980s	1990s	1950–1999
Chum	128,673	162,074	71,527	107,140	44,736	102,830
Coho	6,144	15,800	6,030	2,033	622	6,125
Pink	22,529	65,152	79,544	76,866	48,488	58,516
Sockeye	0	5	0	9	1	3

Source: DFO 2007.

Table 8.4. Department of Fisheries and Oceans escapement data for Haida Gwaii for the years 1950 to 1999

Species	Mean annual escapement					
	1950s	1960s	1970s	1980s	1990s	1950–1999
Chinook	1,813	4,472	820	1,265	2,100	2,094
Chum	479,893	425,179	259,464	348,827	272,523	357,178
Coho	72,014	113,715	104,691	52,926	29,710	74,612
Pink	592,027	678,181	580,647	611,091	1,068,089	706,007
Sockeye	30,980	53,145	42,679	33,919	17,369	35,618

Source: DFO 2007.

of a locally focused salmon fishery. Intensive site surveys have identified at least sixty fish trap sites within Gwaii Haanas, including wood stake weirs (n = 26), rock wall traps (n = 22), and complexes including both wood stake alignments and rock walls (n = 12) (Acheson 1998; Eldridge, Mackie, and Wilson 1993; Mackie and Wilson 1994; Orchard et al. 2010; Zacharias and Wanagun 1993). This likely represents a fraction of the fish traps that existed in the late precontact period, as historic logging often involved floating and skidding logs out along creek beds and undoubtedly had an impact on weirs and stone walls (Mackie and Wilson 1994). Several rock wall sites in the intertidal zone (n = 7) are not directly associated with streams, although such fish traps could have targeted salmon moving through the outer islands and channels on their way to spawning streams (e.g., Langdon 2006). Most of the identified fish traps (n = 53, 88 percent) are directly associated with salmon-bearing streams and likely targeted salmon (Orchard et al. 2010). A preliminary program of dating these fish trap features has produced a series of twenty-four dates on wood stakes and other associated organic remains. All these dates fall within the past 1700 calibrated years BP, with the majority of the samples (n = 21, 88 percent) postdating 1200 cal BP (Orchard et al. 2010). The abundance of fish trap sites in Gwaii Haanas suggests an investment in intensive harvest of local salmon, while preliminary dating results support the rise of this intensive fishery within the Late Holocene, particularly within the past 1,200 years (Orchard 2009; Orchard and Clark 2005).

Site surveys in Gwaii Haanas have also identified a number of smaller sites, several of which are directly associated with fish trap features. Acheson (1998) excavated five small midden sites, only one of which was located on protected waters near a salmon stream. The vertebrate assemblage at this site (FaTs-35) was dominated by fish (92.5 percent of elements by class), of which 98 percent were salmon (Wigen 1990). This dominance, along with the presence of salmon teeth typical of spawning chum salmon, led Wigen (1990:16) to identify the site as a temporary camp for harvesting salmon from the local creek. This short-term salmon harvesting camp further strengthens the argument for a local salmon fishery in Gwaii Haanas. Unfortunately, the site has not been dated, but dating this and other similar deposits will improve our understanding of the salmon fishery in Gwaii Haanas.

The emerging picture of Late Holocene salmon utilization in Gwaii Haanas fits well with the limited ethnographic descriptions available. According to Curtis (1916), Dawson (1880), and Murdock (1934), chum salmon comprised a significant portion of ethnographic Haida subsistence. Perhaps the best description is by Dawson (1880:110B), who describes a large, hook-nosed variety of salmon, almost certainly chum, that

> ascend even very small streams when these are in flood with the autumn rains, and being easily caught and large, they constitute the great salmon harvest of the Haidas. They are generally either speared in the estuaries of the streams or trapped in fish-wiers [sic] made of split sticks, which are ranged across the brooks. The various "rivers" are the property of the several families or subdivisions of the tribes, and at the salmon fishing season the inhabitants are scattered from the main villages; each little party camped or living in temporary houses of slight construction in the vicinity of the streams they own.

Harrison (1925) also indicates that salmon were taken using wood stake weirs and basket traps, while Curtis (1916:131) further notes that "on account of its keeping qualities, dog-salmon [chum] was the principal storage food, and was used throughout the winter whenever the weather did not permit fishing." These descriptions indicate that ethnographically chum salmon were the primary salmon utilized by the Haida, that chum were harvested locally using weirs, and that temporary camps were established near salmon streams during the harvest season. This matches very closely with the archaeological picture described above.

Explaining Late Salmon Intensification in Gwaii Haanas

It seems clear, based on the wide range of evidence outlined above, that a locally focused salmon fishery existed in Gwaii Haanas throughout the millennium preceding European contact, and that it was sufficiently intensive to contribute a significant portion of the Haida diet. It is particularly notable that the increase in the relative abundance of salmon remains in village middens after 1200 cal BP (Orchard 2009; Orchard and Clark 2005) corresponds reasonably well with the appearance of fish weir technology after 1700 cal BP and the prominence of this technology after 1200 cal BP (Orchard et al. 2010). These dates contrast sharply with a variety of much earlier dates elsewhere on the Northwest Coast for both the prominence of salmon in faunal assemblages (Ames and Maschner 1999; Butler and O'Connor 2004; Cannon 1991, 1996; Cannon and Yang 2006; Coupland, Stewart, and Patton 2010; Matson and Coupland 1995) and for the appearance of weir technology (Eldridge and Acheson 1992; Moss and Erlandson 1998; Moss, Erlandson, and Stuckenrath 1990). In general, these results suggest that a proximal explanation for the late increase in salmon use evident in village middens in Gwaii Haanas was a late adoption of weir technology that facilitated this more intensive salmon harvest.

This argument, however, fails to consider the more complex factors that contributed to the adoption of weir technology itself. It is simply not possible, given current data, to determine conclusively why this economic shift occurred. Undoubtedly, it resulted from the complex interrelationship between subsistence, social organization, technology, economic organization, and broader patterns in population and environment. Although we do not yet have sufficient data to interpret these relationships, it is worthwhile to briefly consider some contributing factors that should be further explored in future research.

The lack of major salmon rivers and the ubiquity of small streams with more limited salmon resources within Haida Gwaii may help explain the delay in the intensification of salmon use in this area compared to the much earlier dates reported elsewhere on the Northwest Coast. These more limited salmon resources undoubtedly led my colleagues to question the potential for a locally based salmon economy in Gwaii Haanas. The prominence of very rocky, high-energy coastal environments in the region appears much more conducive to a fishery focused on taxa that thrive in such conditions, such as the rockfish that dominate assemblages predating the rise of salmon use (Acheson 1998; Orchard 2009; Orchard and Clark 2005). Local ecological conditions cannot entirely explain this pattern, however, as such outer coastal areas of the Northwest Coast, including Haida Gwaii, western Vancouver Island (Monks 2006; Streeter 2002), and southeast Alaska (Maschner 1991, 1997; Moss 1989; Moss and Erlandson 1998; Moss, Erlandson, and Stuckenrath 1990), did ultimately adopt economies that intensively used salmon.

Alternately, changes in the local abundance of relevant taxa, namely salmon and rockfish, may have contributed to the economic changes outlined here. Research in southcentral Alaska, for example, has identified a marked decline in sockeye salmon productivity between 100 BC and AD 800, and a period of consistent abundance from AD 1200 to 1900 (Finney et al. 2002; Gregory-Eaves et al. 2003). The intervening period of increasing salmon abundance between AD 800 and 1200 has been linked to population increases and increased use of salmon by Kodiak Islanders (Finney et al. 2002; Knecht 1995). The increasing focus on salmon use in Gwaii Haanas may also have been driven by declining returns on previously important resources. This is particularly true for an economy initially focused on rockfish (Orchard 2009; Orchard and Clark 2005), which are nonmigratory fish that commonly reach ages of fifty to two hundred years (Love, Yoklavich, and Thorsteinson 2002; Munk 2001). These life-history traits make rockfish highly susceptible to overexploitation (Berkeley et al. 2004), and McKechnie (2007) and McMillan et al. (2008) have documented prehistoric overharvesting and subsequent depression of local rockfish populations in Barkley Sound on western Vancouver Island. While patterns in the local abundance of prey species have thus been implicated in cultural developments elsewhere on the coast, it is not yet possible with existing data to assess the role of such local environmental changes in the economic shifts described for Late Holocene Haida Gwaii.

Demographic trends may also have contributed to changing economic strategies in Gwaii Haanas. Elsewhere on the Northwest Coast, major cultural and economic shifts, including intensification of salmon use, have been argued to result largely from population pressure (Croes and Hackenberger 1988). Acheson (1998) noted that in the Kunghit region of Gwaii Haanas the intensification of salmon appeared to coincide with an increase in house floor deposits. A storage-based economy has been widely argued to correlate with increased sedentism, both requiring and facilitating population increases (e.g., Rowley-Conwy and Zvelebil 1989). Although the increase in house floor deposits on Gwaii Haanas may indicate increasing sedentism, it may also reflect population increase, which in turn would benefit from the intensification of the salmon fishery.

Finally, increasing intercultural interaction across the Northwest Coast may have helped drive the shift to intensified salmon use on the outer coast. A variety of archaeological evidence has led MacDonald (1969; MacDonald and Inglis 1981) and others (Ames 2003; Croes 1997; Fladmark 1989; Sutherland 2001) to argue for the development of a North Coast Interaction Sphere, beginning as early as 4000 BP. For Haida Gwaii, Christensen and Stafford (2005; cf. Fedje and Mackie 2005) find no evidence for trade with the mainland prior to the Graham Tradition, which began ca. 5000 BP. Mackie and Acheson (2005:302) emphasize the geographic isolation of the archipelago and suggest that "Haida culture history can be characterised by long-term in situ development, with important but intermittent external influences from their neighbours." Interestingly, several of the excavated Late Holocene sites from Gwaii Haanas have produced artifacts, including small amber beads, ground nephrite, and dentalium beads, which suggest possible trade with the mainland (Acheson 1998; Orchard 2009). Although some level of interaction and trade may have existed much earlier (Carlson 1994), increasing interaction between Haida Gwaii and the mainland during the Late Holocene may have contributed to the adoption of a salmon-focused economy in Haida Gwaii mirroring that which had existed for a much longer period among mainland groups (Coupland, Stewart, and Patton 2010). This fits with suggestions that the Late Graham Tradition (ca. 2000 BP to contact) was characterized by the rise of extensive trade (Fedje and Mackie 2005; Fedje et al. 2008).

Conclusions

Late Holocene fisheries in Gwaii Haanas are characterized by variable use of several prominent fish taxa, including herring, salmon, rockfish, and halibut (Acheson 1998; Orchard 2009). Cross-cutting this regional pattern is a temporal shift from earlier, more generalized economies with a slight focus on rockfish prior to ca. 1200 BP to later specialized economies focused on salmon after 800 BP, with a transitional period in between (Orchard 2009; Orchard and Clark 2005). The late prominence of salmon is notable given the more general importance of salmon across the Northwest Coast at various times during the Holocene.

Radiographic and metric analyses of a sample of almost 2,000 salmon vertebrae from Late Holocene sites in Gwaii Haanas, and the analysis of a small selection of vertebrae for stable isotopes of nitrogen and carbon, suggest that archaeological salmon assemblages from Gwaii Haanas are dominated by chum salmon and also contain pink and possibly coho salmon, while sockeye and chinook appear to be absent. This species composition matches very well with salmon populations spawning in Gwaii Haanas as recorded in

DFO escapement surveys. Surveys also point to an average of more than one million salmon spawning in Haida Gwaii in the second half of the twentieth century, after significant impacts from commercial fishing, mining, and logging. Together, these data suggest that the rise of an economy characterized by intensive salmon harvesting after 1200 BP was focused on local salmon populations. It is impossible to rule out the possibility that some salmon were obtained from more distant locales or through alternate means, but it appears that many, if not all, of the salmon represented in archaeological assemblages from Gwaii Haanas were obtained through a focused harvest of locally spawning populations.

This interpretation is further supported by the presence of numerous fish trap sites and salmon harvesting sites in Gwaii Haanas and is consistent with ethnographic descriptions of Haida salmon fishing. In fact, the rise of a salmon-focused economy in the Late Holocene, much later than similar economies described elsewhere on the Northwest Coast (Butler and O'Connor 2004; Cannon and Yang 2006; Coupland, Stewart, and Patton 2010; Matson and Coupland 1995), was likely based on a comparably late adoption of mass harvest technologies, namely fish weirs. Dates from a preliminary sample of twenty-four wood stakes in Gwaii Haanas are all more recent than 1700 cal BP, with the majority postdating 1200 cal BP (Orchard et al. 2010). At this point, it is not possible to interpret the more fundamental factors responsible for this economic shift. Likely, these changes resulted from a combination of local and regional cultural and demographic factors. Increasing human population in Haida Gwaii may have stimulated the need for an intensified focus on a storable resource (after Croes and Hackenberger 1988), while increasing intercultural interaction on the north coast may have exposed the Haida to the possibilities of a salmon-focused economy as it existed among the Tsimshian and other mainland groups. Currently available data do not, however, facilitate an assessment of the complex and interrelated changes that must have occurred in subsistence, social organization, technology, and economic organization, along with their possible links to population and environment.

Considerable work remains to be done to further flesh out these arguments. Clearly, a locally focused, intensive, salmon-harvesting economy developed in Gwaii Haanas after ca. 1200 cal BP. This represents a significant advance in our understanding of economic patterns in Haida Gwaii and on the Northwest Coast more generally. These results also provide further evidence of the widely variable timing and nature of fisheries practices and cultural developments across the Northwest Coast, and contribute to our understanding of the variable trajectories towards the rise of the ethnographically described cultures.

Acknowledgments

I would like to thank Aubrey Cannon and Madonna Moss for the invitation to participate in the North Pacific Fisheries session at the 2008 SAA Conference. Financial support during the preparation of this paper was provided by the Social Science and Humanities Research Council of Canada. The recent archaeological research in Gwaii Haanas was made possible by the Council of the Haida Nation, the Archipelago Management Board, Parks Canada, Gwaii Haanas Staff, Marty Magne, Daryl Fedje, Ian Sumpter, Alan Davidson, Quentin Mackie, GHEAP field crews, Susan Crockford and Becky Wigen, the British Columbia

Heritage Trust, SSHRC, the University of Toronto, and the University of Victoria. Gord Mawdsley at Sunnybrook Health Sciences Centre in Toronto provided access to radiography equipment and helped with the analysis of salmon vertebrae. Meghan Burchell, Aubrey Cannon, Terence Clark, and Madonna Moss provided useful comments on earlier drafts of this paper.

References Cited

Acheson, Steven. *In the Wake of the* Ya'áats' Xaatgáay *["Iron People"]: A Study of Changing Settlement Strategies among the Kunghit Haida*. BAR International Series 711. Oxford: British Archaeological Reports, 1998.

Ames, Kenneth M. "The Northwest Coast." *Evolutionary Anthropology* 12 (2003): 19–33.

Ames, Kenneth M., and Herbert D. G. Maschner. *People of the Northwest Coast: Their Archaeology and Prehistory*. London: Thames and Hudson, 1999.

Berkeley, Steven A., Mark A. Hixon, Ralph J. Larson, and Milton S. Love. "Fisheries Sustainability via Protection of Age Structure and Spatial Distribution of Fish Populations." *Fisheries* 29 (2004): 23–32.

Butler, Virginia L. "Natural Versus Cultural Salmonid Remains: Origin of The Dalles Roadcut Bones, Columbia River, Oregon, U.S.A." *Journal of Archaeological Science* 20 (1993): 1–24.

Butler, Virginia L., and Sarah K. Campbell. "Resource Intensification and Resource Depression in the Pacific Northwest of North America: A Zooarchaeological Review." *Journal of World Prehistory* 18 (2004): 327–405.

Butler, Virginia L., and Jim E. O'Connor. "9000 Years of Salmon Fishing on the Columbia River, North America." *Quaternary Research* 62 (2004): 1–8.

Cannon, Aubrey. *The Economic Prehistory of Namu*. Burnaby, BC: Department of Archaeology, Simon Fraser University, 1991.

——."The Early Namu Archaeofauna." In *Early Human Occupation in British Columbia*, Roy L. Carlson and Luke Dalla Bona, eds., pp. 103–110. Vancouver: UBC Press, 1996.

——."Faunal Remains as Economic Indicators on the Pacific Northwest Coast." In *Animal Bones, Human Societies*, Peter Rowley-Conwy, ed., pp. 49–57. Oxford: Oxbow Books, 2000a.

——."Assessing Variability in Northwest Coast Salmon and Herring Fisheries: Bucket-Auger Sampling of Shell Midden Sites on the Central Coast of British Columbia." *Journal of Archaeological Science* 27 (2000b): 725–737.

Cannon, Aubrey, and Dongya Y. Yang. "Early Storage and Sedentism on the Pacific Northwest Coast: Ancient DNA Analysis of Salmon Remains from Namu, British Columbia." *American Antiquity* 71 (2006): 123–140.

Carlson, Roy L. "Trade and Exchange in Prehistoric British Columbia." In *Prehistoric Exchange Systems in North America*, Timothy G. Baugh and Jonathon E. Ericson, eds., pp. 307–361. New York: Plenum, 1994.

Carrothers, W. A. *The British Columbia Fisheries*. Political Economy Series, No. 10. Toronto: University of Toronto Press, 1941.

Casteel, Richard W. "Comparison of Column and Whole Unit Samples for Recovering Fish Remains." *World Archaeology* 8 (1976): 192–196.

Christensen, Tina, and Jim Stafford. "Raised Beach Archaeology in Northern Haida Gwaii: Preliminary Results from the Cohoe Creek Site." In *Haida Gwaii: Human History and Environment from the Time of Loon to the Time of the Iron People*, Daryl W. Fedje and Rolf W. Mathewes, eds., pp. 245–273. Vancouver: UBC Press, 2005.

Coupland, Gary, Kathlyn Stewart, and Katherine Patton. "Do You Never Get Tired of Salmon? Evidence for Extreme Salmon Specialization at Prince Rupert Harbour, British Columbia." *Journal of Anthropological Archaeology* 29 (2010): 189–207.

Croes, Dale R. "The North-Central Cultural Dichotomy on the Northwest Coast of North America: Its Evolution as Suggested by Wet-Site Basketry and Wooden Fish-Hooks." *Antiquity* 71 (1997): 594–615.

Croes, Dale R., and Steven Hackenberger. "Hoko River Archaeological Complex: Modeling Prehistoric Northwest Coast Economic Evolution." In *Prehistoric Economies of the Pacific Northwest Coast*, B. L. Isaac, ed., pp. 19–85. Research in Economic Anthropology Supplement 3. Greenwich, CT: JAI Press, 1988.

Curtis, Edward S. *The North American Indian*, vol. 11. New York: Johnson Reprint Corporation, 1916.

Dalzell, Kathleen E. *The Queen Charlotte Islands, vol. 1: 1774–1968*. Madeira Park, BC: Harbour Publishing, 1968.

Dawson, George M. *Report on the Queen Charlotte Islands, 1878*. Geological Survey of Canada Report of Progress for 1878–79, vol. 4. Montreal: Geological Survey of Canada, 1880.

DeNiro, Michael J. "Postmortem Preservation and Alteration of *In Vivo* Bone Collagen Isotope Ratios in Relation to Palaeodietary Reconstruction." *Nature* 317 (1985): 806–809.

Department of Fisheries and Oceans Canada (DFO). Salmon Escapement Figures for Statistical Areas 1 and 2. http://www.pac.dfo-mpo.gc.ca/northcoast/Areas/ (accessed August 6, 2007).

Donald, Leland. "The Northwest Coast as a Study Area: Natural, Prehistoric, and Ethnographic Issues." In *Emerging from the Mist: Studies in Northwest Coast Culture History*, R. G. Matson, Gary Coupland, and Quentin Mackie, eds., pp. 289–327. Vancouver: UBC Press, 2003.

Eldridge, Morley, and Steven Acheson. "The Antiquity of Fish Weirs on the Southern Coast: A Response to Moss, Erlandson, and Stuckenrath." *Canadian Journal of Archaeology* 16 (1992): 112–116.

Eldridge, Morley, Alexander Mackie, and Bert Wilson. Archaeological Inventory of Gwaii Haanas National Park Reserve, 1992. MS on file, Parks Canada, Victoria, 1993.

Fedje, Daryl W., and Quentin Mackie. "Overview of Cultural History." In *Haida Gwaii: Human History and Environment from the Time of Loon to the Time of the Iron People*, Daryl W. Fedje and Rolf W. Mathewes, eds., pp. 154–162. Vancouver: UBC Press, 2005.

Fedje, Daryl, Quentin Mackie, Duncan McLaren, and Tina Christensen. "A Projectile Point Sequence for Haida Gwaii." In *Projectile Point Sequences in Northwestern North America*, Roy L. Carlson and Martin P. R. Magne, eds., pp. 21–43. Burnaby, BC: Archaeology Press, Simon Fraser University, 2008.

Finney, Bruce P., Irene Gregory-Eaves, Marianne S. V. Douglas, and John P. Smol. "Fisheries Productivity in the Northeastern Pacific Ocean over the Past 2,200 Years." *Nature* 416 (2002): 729–733.

Fladmark, Knut R. "The Native Culture History of the Queen Charlotte Islands." In *The Outer Shores*, Geoffrey G. E. Scudder and Nicholas Gessler, eds., pp. 199–221. Skidegate, BC: Queen Charlotte Islands Museum Press, 1989.

Gregory-Eaves, Irene, John P. Smol, Marianne S. V. Douglas, and Bruce P. Finney. "Diatoms and Sockeye Salmon (*Oncorhynchus nerka*) Population Dynamics: Reconstructions of Salmon-derived Nutrients over the Past 2,200 Years in Two Lakes from Kodiak Island, Alaska." *Journal of Paleolimnology* 30 (2003): 35–53.

Hanson, Diane K. "Late Prehistoric Subsistence in the Strait of Georgia Region of the Northwest Coast." PhD diss., Simon Fraser University. Ann Arbor, MI: University Microfilms, 1991.

Harrison, Charles, Rev. *Ancient Warriors of the North Pacific: The Haidas, Their Laws, Customs and Legends, with Some Historical Account of the Queen Charlotte Islands.* London: H. F. and G. Witherby, 1925.

Hart, John Lawson. *Pacific Fishes of Canada.* Ottawa: Fisheries Research Board of Canada, 1973.

Hewes, Gordon W. "Indian Fisheries Productivity in Pre-contact Times in the Pacific Salmon Area." *Northwest Anthropological Research Notes* 7 (1973): 133–155.

Knecht, Richard. "The Late Prehistory of the Alutiiq People: Culture Change on the Kodiak Archipelago from 1200–1750 AD." PhD diss., Bryn Mawr College. Ann Arbor, MI: University Microfilms, 1995.

Kopperl, Robert E. "Herring Use in Southern Puget Sound: Analysis of Fish Remains at 45-KI-437." *Northwest Anthropological Research Notes* 35 (2001): 1–20.

Krishka, Brian. Summary of Aquatic Resource Information for Gwaii Haanas: A Compilation and Synthesis of Aquatic Inventories for Lakes and Streams. MS on file, Gwaii Haanas National Park Reserve, Skidegate, BC, 1997.

Langdon, Stephen J. "Tidal Pulse Fishing: Selective Traditional Tlingit Salmon Fishing Techniques on the West Coast of the Prince of Wales Archipelago." In *Traditional Ecological Knowledge and Natural Resource Management*, Charles R. Menzies, ed., pp. 21–46. Lincoln: University of Nebraska Press, 2006.

Love, Milton S., Mary Yoklavich, and Lyman Thorsteinson. *The Rockfishes of the Northeast Pacific.* Berkeley: University of California Press, 2002.

MacDonald, George F. "Preliminary Culture Sequence from the Coast Tsimshian Area, British Columbia." *Northwest Anthropological Research Notes* 3 (1969): 240–254.

MacDonald, George F., and Richard I. Inglis. "An Overview of the North Coast Prehistory Project (1966–1980)." *BC Studies* 48 (1981): 37–63.

Mackie, Alexander P., and Bert Wilson. Archaeological Inventory of Gwaii Haanas 1993. Final Report Prepared for Archaeological Services, Western Region, Parks Canada and Gwaii Haanas Archipelago Management Board. MS on file, Parks Canada, Victoria, 1994.

Mackie, Quentin, and Steven Acheson. "The Graham Tradition." In *Haida Gwaii: Human History and Environment from the Time of Loon to the Time of the Iron People*, Daryl W. Fedje and Rolf W. Mathewes, eds., pp. 274–302. Vancouver: UBC Press, 2005.

Maschner, Herbert D. G. "The Emergence of Cultural Complexity on the Northern Northwest Coast." *Antiquity* 65 (1991): 924–934.

———."Settlement and Subsistence in the Later Prehistory of Tebenkof Bay, Kuiu Island, Southeast Alaska." *Arctic Anthropology* 34 (1997): 74–99.

Matson, R. G., and Gary Coupland. *The Prehistory of the Northwest Coast.* San Diego: Academic Press, 1995.

McKechnie, Iain. "Investigating the Complexities of Sustainable Fishing at a Prehistoric Village in Western Vancouver Island, British Columbia, Canada." *Journal for Nature Conservation* 15 (2007): 208–222.

McMillan, Alan D., Iain McKechnie, Denis E. St. Claire, and S. Gay Frederick. "Exploring Variability in Maritime Resource Use on the Northwest Coast: a Case Study from Barkley Sound, Western Vancouver Island." *Canadian Journal of Archaeology* 32 (2008): 214–238.

Monks, Gregory G. "The Fauna from Ma'acoah (DfSi-5), Vancouver Island, British Columbia: An Interpretive Summary." *Canadian Journal of Archaeology* 30 (2006): 272–301.

Moss, Madonna L. "Archaeology and Cultural Ecology of the Prehistoric Angoon Tlingit." PhD diss., University of California, Santa Barbara. Ann Arbor, MI: University Microfilms, 1989.

Moss, Madonna L., and Jon M. Erlandson. "A Comparative Chronology of Northwest Coast Fishing Features." In *Hidden Dimensions: The Cultural Significance of Wetland Archaeology*, Kathryn Bernick, ed., pp.180–198. Vancouver: UBC Press, 1998.

Moss, Madonna L., Jon M. Erlandson, and Robert Stuckenrath. "Wood Stake Weirs and Salmon Fishing on the Northwest Coast: Evidence from Southeast Alaska." *Canadian Journal of Archaeology* 14 (1990): 143–158.

Munk, Kristen M. "Maximum Ages of Groundfishes in Waters off Alaska and British Columbia and Considerations of Age Determination." *Alaska Fishery Research Bulletin* 8 (2001): 12–21.

Murdock, George P. "The Haidas of British Columbia." In *Our Primitive Contemporaries*, G. P. Murdock, pp. 221–263. New York: MacMillan Company, 1934.

Northcote, T. G., and D. Y. Atagi. "Pacific Salmon Abundance Trends in the Fraser River Watershed Compared with Other British Columbia Systems." In *Pacific Salmon & Their Ecosystems: Status and Future Options*, Deanna J. Stouder, Peter A. Bisson, and Robert J. Naiman, eds., pp. 199–219. New York: International Thomson Publishing, 1997.

Orchard, Trevor J. "Otters and Urchins: Continuity and Change in Haida Economy during the Late Holocene and Maritime Fur Trade Periods." PhD diss., University of Toronto. Ann Arbor, MI: University Microfilms, 2007.

———. *Otters and Urchins: Continuity and Change in Haida Economy during the Late Holocene and Maritime Fur Trade Periods*. BAR International Series 2027. Oxford: Archaeopress, 2009.

Orchard, Trevor J., and Terence Clark. "Multidimensional Scaling of Northwest Coast Faunal Assemblages: A Case Study from Southern Haida Gwaii, British Columbia." *Canadian Journal of Archaeology* 29 (2005): 88–112.

Orchard, Trevor J., Nicole Smith, Iain McKechnie, and Daryl Fedje. "Terrestrial, Aquatic and Intertidal Archaeological Resources in Gwaii Haanas: Towards a More Complete Picture of Late Holocene Human Resource and Landscape Use." Paper presented at the Canadian Archaeological Association Conference, Calgary, Alberta, 2010.

Orchard, Trevor J., and Rebecca J. Wigen. "Halibut Use on the Northwest Coast Reconciling Ethnographic, Ethnohistoric, and Archaeological Data." Paper presented at the Northwest Anthropological Conference, Victoria, BC, April 2008.

Rowley-Conwy, Peter, and Marek Zvelebil. "Saving It for Later: Storage by Prehistoric Hunter-Gatherers in Europe." In *Bad Year Economics: Cultural Responses to Risk and Uncertainty*, Paul Halstead and John O'Shea, eds., pp. 40–56. Cambridge: Cambridge University Press, 1989.

Satterfield, Franklin R., IV, and Bruce P. Finney. "Stable Isotope Analysis of Pacific Salmon: Insight into Trophic Status and Oceanographic Conditions over the Last 30 Years." *Progress in Oceanography* 53 (2002): 231–246.

Selbie, Daniel T., Bert A. Lewis, John P. Smol, and Bruce P. Finney. "Long-Term Population Dynamics of the Endangered Snake River Sockeye Salmon: Evidence of Past Influences on Stock Decline and Impediments to Recovery." *Transactions of the American Fisheries Society* 136 (2007): 800–821.

Streeter, Ian M. "Seasonal Implications of Rockfish Exploitation in the Toquaht Area, British Columbia." Master's thesis, University of Manitoba, 2002.

Sutherland, Patricia D. "Revisiting an Old Concept: The North Coast Interaction Sphere." In *Perspectives on Northern Northwest Coast Prehistory*, Jerome S. Cybulski, ed., pp. 49–59.

Archaeological Survey of Canada, Mercury Series Paper 160. Hull, Quebec: Canadian Museum of Civilization, 2001.

Suttles, Wayne, ed. *Northwest Coast*. Handbook of North American Indians, vol. 7. Washington, DC: Smithsonian Institution, 1990.

Szpak, Paul, Trevor J. Orchard, and Darren R. Gröcke. "A Late Holocene Vertebrate Food Web from Southern Haida Gwaii (Queen Charlotte Islands, British Columbia)." *Journal of Archaeological Science* 36 (2009): 2734–2741.

Wigen, Rebecca J. Identification and Analysis of Vertebrate Fauna from Eighteen Archaeological Sites on the Southern Queen Charlotte Islands. Report to British Columbia Heritage Trust. MS on file, Cultural Resource Library, Ministry of Sustainable Resource Management, Victoria, BC, 1990.

Wigen, Rebecca J., and Barbara R. Stucki. "Taphonomy and Stratigraphy in the Interpretation of Economic Patterns at Hoko River Rockshelter." In *Prehistoric Economies of the Pacific Northwest Coast*, B. L. Isaac, ed., pp. 87–146. Research in Economic Anthropology Supplement 3. Greenwich, CT: JAI Press, 1988.

Zacharias, Sandra K., and Wanagun (Richard S. Wilson). Gwaia Haanas 91 Archaeology: 1991 Archaeological Resource Inventory of Southeastern Moresby Island within the Gwaii Haanas/ South Moresby National Park Reserve. MS on file, Parks Canada, Victoria, 1993.

CHAPTER 9

Locational Optimization and Faunal Remains in Northern Barkley Sound, Western Vancouver Island, British Columbia

Gregory G. Monks, *Department of Anthropology, University of Manitoba*

W hy are sites located where they are? The reasons are many and include the availability of suitable terrain, fresh water, subsistence and other resources, and protection (or exposure) to specific climatic conditions and to features of the social environment, to name a few. On the Northwest Coast, consideration of this subject has generally focused on the availability of subsistence resources within a cultural ecology conceptual framework. Dominating characteristics of the environment, particularly the abundance of salmon, and variability in the distribution of this and other subsistence resources are generally viewed as primary determinants of cultural adaptations, including settlement location.

The lip service that cultural ecology pays to the assumed reciprocal relationship between culture and environment tends to mask the largely deterministic application of this theoretical framework. In Northwest Coast archaeological literature, this determinism is reflected in analyses that adopt Suttles's (1962, 1968) model of environmental variation. To be sure, variation in the availability of subsistence resources is an empirically real phenomenon, and cultural groups did indeed adapt to this variability in a number of ways. But two critical elements of this conceptualization are missing in most applications. One is the corresponding effect of culture on environment, which the theoretical framework of cultural ecology assumes. This reciprocal effect has been conveniently overlooked until recently (Kuhnlein and Turner 1991; Turner 1999; Williams 2006). The second missing element is an avenue of inquiry that enables researchers to understand how and in what ways the variation in the environment is translated into specific cultural formations (but see Donald and Mitchell 1975, 1994). Further, recognizing these specific formations in the archaeological record and setting them in the context of environmental variation is difficult for the archaeologist.

This paper addresses the problem of identifying in the archaeological record how specific aspects of a once-extant culture are related to specific aspects of environmental variation. This aim is pursued using the concept of locational optimization from optimal foraging theory (OFT) and applying it to faunal assemblages, mainly fish, from two sites in northern Barkley Sound, western Vancouver Island, British Columbia (Figure 9.1), excavated as part of the Toquaht Archaeological Project. Site location is here taken as one aspect of cultural adaptation that addressed environmental variation in subsistence resource availability, and the faunal data are taken as supporting (or not supporting) the reasons why sites were located where they were.

Background

Drucker (1951:37, 49) indicated that local groups who occupied "outside" sites (those exposed to the open ocean) endeavored to gain access to "inside" sites (protected inner coast locations) that controlled salmon streams. Whaling was reported as having emerged as a matter of survival by local groups on the outside of Nootka Island and on the outside of Esperanza Inlet who had no access to salmon streams. From these groups, whaling spread among the Nuu-chah-nulth (Drucker 1951:49). The principal social, economic, and political unit of the Nuu-chah-nulth was the local group, a set of patrilineally related males, along with their wives, children, and associates, who shared common descent from a known ancestor, i.e., a lineage. Some lineages shared a common winter village with other lineages, in which case Drucker identified them as tribes. In turn, some tribes were organized into confederacies, of which Drucker identified four (Drucker 1951:220). Although Drucker did not address the matter, it follows that larger social groupings would have broader access to a range of territories and resources than smaller groupings. This interpretation is supported by the research of Donald and Mitchell (1994), who showed that complexity of sociopolitical organization among the Nuu-chah-nulth varied directly with salmon escapements. Drucker also implied that some groups were exclusively "inside" and some were exclusively "outside." While this situation may have been true, it is more likely to have applied to local groups than to tribes or confederacies, and the situation among the Toquaht, to whom this chapter now turns, was not as Drucker described.

The Toquaht were the original Nuu-chah-nulth occupants of Barkley Sound (St. Claire 1991:53), and their territory extended from the mouth of Ucluelet Inlet on the "outside" to Lyell Point on the "inside" (Figure 9.2). They were an autonomous local group that, despite massive population decline after European contact, never amalgamated into a larger social unit (St. Claire 1991:55). The Toquaht were therefore the smallest social unit that Drucker defined, but they had access to both "outside" and "inside" territory and the resources therein. Earliest occupancy of Toquaht territory began approximately 4,000 to 3,800 years

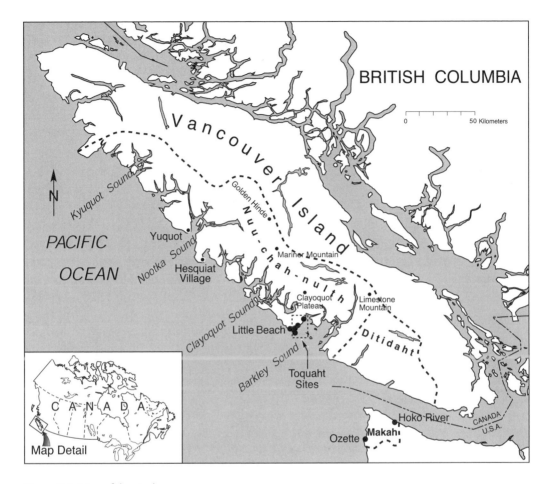

Figure 9.1. Map of the study area.

ago (McMillan 1999:71, Table 4) with a material culture that most closely resembles that of the Locarno Beach period in the Strait of Georgia. Approximately 2,000 years ago, the chipped stone assemblage disappeared and there appeared a material culture described as the West Coast Culture Type (Mitchell 1990), consistent with ethnographic Nuu-chah-nulth assemblages (McMillan 1999:71).

The Toquaht Archaeological Project conducted excavations and survey in Toquaht territory between 1991 and 1996. Excavations at five sites revealed that, over the span of occupation, one original site, DfSi-4, known as Ch'uumat'a, showed the earliest evidence of occupation beginning about ca. 4000–3800 BP and continuing until approximately 720 BP (McMillan 1999, Table 4). Prior to the cessation of major occupation at this site, a second site, DfSj-23, known as T'ukw'aa, was established facing the open ocean. This site was also large, and one portion exhibits front and rear terraces. Early occupation of this site began about 1200 BP and continued into the historic period. About 800 BP the rocky pinnacle adjacent to the village site (DfSj-23B) began use as a defensive location. In the nineteenth century, these sites were described as a spring and summer village of the Toquaht where fish, seals, and whales were taken (McMillan 1999:66–67).

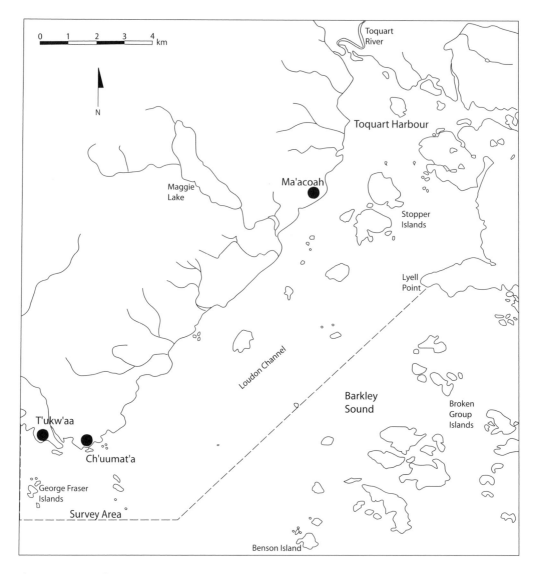

Figure 9.2. Map of Toquaht traditional territory showing selected site locations.

Approximately 600 BP the "inside" site of Ma'acoah (DfSi-5) was established.* This site was identified in oral history and in nineteenth-century Euro-Canadian documents as a winter village and fishing station (McMillan 1999:63). Two smaller sites, DfSj-29 and DfSj-30 on the offshore George Fraser Islands, were also tested and revealed occupation histories spanning the past 400 years. These sites are thought to be lookout sites for monitoring whale movements, but they may also have been distant early warning locations in times of hostility.

There was, then, an apparent proliferation of sites in Toquaht territory. Early occupants who lived at Ch'uumat'a and who possessed a material culture similar to the Locarno Beach period material in the Strait of Georgia appear either to have been replaced around 2000 BP

---

* An earlier date of 1,840 ± 80 BP was obtained from clay at the base of the site (McMillan 1999, Table 4), but the cultural origin of this date is questionable (McMillan, 2003, pers. comm.).

by Wakashan-speaking people or to have taken up a material culture consistent with the ethnographic Nuu-chah-nulth (McMillan 1999; Mitchell 1990). About 800 years after the transition to the West Coast Culture Type material culture, the second site of T'ukw'aa was established, creating two major habitation sites in the outer part of Toquaht territory. About 400 years later, major occupation at Ch'uumat'a ceased, and the defensive location adjacent to T'ukw'aa came into use. Several hundred years later, about 600 BP, the inner site of Ma'acoah came into use, creating "inside" and "outside" focal sites for the Toquaht. Missing in this sketch is the real possibility of a major habitation site at the mouth of the Maggie River midway between the inner and outer sites. Excavation, analysis, and dating of this site will be important for a fuller understanding of the evolution of Toquaht settlement. Occupation on offshore islands involved special purpose satellite sites apparently related to observation of the seascape off Toquaht territory. The picture of settlement history that currently exists is one of site proliferation from an original location, Ch'uumat'a, that continued to be used despite a change in material culture and possibly in occupants. The more recent occupants of that site established a second site, T'ukw'aa, that was large, terraced, and distinctly "outside." Sometime thereafter, the original site fell into relative disuse and the defensive location at the outside site was established. Shortly thereafter, a major "inside" site was founded and used and, still later, offshore observation sites came into use.

Why did these changes in settlement locations take place? Two approaches to this question have been presented. First, Streeter's (2002) analysis of rockfish seasonality indicated that the abundance of rockfish in Toquaht faunal assemblages declined markedly in the later prehistoric period at the same time as salmon abundance markedly increased. This finding has been verified at Ma'acoah (Monks 2006), at Ts'ishaa (DfSi-16) in the Broken Group of central Barkley Sound (Frederick and Crockford 2003; McKechnie 2005), and, as described below, at T'ukw'aa. Rahn (2002) presented the second approach to this question by using nearest neighbor analysis to show a transition from three focal settlement areas during the archaeological past to two focal settlement areas in the ethnographic present. The spatial distribution of the three focal settlement areas in the archaeological past appeared to be based on considerations of social conceptions or appropriate inter-site distance, whereas the two focal settlement areas during the ethnographic period were based on access to subsistence resources. These two studies suggest that the shift from rockfish to salmon in the Toquaht faunal assemblages occurs in the context of the shift from three focal settlements to two. The studies also suggest that other approaches may shed light on the question of why this shift occurred and why "inside" and "outside" focal areas characterize the ethnographic period.

The approach taken here is that of locational optimization. Set within optimal foraging theory, this approach hypothesizes that groups will attempt to maximize their access to, hence maximize the productivity of, resource patches through strategic placement of habitation sites that minimize costs in terms of travel time and effort, search time, and handling time (Smith 1981:634). The majority of the faunal remains in all Toquaht assemblages derive from marine environments, so travel is assumed to have occurred by watercraft. This assumption permits the use of distance as a proxy for travel cost because the same surface is being traversed. Thus, locational optimization predicts that the faunal resources brought to a site will tend to come from proximate patches rather than distant patches and that faunal assemblages will therefore tend to reflect locally exploited resource patches.

The following analysis compares the faunal assemblages from T'ukw'aa (DfSj-23A), an "outside" site, and Ma'acoah (DfSi-5), an "inside" site, both of which constitute the focal points of the ethnographic period settlement pattern in Rahn's analysis.

The different environments in which the sites are located should produce different faunal signatures if the locational optimization model applies. The inside site, Ma'acoah, should exhibit fish remains dominated by salmon, herring, and small flatfish, and T'ukw'aa should exhibit more remains of halibut. Among the sea mammals, sea otters, fur seals, and other large pinnipeds, as well as whales and delphinids, should be more abundant at the "outside" site whereas the "inside" site should contain more harbor seals. Among the shellfish, California mussels and large acorn barnacles should be more abundant at T'ukw'aa, and butter clams, bent nose clams, and bay mussels should be more common at Ma'acoah. The birds and land mammals are more difficult to characterize due to their mobility, in the former case, and their ubiquity, in the latter case. The only exception where birds are concerned might be albatross, although the now-extinct *Phoebastria albatrus* likely was available at both sites and the pelagic albatross *P. nigripes* is normally found far offshore and therefore largely out of reach of "outside" site occupants.

Optimal foraging theory assumes that economic rationality is of primary importance. As originally formulated, it also assumes that the forager is an individual, and that each individual is in competition with all other individuals. These assumptions present problems when applying OFT to most human groups. At some level, inter-individual competition occurs, but it does so within a social structure, and it occurs also in direct and indirect ways. As well, groups compete directly and indirectly with each other, and the members of each group perform a range of tasks to that end.

The issue of groups and optimization theory has been addressed by Borgerhof Muldur (1991), among others, with the intent of harmonizing OFT with the human practice of group behavior. Her formulations continue to focus on the individual and optimization, whereas the individual is not always at liberty, especially in traditional societies, to operate as a free agent in all circumstances. Thus, a consideration of the group as the forager is required.

The internal dynamics of foraging groups are highly variable and must be discussed on a case-by-case basis. Binford's (1980) formulation focused on groups in general and their behavior with respect to obtaining food and other resources. Binford (1980) provided a model in which hunter-gatherer groups may be positioned along a scale that is characterized at one end by foragers who move opportunistically through their environment, taking resources as they are encountered. It should be noted that the forager in Binford's definition is much more specific than the forager of OFT. At the other end of Binford's scale are collectors who strategically plan to move to specific places at specific times of year with specific task groups in order to acquire specific resources. This model has strongly influenced North American archaeologists, but it contains an inherent problem: assigning a particular group to a particular point on the scale. Is a specific group more collector than forager? Are they more collector than another group? According to what criteria are these decisions made?

An alternative to Binford's model is the concept of multiple layers rather than a single scale. Some items are widely distributed and are easily obtained in a number of places. Other items are more restricted in their distribution and are less easily obtained. The availability of resources is not just a spatial issue—it is also an issue of seasonal availability and the ease with which they can be obtained with the available technology. Similarly, the importance to a group of different resources may vary. A result of these considerations is the proposal made here that one might identify a foraging layer that consists of those resources that are ubiquitous in time and/or space. Some of these may be culturally important while others are relatively unimportant. In Barkley Sound, for example, clams and rockfish are widely distributed and they are important whereas sculpins are widely distributed but unimportant. One can also identify a collecting layer that consists of those important resources that are not widely available in time and/or space. The important resources in this layer are obtained through scheduling decisions involving where to be at what time with what task group. Some of the important resources will be taken in abundance (e.g., salmon, herring), whereas other important resources are not available in such abundance (e.g., halibut). Resources that are widely available and important will have relatively little effect on where sites are located, especially those sites identified by Binford as residential sites. Specific purpose extractive sites support such residential sites, and the faunal remains at each site type should reflect this relationship; the former should contain a more limited range of fauna than the latter. Sometimes, however, special purpose extraction also occurs from residential sites, e.g., whaling and halibut fishing, in which case the residential site faunal assemblage will reflect both the common, everyday resources obtained from the site as well as special resources that can only be obtained from that location.

Locational optimization, then, may be better reflected in special purpose camps than in residential sites. Residential camps located proximate to a number of potential resource patches that may or may not be supported by special purpose extractive camps will show a diversity of faunal remains, and it is only the special purpose camps that will exhibit a limited range of fauna in large numbers (e.g., salmon fishing stations, bison drive sites). Thus the choice of where to locate a residential site may not be influenced by the ubiquitous important and unimportant (foraged) resources; instead, it will likely be a few important and unevenly distributed (collected) resources that, in combination with the important foraged resources, have the greatest influence on the choice of where to locate a residential site.

None of the foregoing discussion considers anything apart from economic maximization of benefit over cost. Marshall (1993), however, included a mix of topographic, social, and resource variables in her analysis of site locations in Nootka Sound. Her analysis, aimed at tracing Nuu-chah-nulth political continuity from the past to present, identified such variables as inner versus outer coast setting, presence of whaling, availability of salmon, seasonal occupation, and political aggregation as factors that influenced site size and location (Marshall 1993:143–146). Her study did not attempt to demonstrate specifically which variables had what effect on site size, location, function, or intensity of occupation, but the broad mix of relevant variables that she identified contains within it the seeds of further analyses. Among other things, Marshall classified sites on the basis of size. Type II sites are large, show surface features, were placed so as to serve as sociopolitical centers, were occupied over long periods, evidenced a great variety of subsistence procurement activities, and emphasized manufacturing tasks (Marshall 1993:145). Into this type fall both DfSj-23A (T'ukw'aa) and DfSi-5 (Ma'acoah). The faunal assemblages of both sites should exhibit a variety of subsistence activities and, as OFT predicts, should emphasize a mix of locally available taxa.

Maschner and Stein (1995) determined that Tlingit site locations were influenced by topographic variables such as proximity of fresh water, slope of the shore, protectedness of the beach, and exposure to sunlight. The size of sites and, by extension, the functions they served, the numbers of people who occupied sites at one time, and the length of occupation could not be determined (Maschner and Stein 1995:932).

Mackie's (2003) site location analysis considered all shell middens as equal and defined midden zones when site buffer areas overlapped. His analysis led him to reject the idea that large, central midden zones represent home bases in a logistically organized mobility system because "[i]f these home bases are situated relative to resources (as is central to the home-base concept), then it follows that the resources themselves must be arranged in a mutually interdependent way across this large study area" (Mackie 2003:278). Such a rejection is suspect, however, because large, central midden zones are analytical constructs, not necessarily emically defined home bases. Also, the operational assumption that all middens are equal is not borne out in ethnographic or archaeological reality. Mackie went on to say, though, that "there is reason to believe that specific single site locations are influenced by the local environmental (not resource) variables identified in Tebenkof Bay. Therefore, base camp sites might be central relative to resources, such sites might indeed be larger than average and contribute to a settlement hierarchy, but it is difficult to equate the behavior and constraints underlying local logistical mobility with regional interdependent centrality" (Mackie 2003:278–279). The distinction between operationally defined analytic units (midden zones) and individual sites, with their particular locations and functions, is recognized beyond the location-allocation analysis. It further appears that large, central sites may have been logistically situated in order to provide convenient access to a range of resources. The difficulty Mackie notes in equating logistical mobility with regional interdependent centrality results, at least in part, from the differences in scale that each analysis addresses.

The larger point is the clear need for much more work on the reasons why different types of sites are located where they are. In particular, aspects of social distance as well as personal agency and situational contingency likely contribute to the choice of site location and the faunal remains that will be found there. The analysis presented in this paper examines resource variables relevant to two base camps within the conceptual framework of locational optimization in order to identify which resources may have influenced the locations of these sites. This analysis, like Mackie's, attempts "to identify a scale in which habitual action makes both a significant and recognizable contribution to the archaeological signature" (Mackie 2003:286). Yet clear-cut realization of predictions set forth here should not be expected.

Methods

The measure used to compare the two assemblages will be %NISP. This measure controls different sample sizes and allows the relative abundance of taxa in each assemblage to be compared reliably. This measure is affected by sample size, and %MAU could be used to standardize and eliminate sample size effects. The end result for this analysis would be the same because both samples are large so variation in calculated percentages would be miniscule and because the data will be tested using rank order correlation so the ranked %NISP and the ranked %MAU would be the same. Further, the very uneven abundances of taxa in each assemblage mean that %MAU values will rapidly diminish to three or four decimal places, and thereby make their interpretation ambiguous. The data are presented below by major class. The DfSi-5 data represent the entire faunal assemblage for each class, but the DfSj-23A data represent only the sample that was drawn for analysis. This sample consisted of 40 percent of all level bags for excavation units that were radiocarbon dated, plus one discretionary unit to provide area coverage (see Monks 2000). Remains are presented at different levels of taxonomic identification depending on the identifiability of each element and the comparative value of the taxon. Thus Cetacea are presented for sea mammals, along with northern fur seal, whereas unidentifiable sea mammal values are omitted. Within the presentation of each class are two discussions. The first addresses taxa that both assemblages share, and the second deals with taxa found in only one of the assemblages. Within each class, it is hypothesized that taxa from local patches will be more abundant than taxa from distant patches. It is also expected that abundant taxa and those present in just one assemblage will be local.

The shift from rockfish to salmon is a phenomenon noted at both sites ca. 600 BP and has also been observed at Ts'ishaa in central Barkley Sound. There are thus issues of scale involved in any comparisons between the two assemblages considered here. The larger scale issue of the rockfish-salmon transition is a feature of both assemblages. The large quantities of both taxa in each assemblage may, at first glance, appear to overshadow more subtle variations between the assemblages that reflect the acquisition of resources in locally restricted patches from optimally located sites.

Fish

Table 9.1 presents the absolute and relative amounts of fish that are common to both sites. As reported elsewhere (Monks 2006), herring is the predominant taxon at DfSi-5, accounting for more than 50 percent of the NISP. Herring accounts for less than 12 percent of the NISP at DfSj-23A. Herring school in vast abundance in Loudon Channel during the winter, and they spawn on protected beaches in early spring (Taylor 1983). Such beaches are common in the protected inner locations around DfSi-5 but are more distant from the outer site of DfSj-23A. Other taxa that prefer protected inner locations with low-energy beaches include small flatfish and other small fish. Anchovy is more abundant at DfSi-5, as are starry flounder and pile perch. Rock sole appears in low abundance at both sites. Taxa that prey on small fish, especially herring, are also expected in sheltered inner locations, and dogfish shark is one of these, although they appear only slightly less frequently at DfSj-23A.

Rocky outer shores and high-energy beaches such as those near DfSj-23A, on the other hand, should support a different fauna, including rockfish, cabezon, lingcod, halibut, and bluefin tuna. Table 9.1 shows that these taxa appear more abundantly at DfSj-23A than at DfSi-5. One would expect that salmon would be more abundant at DfSi-5 because of its

Table 9.1. Fish taxa common to both assemblages

Taxon	Common Name	DfSi-5 NISP	DfSi-5 %NISP	DfSj-23A NISP	DfSj-23A %NISP
Anoplopoma	Sablefish	1	0.0	2	0.0
Clupea	Herring	2,834	52.0	1,644	11.5
Engraulis	Anchovy	342	6.3	37	0.3
Gadus	Pacific Cod	34	0.6	109	0.6
Hemilepidotus	Red Irish Lord	7	0.1	8	0.1
Hexagrammos	Kelp Greenling	34	0.6	719	5.0
Hippoglossus	Pacific Halibut	11	0.2	637	4.4
Hydrolagus	Ratfish	19	0.3	46	0.3
Lepidopsetta	Rock Sole	2	0.0	4	0.0
Merluccius	Pacific Hake	1	0.0	133	0.9
Oncorhynchus	Salmon	950	17.4	4,421	30.8
Ophiodon	Ling Cod	82	1.5	664	4.6
Platichthys	Starry Flounder	50	0.9	10	0.1
Porichthys	Midshipman	67	1.2	138	1.0
Rhacochilus	Pile Perch	168	3.1	151	1.1
Scorpaenichthys	Cabezon	50	0.9	569	4.0
Sebastes	Rockfish	233	4.3	3,186	22.2
Squalus	Dogfish	496	9.1	1,114	7.8
Theragra	Walleye Pollock	2	0.0	9	0.1
Thunnus	Bluefin Tuna	20	0.4	186	1.3
	Total	5,402	99.2	13,785	96.0

proximity to the Toquart River, but, in fact, salmon is the most abundant taxon at DfSj-23A, where its relative abundance is almost twice that of DfSi-5. This deviation from optimization theory may have several explanations. First, the emphasis on salmon was so great after its rise to prominence that it was widely transported in abundance, thereby appearing as stored foods in sites distant from the patch where it was obtained. Interestingly, eleven salmon dentaries were found at DfSj-23A, and several of them exhibited spawning teeth of male chum salmon, but no such evidence was observed in the DfSi-5 assemblage. It might be, then, that salmon was of such importance as to be transported to outer coast sites, but, whether preserved or fresh, the heads remained on at least some fish. Alternatively, some salmon may have been caught in saltwater before they ascended the Toquart River. Second, the abundance of herring at DfSi-5 may obscure the relative abundance of salmon and the proximity to the patch of its acquisition. Third, the function of each site may have influenced the distribution of resources from the patches where they were obtained. Fourth, site functions may have changed during the past 1,200 years, thereby obscuring trends hidden by this whole-assemblage analysis.

This anomaly may represent a macro-level trend that differs from and partially obscures micro-level trends of equal importance. Further support for this latter interpretation lies in the statistical comparison of these data. The fish remains of the two assemblages are expected to differ with the different resource patches optimally exploited from each site location. Thus the substantive hypothesis states that the rank order of taxa in one assemblage will differ from that in the other ($r_s = 0$), and the null hypothesis states that rank orders of taxa in each assemblage will be the same ($r_s = 1$). Spearman's rank order correlation coefficient is 0.7053, which exceeds the critical value of 0.450 at $\alpha \leq 0.05$ for a two-tailed test (n = 20). The coefficient is positive and quite strong and does not allow the null hypothesis to be rejected. Storage and transportation of food resources, the general accessibility of food resource patches, and the ubiquity of many food resources are likely homogenizing factors in this comparison. Nevertheless, the preceding discussion makes clear that a few very important differences between the fish assemblages do conform to the expectation of local patch exploitation based on locational optimization. Further, the method used here promotes a picture of similarity by removing all taxa that do not appear in both assemblages. It is to those taxa that I now turn.

Transportation of stored foods may homogenize assemblages, especially if those stored foods are culturally important and taken in large numbers. As a partial test for the effect of stored food transport, a rank order correlation test was applied to the data in Table 9.1 after herring and salmon were removed. These two taxa are known ethnographically to be highly important and to be stored for later consumption, often transported to other sites. Again, the theoretically derived substantive hypothesis states that the taxa in the two faunal assemblages have different rank orders ($r_s = 0$), and the null hypothesis states that their rank orders are the same ($r_s = 1$). The Spearman's rank order correlation coefficient is 0.6037(n = 18), which exceeds the critical value of 0.475 for a two-tailed test at $\alpha \leq .05$. The null hypothesis still cannot be rejected, so the substantive hypothesis is unsupported. The two assemblages do not differ even when stored and transported resources are removed. There is an embedded problem in removing either or both salmon and herring because some of each may actually have been taken at patches near one or both sites. As well, the Toquaht were a single local group, so kinship affiliation is likely to have enabled occupants of DfSj-23A to take herring at inside locations and salmon on Toquart River. Dried and stored salmon and herring taken near, and processed at, DfSi-5 could have been transported to DfSj-23A, where they were given to kin who returned the favor with "outside" products or access to "outside" resource patches.

Another line of reasoning was also evaluated. If salmon were processed at DfSi-5 and distributed as stored fish to DfSj-23A, it might be expected that the ratio of cranial to vertebral elements would be more even at the former site compared to the latter. Hoffman, Czederpiltz, and Partlow (2000:700) reviewed the literature on cultural and natural taphonomic processes that affect the preservation of salmon skeletal parts. Some of the variables they identify can be controlled in the Toquaht case: namely, (a) the processing method, which is likely the same for all local group members, (b) the chemistry of the soil in which the bones were deposited, and (c) the age of the deposits in which the bones were found. Even though the density of salmon bones has been measured (Butler and Chatters 1994), neither their attrition through taphonomic processes (e.g., cooking, weathering, burial) nor the differential attrition of cranial versus postcranial elements

Table 9.2. Fish taxa found only in one assemblage

Taxon	Common Name	DfSi-5 NISP	DfSi-5 %NISP	DfSj-23A NISP	DfSj-23A %NISP
Atheresthes	Arrowthooth Flounder	22	0.4	0	0.0
Cottidae	Sculpins	0	0.0	19	0.1
Embiotoca	Seaperch	0	0.0	2	0.0
Enophrys	Buffalo Sculpin	9	0.2	0	0.0
Eopsetta	Petrale Sole	0	0.0	4	0.0
Leptocottus	Staghorn Sculpin	1	0.0	0	0.0
Microstomus	Dover Sole	1	0.0	0	0.0
Myoxocephalus	Sculpin	9	0.2	0	0.0
Parophrys	English Sole	2	0.0	0	0.0
Embiotocidae	Surfperch	0	0.0	263	1.8
Pleuronectidae	Flatfish	0	0.0	259	1.8
	Total	45	0.8	549	3.8

has been adequately studied.[*] With these cautions in mind, the ratio of vertebrae to cranial elements was calculated for both DfSi-5 and DfSj-23A. At the former site, the ratio was 77:1, and at the latter site 80:1. To evaluate the possibility that Toquaht butchering practices included more than simply the cranium, a further calculation was made for both sites of the ratio of axial elements to all cranial, pectoral, and pelvic elements. At DfSi-5 the ratio was 36:1, and at DfSj-23A it was 39:1. These results do not reliably indicate whether the ratios are signatures of processing or of storage, but they do suggest that the same signature appears in both sites. The similarity of these ratios at both sites may indicate that the same processes affected the entry of salmon remains into both assemblages. Though DfSi-5 may be more optimally located than DfSj-23A for salmon acquisition, the present evidence does not indicate that salmon were processed "inside" and transported "outside."

Table 9.2 presents the fish taxa found only in one site. The frequencies of most taxa are low and provide little evidence of patch exploitation from optimally located sites. Perch and flatfish, although they appear in modest numbers at DfSj-23A, are difficult to evaluate for their importance because of the family-level identification. Without knowing if the perch are surf perch or if the flatfish are halibut (which observation suggests that many are but fragmentation precludes definite confirmation), the patch from which they were taken cannot be identified with confidence and the optimal location hypothesis cannot be evaluated.

---

[*] Ben Collins, formerly an undergraduate at the University of Manitoba and now a master's student at the University of Alberta, undertook an experiment in which selected cranial elements and vertebral elements of two salmon were immersed in four solutions of differing pH. His results suggested that the cranial elements lost more density than the vertebral elements over the same period of time (Collins 2010).

Table 9.3. Shellfish taxa common to both assemblages

Taxon	Common Name	DfSi-5 NISP	DfSi-5 %NISP	DfSj-23A NISP	DfSj-23A %NISP
Astraea	Turban Shell	288	16.3	75	0.3
Haliotus	Abalone	1	0.1	5	0.0
Margarites	Smooth Margarite	7	0.4	38	0.1
Polinices	Moon Snail	25	1.4	13	0.0
Nucella	Whelk	5	0.3	146	0.6
Clinocardium	Basket Cockle	4	0.2	57	0.2
Protothaca	Littleneck Clam	131	7.4	7,543	28.9
Saxidomus	Butter Clam	1,213	68.8	4,970	19.1
Tresus	Horse Clam	20	1.1	598	2.3
Semibalanus	Acorn Barnacle	8	0.5	645	2.5
Balanus nubilus	Giant Acorn Barnacle	4	0.2	867	3.3
Mytilus	Bay Mussel	56	3.2	2	0.0
	Total	1,762	99.9	14,959	57.3

Shellfish

The shellfish common to both assemblages are presented in Table 9.3. Bivalves were counted by the number of umbos, and these were not matched within each taxon. Gastropods were counted by columella fragments, which were also not matched within each taxon. Barnacles were counted by numbers of body plates and bases, and sea urchins were counted by test fragments and spine fragments. A major difference in the assemblages is in the proportions of shared taxa. Almost all (97 percent) of the DfSi-5 shellfish assemblage are the same taxa represented in the DfSj-23A assemblage, but only 57 percent of the taxa in that assemblage is represented at Ma'acoah. Thus a far wider range of taxa were exploited from the outer coast location.

The predicted exploitation of local resource patches should be observed more clearly for shellfish than for fish because shellfish are less mobile, their ecological requirements more specific, and discarded shells are unlikely to be transported in the way that some elements of stored fish are. Protected inner coast taxa that prefer low-energy, fine-grained foreshore predominate in the DfSi-5 assemblage (turban shell, butter clam, moon snail, bay mussel), and taxa that prefer high-energy, coarse-grained, and rocky foreshores are more abundant in the DfSj-23A assemblage (littleneck clam, small and giant barnacles, horse clam). The assemblages are expected to differ in the rank order of their taxa ($r_s = 0$), so the null hypothesis states that the rank orders of the taxa in one assemblage will enable prediction of rank order of the taxa in the other assemblage ($r_s = 1$). The Spearman's rank order correlation coefficient is 0.29, which does not exceed the critical value of 0.591 at the $\alpha \leq 0.05$ significance level for a two-tailed test (n = 12). This result permits the null hypothesis to be rejected, so the substantive hypothesis of no relationship between the assemblages is supported. Shellfish, therefore, may differ in important ways between the two sites, as one might expect given the changes in wave energy as one moves from exposed outer coast to protected inner coast and given the fairly specific substrate and salinity requirements of a number of mollusks. As well, many taxa are not found at Ma'acoah.

The taxa found in only one assemblage are also illuminating, and these are presented in Table 9.4. Two points are immediately apparent. The only taxon present in DfSi-5 is jingle or rock oyster, of which there is one. This taxon prefers hard substrates in sheltered waters, an econiche found near DfSi-5 but not near DfSj-23A. More importantly, twenty-nine taxa in DfSj-23A are not represented in DfSi-5. There is thus a far greater taxonomic richness in the outer coast site, suggesting a greater diversity of patches, or greater diversity within patches that are close to DfSj-23A. Rocky, high-energy, outer-coast littoral taxa are represented by some of the most abundant shellfish remains at DfSj-23A. In particular, purple sea urchin and California mussel are abundant in this assemblage, and the presence of whale barnacle reflects the acquisition of whales from an open ocean patch. Lesser amounts of high- and medium-energy littoral taxa (sculptured rock shell, chiton, basket cockle, limpet, rock scallop, black top shell, red turban shell) are also found. By contrast, only a few taxa (*Macoma*, bodega clam) prefer quiet water with fine substrates. Both shellfish assemblages share a limited number of important food species, particularly butter clam and littleneck clam, followed by relatively minor taxa such as turban shells, horse clam, bay mussel, barnacles, and basket cockle. The species richness of DfSj-23A distinguishes the two assemblages.

Another important difference is the access to shellfish patches available from outer and inner coast locations. A gradient of substrate types are spatially distributed between these extremes providing different habitats for shellfish taxa that often have fairly narrow econiches. Thus the site located among high-energy patches displays a shellfish assemblage that reflects the local variety of patches, as does the site in the protected inner coast location. The much greater range of taxa exploited from the outer coast site and the considerable importance of purple sea urchin and California mussel are also significant. This greater range of taxa likely results from the high contrast in energy between exposed and sheltered microenvironments. Islands and islets, of which there are a number, offer exposed and lee shores as well as gradations in between. On exposed shores, the vertical distribution of littoral zones is great, from the bottom of the undertow range at low tide to the height of the surf zone at high tide. The Vancouver Island shore also offers coves and bays with varying degrees of exposure that permit beaches with different substrate coarseness to develop. Thus the variety of microenvironments on the outer coast may account for the greater range of shellfish taxa at DfSj-23A than at DfSi-5.

Sea Mammals

The sea mammals common to both assemblages are shown in Table 9.5. All sea mammal taxa identified at DfSi-5 are also found at DfSj-23A. As shown below, very few taxa are present only at DfSj-23A. The hypothesis of specific patch exploitation from different site locations is not borne out by the sea mammal data. Whale remains, which one would expect to be more numerous in an outer coast assemblage, are only slightly more relatively abundant there. The whale remains from DfSi-5 are almost entirely unidentifiable fragments, the exception being one rib that was excavated in several sections. The whale

Table 9.4. Shellfish taxa found only in one assemblage

Taxon	Common Name	DfSi-5 NISP	DfSi-5 %NISP	DfSj-23A NISP	DfSj-23A %NISP
Pododesmus	Jingle Shell	1	0.1	0	0.0
Strongylocentrotus	Purple Sea Urchin	0	0.0	6,136	23.5
Mytilus	California Mussel	0	0.0	3,760	14.4
Ocenebra	Sculptured Rock Shell	0	0.0	383	1.5
Katharina	Chiton	0	0.0	237	0.9
Tegula	Black Top Shell	0	0.0	219	0.8
Macoma	Bent-Nose Clam	0	0.0	139	0.5
Glycymeris	Ark Shell	0	0.0	60	0.2
Nassarius	Channeled Dog Whelk	0	0.0	54	0.2
Lottidae	Limpet	0	0.0	43	0.2
Searlesia	Spindle Shell	0	0.0	34	0.1
Hinnites	Purple-Hinged Rock Scallop	0	0.0	10	0.0
Macoma	Polluted Macoma	0	0.0	6	0.0
Coronulidae	Whale Barnacle	0	0.0	6	0.0
Diodora	Rough Keyhole Limpet	0	0.0	5	0.0
Lottia	Fingered Limpet	0	0.0	4	0.0
Hipponix	Flat Hoof-Shell	0	0.0	4	0.0
Lottia	Shield Limpet	0	0.0	4	0.0
Tegula	Dusky Turban	0	0.0	3	0.0
Ostrea	Native Oyster	0	0.0	3	0.0
Tellina	Bodega Tellin	0	0.0	1	0.0
Tellina	Button Tellin	0	0.0	1	0.0
Macoma	Inconspicuous Macoma	0	0.0	1	0.0
Fusitriton	Oregon Triton	0	0.0	1	0.0
Gari	Sunset Shell	0	0.0	1	0.0
Bittium	Threaded Bittium	0	0.0	1	0.0
	Total	1	0.1	11,116	42.6

remains at DfSj-23A, on the other hand, contain a number of pieces identifiable to element, portion, and species (Monks 2001, 2003; Monks, McMillan, and St. Claire 2001), even though there is also a large quantity of unidentifiable fragments in this assemblage. Northern fur seal was expected to be more abundant at DfSj-23A, and it is only slightly so. California sea lion is about equally distributed, as is northern elephant seal. Dolphin/porpoise is relatively more abundant in the outer coast assemblage, as predicted, but generic seal remains, which are the most abundant taxon in each assemblage, are more common in the sheltered inner coast site, as is harbor seal. There appears to be little difference between the two assemblages, and each is represented by all, or almost all, of the remains from this class. The theoretically described substantive hypothesis of no relationship between assemblages ($r_s = 0$) is evaluated against the null hypothesis that the rank order of taxa in both assemblages is the same ($r_s = 1$). The Spearman's rank order correlation coefficient is

Table 9.5. Sea mammal taxa common to both assemblages

Taxon	Common Name	DfSi-5 NISP	DfSi-5 %NISP	DfSj-23A NISP	DfSj-23A %NISP
Cetacea	Whale	126	23.6	315	28.4
Delphinidae	Dolphin/Porpoise	2	0.4	706	6.4
Pinnipedia	Seals/Sea Lion	274	51.2	3,708	33.9
Callorhinus	Northern Fur Seal	84	15.7	2,070	18.9
Eumetopias	Northern Sea Lion	2	0.4	114	1.0
Mirounga	Northern Elephant Seal	1	0.2	5	0.0
Phoca	Harbour Seal	32	6.0	81	0.7
Zalophus	California Sea Lion	14	2.6	228	2.1
	Total	535	100.0	10,027	91.6

0.8155, which exceeds the critical value of 0.738 for $\alpha \leq 0.05$ for a two-tailed test (n = 8). The null hypothesis cannot be rejected, so the substantive hypothesis cannot be supported. Sea mammal taxa appear in similar ranked proportions in each assemblage.

The taxa found in only one assemblage are listed in Table 9.6. As already stated, all are found at DfSj-23A, and the list is short. Most noticeable is the family-level identification of eared seals (Otariidae), which includes *Zalophus*, *Callorhinus*, and *Eumetopias*, that are present in DfSi-5. Whether these figures result from investigator variability or fragmentation, they suggest that the relative frequencies of northern fur seal and northern sea lion are slightly increased in relation to the recoveries at DfSi-5, and the relative frequencies for California sea lion are likely even more similar. The presence of Dall's porpoise and harbor porpoise in DfSj-23A simply augments the situation seen in Table 9.5 and points to the greater attention paid to their acquisition from the outer coast site. The value for family-level identification of earless seal (Phocidae) changes nothing in Table 9.5, and the minimal presence of sea otter bones in DfSj-23A indicates their low importance in the Aboriginal food quest.

The predicted optimization of site location is not observed in these data. The most likely explanation is that these aquatic taxa are highly mobile and probably preyed upon herring, which, as noted earlier, school in vast abundance in Loudon Channel. All these predators can be accessed relatively easily from both sites throughout the winter, weather permitting. It has been argued elsewhere (Monks 1987) that herring provide the basis for extensive food chains upon which Aboriginal peoples capitalized, and that humpback whales were formerly year-round residents in Barkley Sound (Monks, McMillan, and St. Claire 2001). Thus the similarity in sea mammal assemblages may result from the "herring patch" that consisted of schooling herring in Loudon Channel throughout the winter and spawning herring that used sheltered inner beaches in the early spring. This transient and morphologically variable patch attracted a series of predators that were preyed upon in turn by residents at both sites. This overarching generalization should not obscure the lower level distinctions between the sites in terms of the delphinids that appear to be more sought from the outer coast site and the harbor seals that prefer calmer water and appear more frequently in the inner coast site.

Table 9.6. Sea mammal taxa found only in one assemblage

Taxon	Common Name	DfSi-5 NISP	DfSi-5 %NISP	DfSj-23A NISP	DfSj-23A %NISP
Otariidae	Eared Seal	0	0.0	597	5.5
Phocoenoides	Dall's Porpoise	0	0.0	200	1.8
Phocoena	Harbour Porpoise	0	0.0	101	0.9
Phocidae	Earless Seal	0	0.0	18	0.2
Enhydra	Sea Otter	0	0.0	8	0.1
	Total	0	0.0	924	8.4

Discussion

The predictions of optimal site location in order to exploit different patches most effectively receive limited support. Shellfish differ most between the assemblages, suggesting that access to a greater number and diversity of shellfish patches may have influenced the choice of site location. Shellfish are far less mobile than are fish or sea mammals, so it may not be surprising that shellfish assemblages differ the most. Fish and sea mammals are more ubiquitous due to their mobility and their ability to inhabit a range of econiches. Also, human groups traveling over water can move easily to a variety of patches, and they can easily transport the products of a number of patches to sites of their choosing. Finally, site locations may be selected for reasons other than simple economic expediency. DfSj-23A, for example, is adjacent to the defensive location of DfSj-23B, and the site also commands the entrance to Ucluelet Inlet, which allowed the occupants to control those groups living in the inlet (St. Claire 1991:53). Similarly, DfSi-5 is reported to have been protected by a palisade in the Late Prehistoric period (McMillan 1999:63; St. Claire 1991:163). Economic optimization models do not address such social and demographic factors. Economic optimization models also ignore the growth of population, the elaboration of social structures, concepts of ownership or stewardship, and costly signaling. Finally, the vagaries of human decision-making in past contexts, i.e., agency and contingency, cannot be addressed by economic optimization models.

The analysis presented here indicates that locational optimization may not be clearly reflected in faunal assemblages. Further, it suggests a danger in seeking "prime mover" or key patch(es). The data show that specific locally available shellfish patches were exploited from each site, that fish and sea mammal patches were more ubiquitous than shellfish patches, and that the products of patches were distributed beyond the sites at which they were acquired. It may be, in fact, that stored shellfish were also transported, but no faunal signature of it remains. This analysis suggests that both specific and ubiquitous patches, in combination, provided the suite of subsistence resources that heavily influenced site locations.

Does this mean that Rahn's analysis was incorrect? The answer is no, because his analysis was based on a theoretical availability of patches, whereas the analysis presented here represents actual data that derive from choices and decisions about which of the theoretical patches could be exploited from each location. As well, the coordination of patch exploitation from each site location was undertaken as part of a larger corporate venture by the Toquaht local group. This venture involved social relations between related kin groups that occupied the inner and outer sites, the exchange of resources along kinship lines, the shared access to owned resource locations, and a mutual recognition that cooperation within the group as a whole maximized the reproductive advantages of the group members. The sites discussed here were located so as to have easily accessible patches that provided a variety of food resources on a yearly basis, large quantities of essential food resources, and access to specific resource extraction locations unique to each setting. From each location, resources more easily accessible from that location were exploited, in addition to more ubiquitous resources. In addition, the residential function of both sites and the effects of storage, transportation, and redistribution of resources between them tended to homogenize the archaeologically observed distribution of resources from more specifically located patches.

The locational optimization concept has been found not to work especially well in the Toquaht case. Why? One reason is the small scale of the study. The Toquaht local group territory is not very large and contains resource patches that are present in a number of places. The ~12 km distance between the sites is relatively short by canoe travel, so fish and sea mammal products could easily be shared or exchanged by occupants of each site. The exploited taxa are mobile, especially fish and sea mammals, so their patches are difficult to define. A pinniped rookery is clear enough, but once the animals leave it, the definition of their patch becomes quite diffuse. Another reason is the mobility of the people. Water transportation by canoe allows access to relatively distant patches with considerable economy of effort, thereby further homogenizing the territory. Transportation of food resources away from the patches where they were obtained blurs the signature of patch exploitation. Finally, noneconomic, nonoptimal cultural considerations, as well as unpredictable environmental variation, merit attention as factors that affect where sites are located and what faunal remains are found in them. A cautionary lesson emerges from this analysis: that locational optimization may be more likely to be revealed in faunal assemblages that (a) do not involve storage and transport and (b) are found in heterogeneous environments where patches are restricted and well defined.

The notion that foraging and collecting can be conceptualized as layers rather than as a composite point on a continuum appears to have some merit. Salmon, herring, and, before ca. 600 BP, rockfish may have been essential food resources. They, and the availability of fresh water, comprise the collecting layer and will heavily influence the choices about where a site will be located. Other resources can be reliably foraged no matter where the site is located due either to the ubiquity of the resources and/or to the ease of accessing them. Such resources will comprise the foraging layer and will also include the fortuitously acquired resources that appear in small numbers in faunal assemblages.

Acknowledgments

I wish to thank the Toquaht First Nation, especially Chief Bert Mack, for the opportunity to work on these sites and for financial support in the process. Similarly, thanks go to the Nuu-chah-nulth Tribal Council and the British Columbia Heritage Trust for financial support. Two grants from the Social Sciences and Humanities Research Council (410–03–0962 and 410–2006–1068), as well as support from the University of Manitoba, are also gratefully acknowledged. Alan McMillan and Denis St. Claire kindly invited me to be the faunal analyst on the Toquaht Archaeological Project. Faunal identifications were accomplished with comparative collections at the Royal British Columbia Museum and Pacific Identifications. Many people assisted with the identifications, but particular thanks go to Megan Caldwell, Becky Wigen, Susan Crockford, and Gay Frederick for the DfSj-23A fish; to Adam Allentuck, Tavis Allard, and Megan Caldwell for the DfSj-23A sea mammals; and to Allison Machovec for the DfSj-23A shellfish. Gratitude is also expressed to Madonna Moss and Aubrey Cannon for inviting me to participate in this volume and for offering constructive criticism on an earlier draft. All errors and omissions are the author's sole responsibility.

References Cited

Binford, L. R. "Willow Smoke and Dog's Tails: Hunter-Gatherer Settlement Systems and Archaeological Site Formation." *American Antiquity* 45 (1980): 4–20.

Borgerhof Muldur, M. "Human Behavioral Ecology." In *Behavioral Ecology: An Evolutionary Approach*, 3rd ed., J. R. Krebs and N. B. Davies, eds., pp. 69–98. Oxford: Blackwell Scientific Publications, 1991.

Butler, Virginia L., and J. C. Chatters. "The Role of Bone Density in Structuring Prehistoric Salmon Bone Assemblages." *Journal of Archaeological Science* 21 (1994): 413–424.

Collins, Benjamin R. "Element Survivability of *Salmo salar*." *Journal of Taphonomy* 8(4) (2010): 291–300.

Donald, L., and D. H. Mitchell. "Some Correlates of Local Group Rank among the Southern Kwakiutl." *Ethnology* XIV (1975): 325–346.

———. "Nature and Culture on the Northwest Coast of North America: The Case of the Wakashan Salmon Resource." In *Key Issues in Hunter-Gatherer Research*, E.S.J. Burch and L. J. Ellanna, eds., pp. 95–117. Oxford: Berg, 1994.

Drucker, Philip. *The Northern and Central Nootkan Tribes*. Washington, DC: Bureau of American Ethnology Bulletin 144, 1951.

Frederick, S. G., and S. J. Crockford. "Appendix D: Analysis of the Vertebrate Fauna from Ts'ishaa Village, DfSi-16, Benson Island." In *Ts'ishaa: Archaeology and Ethnography of a Nuu-chah-nulth Origin Site in Barkley Sound*, A. D. McMillan and D. E. St. Claire, eds., pp. 173–205. Victoria: Parks Canada and Tseshaht Nation, 2003.

Hoffman, B. W., J.M.C. Czederpiltz, and M. A. Partlow. "Heads or Tails: The Zooarchaeology of Aleut Salmon Storage on Unimak Island, Alaska." *Journal of Archaeological Science* 27 (2000): 699–708.

Kuhnlein, H. V., and N. J. Turner. *Traditional Plant Foods of Canadian Indigenous People*, vol. 8. Amsterdam: Gordon and Breach, 1991.

Mackie, Q. "Location-Allocation Modelling of the Shell Midden Distribution on the West Coast of Vancouver Island." In *Emerging from the Mist: Studies in Northwest Coast Culture History*, R. G. Matson, G. Coupland, and Q. Mackie, eds., pp. 260–288. Vancouver: UBC Press, 2003.

Marshall, Y. A. "Political History of the Nuu-Chah-Nulth People: A Case Study of the Mowachaht and Muchalaht Tribes." PhD diss., Simon Fraser University, 1993.

Maschner, H. D. G., and J. Stein. "Multivariate Approaches to Site Location on the Northwest Coast of North America." *Antiquity* 69 (1995): 61–73.

McKechnie, I. "Five Thousand Years of Fishing at a Shell Midden in the Broken Group Islands, Barkley Sound, British Columbia." Master's thesis, Simon Fraser University, 2005.

McMillan, A. D. *Since the Time of the Transformers: The Ancient Heritage of the Nuu-Chah-Nulth, Ditidaht, and Makah.* Vancouver: UBC Press, 1999.

Mitchell, D. H. "Prehistory of the Coasts of Southern British Columbia and Northern Washington." In *Northwest Coast,* Handbook of North American Indians, vol. 7, W. P. Suttles, ed., pp. 340–358. Washington DC: Smithsonian Institution, 1990.

Monks, G. G. "Prey as Bait: The Deep Bay Example." *Canadian Journal of Archaeology* 11 (1987): 119–142.

———. "How Much Is Enough? An Approach to Sampling Ichthyofaunas." In *Studies in Canadian Zooarchaeology: Papers in Honour of Howard G. Savage*, M. Friesen, ed. *Ontario Archaeology* 69 (2000): 65–75.

———. "Quit Blubbering: An Examination of Nuu-chah-nulth Whale Butchery." *International Journal of Osteoarchaeology* 11 (2001): 136–149.

———. "Cultural Taphonomy of Nuu-Chah-Nulth Whalebone Assemblages." In *Emerging from the Mist: Studies in Northwest Coast Culture History*, R. G. Matson, G. Coupland, Q. Mackie, eds., pp. 188–212. Vancouver: UBC Press, 2003.

———. "The Fauna from Ma'acoah (DfSi-5), Vancouver Island, British Columbia: An Interpretive Summary." *Canadian Journal of Archaeology* 30 (2006): 272–301.

Monks, G. G., A. D. McMillan, and D. E. St. Claire. "Nuu-chah-nulth Whaling: Archaeological Insights into Antiquity, Species Preferences and Cultural Importance." *Arctic Anthropology* 38 (2001): 60–81.

Rahn, R. B. "Spatial Analysis of Archaeological Sites in Barkley Sound, Vancouver Island." Master's thesis, University of Manitoba, 2002.

Smith, E. A. "The Application of Optimal Foraging Theory to the Analysis of Hunter-Gatherer Group Size." In *Hunter-Gatherer Foraging Strategies: Ethnographic and Archaeological Analyses*, B. D. Winterhalder and E. A. Smith, eds., pp. 35–65. Chicago: University of Chicago Press, 1981.

St. Claire, D. E. "Barkley Sound Tribal Territories." In *Between Ports Alberni and Renfrew: Notes on West Coast Peoples*, E. Y. Arima, D. E. St. Claire, L. Clamhouse, J. Edgar, C. Jones, and J. Thomas, eds., pp. 13–202. Ottawa: Canadian Ethnology Service, Canadian Museum of Civilization, 1991.

Streeter, I. M. "Seasonal Implications of Rockfish Exploitation in the Toquaht Area, B.C." Master's thesis, University of Manitoba, 2002.

Suttles, W. P. "Variation in Habitat and Culture on the Northwest Coast." Vienna: *Akten des 34 International Amerikanistenkongresses* (1962): 522–537.

———. "Coping with Abundance: Subsistence on the Northwest Coast." In *Man the Hunter*, R. B. Lee and I. Devore, eds., pp. 56–68: Chicago: Aldine, 1968.

Taylor, F. H. C. "The Hydroacoustic Assessment of Herring Distribution and Abundance in Barkley Sound, February 18 to March 12, 1982." Nanaimo: *Canadian Technical Report of Fisheries and Aquatic Sciences,* pp. 1–106, 1983.

Turner, N. J. "'Time to Burn': Traditional Use of Fire to Enhance Resource Production by Aboriginal Peoples in British Columbia." In *Indians, Fire and Land in the Pacific Northwest*, R. Boyd, ed., pp. 185–218. Corvallis: Oregon State University Press, 1999.

Williams, J. *Clam Gardens: Aboriginal Mariculture on Canada's West Coast.* Vancouver: New Star Books, 2006.

CHAPTER 10

Pacific Cod in Southeast Alaska
The "Cousin" of the Fish That Changed the World

Madonna L. Moss, *Department of Anthropology, University of Oregon*

The Pacific cod (*Gadus macrocephalus*) is well represented in many precontact archaeological sites located around the North Pacific Rim, from the Aleutian Islands to the coast of Washington. Despite the abundance of Pacific cod remains archaeologically, relatively little research has been conducted on this species and its importance to the Aboriginal people of the North Pacific (but see Bowers and Moss 2001; Kopperl 2003; Orchard 2003; Souders 1997). For this reason, the potential significance of Pacific cod (and other gadids) to understanding the cultural history of the region has not been broadly recognized. Moreover, the general public is largely unaware of the long-term historical importance of Pacific cod to the region's Indigenous peoples.

The case of Pacific cod presents a vivid contrast to that of salmon, the quintessential North Pacific fish famous around the world. Wissler (1914) referred to the Northwest Coast as the "Salmon Area," and zooarchaeologists have conducted a number of key studies on salmon fishing and use. The intensification of salmon use—capturing large numbers en masse and preserving them for winter storage—is considered foundational to the development of Northwest Coast societies (Butler and Campbell 2006; Cannon 1998; Carlson 1998:31–32; Eldridge and Acheson 1992; Matson 1992:411; Matson and Coupland 1995:173; Moss and Erlandson 1998; Moss, Erlandson, and Stuckenrath 1990). In fact, Northwest Coast archaeologists have been accused of "salmonopea" because of our tendency to attribute all sorts of sociocultural achievements to the economic mainstay of salmon reliance (Monks 1987; see also Stevenson 1998).

The case of Pacific cod also presents a stark contrast to that of Atlantic cod. The Atlantic cod, *Gadus morhua*, is one of the most popular food fishes in the Western world and has been heralded as "the fish that changed the world," not only by popular writer Mark Kurlansky (1998) but by archaeologists working in the North Atlantic region who have documented the Fish Event Horizon in Europe. It is not an exaggeration to say that new archaeological research on Atlantic cod is rewriting the last thousand years of European history (Barrett, Locker, and Roberts 2004a, 2004b; Perdikaris et al. 2007).

The purposes of this paper are to (1) compare Pacific cod to Atlantic cod in an attempt to assess whether attributes of the fish themselves explain the lesser reputation of the former and the popularity of the latter, (2) consider what is known of the biology of Pacific cod in southeast Alaska within the broader ecological context of this species, and (3) assess the ethnographic and archaeological records of Pacific cod use in southeast Alaska. This information provides essential background in an effort to begin to understand the historical ecology of this species and the roles it has played in the economies and cultures of the Tlingit and their ancestors. I close the paper by summarizing some of the key questions that remain in our quest to document the long-term history of Aboriginal use of Pacific cod in southeast Alaska.

The Story of Atlantic Cod

By AD 850, the Norse were involved in long-distance trade of dried marine fish, as evidenced in archaeological sites in Iceland and the Faroe Islands. Until AD 950, almost no marine fish bones are found in Britain's inland archaeological sites. Within a few decades of AD 1000, however, the remains of cod and herring become abundant in inland sites in Britain, particularly in the medieval towns of York, Ipswich, London, and Southampton (Barrett, Locker, and Roberts 2004b:622). This phenomenon, dated to AD 950–1050, is known as the Fish Event Horizon. Cod and other gadids were processed into dried "stockfish" or dried and salted for trade, while herring were salt-cured wet in barrels (Barrett, Locker, and Roberts 2004b:624). In the Norse-occupied parts of the British Isles, "fish middens" of densely packed fish bones showing evidence of preparation for export became common. Some of the marine fish found in British towns probably derived from regional trade but others were the result of long-distance Norse trade. By the early 1100s, the Norwegian king and church began to fully organize the stockfish trade based on the rich winter fisheries of northern Norway (Dugmore et al. 2007:18). By 1250, Iceland was a major center of production of dried fish for export and local consumption. This was to eventually connect Iceland and Norse Scotland into the rapidly expanding urban markets of Europe (Dugmore et al. 2007:18). According to Barrett, Locker, and Roberts (2004b), the Fish Event Horizon marks the break between the so-called Dark Age and later medieval market trade in high-bulk, low-value basic commodities on a large scale across Europe.

Regarding the use of cod on the other coast of the Atlantic, Catherine Carlson (1986) found that cod were the dominant fish in the faunal assemblages from eighteen of the twenty-one sites she studied in the Boothbay region of Maine. In occupations dated over the last 2,500 years, Carlson identified two fishing strategies: "cold season fishing" during winter and spring, and "warm season fishing" during summer and fall. During the cold season (mainly November through February), the Abenaki and their ancestors took mature cod in the marine zone at the mouths of estuaries at depths of three to five fathoms (Carlson 1986:121). During the summer and fall, they focused on flounders and sculpins but took juvenile cod as incidental catches from nearshore and shoal waters (Carlson 1986:120). Both adult and juvenile cod were caught by jigging using baited hooks and lines from oceangoing boats. Because mature cod are available only seasonally, Carlson used them as seasonal indicators. In contrast, juvenile cod are available much of the year and were not considered useful for seasonality determination (Carlson 1986:122). Carlson distinguished between adult and juvenile cod in her assemblages using measurements of thoracic vertebrae.

These seasonal patterns for Boothbay differ from those of the Mikmaw, who focused on open-water fisheries for cod between May and September (Barsh 2002:18). Another factor to consider in the Gulf of Maine is sea level rise and how it may have affected the distribution of fish habitats (Belcher 1989). In the more northerly waters of Nova Scotia and Newfoundland, cod have often been interpreted as evidence of early summer occupation (e.g., Hodgetts et al. 2003; Rojo 1990). Part of this variation is because Atlantic cod spawn at different times and different places (Westrheim 1996), and while some stocks may be "resident" in some estuaries, others migrate significant distances. The biggest cod were taken in deeper, more distant waters.

Bourque, Johnson, and Steneck (2008) reexamined the faunal assemblage from the Turner Farm site on the Gulf of Maine. They found that cod occurred in abundance in the earliest strata of the site, dated to 4350 BP. By 3550 BP, cod declined in relative abundance, and by 1600 BP, they comprised a minor constituent of the assemblage. Bourque, Johnson, and Steneck (2008) interpreted this as a case of "fishing down the food web"—when cod became less available, people shifted to flounders and sculpins. Based on isotope values, these investigators argued that by about 400 BP, cod were no longer caught in the kelp forest ecosystems in the site vicinity, but that fishers traveled to more distant waters for their cod. They argue that Indigenous people had a significant impact on Gulf of Maine ecosystems prior to contact. Yet complications remain because the dating of the most recent component at Turner Farm is problematic, and ca. 400 BP European fishermen were also competing for Atlantic cod. In an earlier paper, Barsh (2002:21) criticized Bourque and colleagues for not considering seasonal differences in the occupations dated to different time periods at the site.

The incursion of Basque, Portuguese, British, French, and Spanish fishermen into North American waters in the 1500s was just the beginning of commercial fishing on this side of the Atlantic. By the mid-1990s, the groundfish and herring fisheries in the Canadian Maritime Provinces and the Gulf of Maine had collapsed. Bourque, Johnson, and Steneck (2008:165) claim that Atlantic cod and virtually all large-bodied fishes are rare and "ecologically extinct" today.

Is Atlantic Cod Superior to Pacific Cod?

This question is one that has been discussed for more than a hundred years. Do the attributes of these fishes help explain the lesser reputation and relative obscurity of Pacific cod versus the popularity of Atlantic or Icelandic cod? Do these attributes help explain the relative lack of archaeological attention on Pacific cod, despite the ubiquity of their bones in sites? Atlantic cod grow larger than Pacific cod, with maximum length of 185 cm for Atlantic cod versus 100 cm for Pacific cod in Canadian waters, and 114 cm for Pacific cod in Russian waters (Westrheim 1996:xiv). Certainly, the larger size of Atlantic cod contributed to efficiencies in processing.

In his 1916 report to the U.S. Commissioner of Fisheries, John N. Cobb wrote, "[t]he fight of the Pacific cod for admission into eastern markets is a typical example of how difficult it is to overcome a prejudice, no matter how insufficiently founded" (Cobb 1916:72). The prejudice against Pacific cod, then and now, is that its value as food is inferior to that of Atlantic cod. The earliest commercial Pacific cod fisheries in Alaska date to the 1860s, but it was not until shore stations became more common in the 1880s that the industry became better established. Those trying to promote Pacific cod have been fighting the battle for more than a hundred years.

Cobb's assessment (1916:72–75) was that the salt cod–producing companies from the East Coast controlled the markets for cod across the United States. They

> did not welcome the intrusion of Pacific cod, and while they were unable to prevent the loss of the greater part of their trade on the Pacific coast, they fought hard for the rest. Dealers and consumers were told in some instances that the fish prepared by this coast's curers were not cod, or that they were a very inferior grade of cod; that the fish would not keep, etc. That these misstatements had a wide dissemination and made a considerable impression is evidenced even to this day in the prejudice which is met with in different sections of the country against Pacific cod. (Cobb 1916:72)

Cobb (1916:72) also explained that some Pacific coast producers "played right into the hands of their trade enemies" because some of the first shipments of salt cod from the West Coast were poorly prepared and shipped east by rail during the warm season. According to Cobb (1916:72–73), the Pacific cod was the only member of the gadid family on the Pacific coast that was large enough for dry salting. While eastern curers rejected Pacific cod, they bought Atlantic hake, cusk, and pollock, which they marketed as "true cod," even though these were considered inferior fish. Regarding the condition of Pacific cod, Cobb (1916:75) admitted that Pacific cod was vulnerable to "reddening" when the salt-cured product was exposed to temperatures higher than 65°F. But Cobb claimed that this bacterial infection was even more common in Atlantic cod cured on the East Coast and in Europe.

Although some information available on the Internet supports the idea that Pacific cod is very similar to Atlantic cod, and that the two species are used interchangeably in the marketplace, it is generally thought that Pacific cod has a slightly higher moisture content and softer texture than Atlantic cod (Seafood Choices Alliance 2008). While large cod can weigh more than 50 pounds, most commercially caught Pacific cod are between 5 and 15 pounds. When cooked, Pacific cod meat is white and flaky with a mild taste. Gregg Hardy (2007), contributing to a discussion list maintained by University of California, Davis, states that after about six to eight months in freezer storage, the texture of the Pacific cod becomes tough. He also writes that storage temperature and how the fish is handled prior to freezing affects the texture of the thawed product. Jon McGraw (2007), from Seafreeze in Seattle, writes that cold storage temperatures should be a steady −10°F or below to increase the shelf life of Pacific cod, and that the condition of the fish is especially vulnerable to fluctuating freezer temperatures. McGraw also states that fish condition depends on whether cod are caught in a pot or by a trawl, and before or during spawning.

The reddening of the salt-cured cod and the hazards of long-term freezer storage are not directly applicable to the precontact period. One concern shared by traditional and commercial cod processors is the presence of "worms." The "sealworm" is the larval form of *Pseudoterranova decipiens* that infests seals as well as both Atlantic and Pacific cod (Westrheim 1996:73). According to Department of Fisheries and Oceans Canada fisheries biologist Alan Sinclair (2008, pers. comm.), cod are vulnerable to parasites in the waters they frequent, whether the Atlantic or Pacific. John Hickman, the plant manager at Sitka Sound Seafoods, explained to me that these "imperfections" can be picked out during processing, but it is a labor-intensive task. Elder Henry Katasse described the Tlingit way of getting rid of worms. First the cod heads were cut off and the fish split lengthwise, and then they were gutted and washed. Then the trunks were laid out, flesh side down, on layers of hemlock boughs. The fish were covered and left like this for three days. Katasse explained:

> Contrary to what people say, it's a known fact that gray cod, the Pacific cod, do have worms. You must remember that this is in the month of March and April. The branches acted as a poultice in getting the worms out which were not visible at the time. We were told [by our grandparents] to shake them off before taking them [the cod] to the smokehouse. (Newton and Moss 2005:14)

Sinclair, who has studied both Atlantic and Pacific cod, maintains that the seasonal condition of cod does not make them more or less vulnerable to parasites. The parasite load in the local waters appears to be the main influence on the condition of the cod.

Regarding seasonal changes in the condition of the flesh, John Hickman confirmed the idea that the flesh of Pacific cod is a bit softer during the spawn than during the winter. Hickman (2008, pers. comm.) said that during spawning, the flesh of the Pacific cod is not as firm and "the belly walls are thinner, with less meat." Today, most commercially caught Pacific cod are taken in the winter. In southeast Alaska, ethnographic evidence suggests that the Tlingit fished Pacific cod during the March–April period, when the fish congregate into shallow waters to spawn.

The idea that Pacific cod is inferior to Atlantic cod is alive and well today. For example, Norfisk, an Icelandic seafood supplier, with offices in Iceland and the United States, makes a number of claims on its website that reiterate the prejudices discussed by Cobb.

Black cod, True cod, Gray cod, Pacific cod, Polar cod, Alaska cod, Red cod, Ling cod... so many fish want to be cod because cod is so popular. But the cod that has earned this great reputation is the North Atlantic Cod, gadus morhua, also knows [sic] as Atlantic Cod. This specie [sic] is the benchmark to which all other white fleshed fish are compared. . . .

Some things you should know about cod:

Atlantic Cod vs. Pacific Cod

Until the 1980s, cod was simply cod, invariably the North Atlantic Cod, or gadus morhua. But in the 80s, the fishing of Pacific cod, gadus macrocephalus, started in earnest, and the fillet of this specie was also marketed as "cod". This is when difficulties with cod species arose. While the actual differences between these two species are few, they are important ones. The flesh of both species looks similar in fact, the Pacific cod is somewhat whiter than the Atlantic in the raw form, while both cook up equally white. The flavour of the two species is similar as well. **It is the texture of the Pacific cod that lessens its value. Sometimes described as 'stringy', 'tough' and 'watery', it has less of the succulence and tenderness of the Atlantic cod.** These attributes mark the Pacific cod as less desirable, and fillets and products of Pacific cod are lower priced than the same products of Atlantic Cod.

The difficulties of both recognizing Atlantic cod from Pacific cod, and understanding the reasons for their different value, has been a recurring issue in the seafood industry. Attracted by the lower price, many buyers choose to minimise these subtle but importance differences, and to use Atlantic and Pacific cod interchangeably. While each cod specie has a legitimate place and value in the market, their differences are obvious enough to the consumer that the issue of consistency is raised every time a substitution is made. (Norfisk [2007], misspellings unchanged, but format adapted and emphasis in bold added)

The Seafood Choices Alliance (2008) website, which is promoting Pacific cod as a "smart choice," phrases the difference between the two cods another way: "[t]he two species look nearly identical, but a keen palate will detect the higher moisture content in the flesh of the Pacific variety, which gives it a considerably softer and flakier texture than its Atlantic counterpart."

The consensus appears to be that Atlantic cod has a quality of texture that is slightly different than that of Pacific cod, at least by today's contemporary standards of taste. Pacific cod has become more important in the marketplace over the last twenty-five years, especially with the decline of Atlantic cod populations. Today almost 90 percent of the cod harvested in the United States is Pacific cod (NOAA 2008). I do not believe that the slight difference in texture, however, is the reason why Pacific coast archaeologists have not focused more attention on the Indigenous use of cod. Such long-term marketing practices do affect popular perceptions about cod, and this may be partly responsible for the underappreciation of the long-term economic importance of this species to the First Peoples of the Pacific coast.

Pacific Cod in the Greater North Pacific Region

Pacific cod is a transoceanic species, found along the continental shelf and upper slope (100–250 m) of the North Pacific from Santa Monica, California, northward to Norton

Sound, Alaska, and on the Asian coast, southward to the northern Yellow Sea (Bakkala et al. 1984). It is a relatively fast-growing species that reaches a maximum age of eighteen years (Alaska Fisheries Science Center 2008). Female cod reach 50 percent maturity at 67 cm (about 5.8 years) and are highly fecund; such a fish produces over a million eggs (Alaska Fisheries Science Center 2008). The earliest documented American commercial use of Pacific cod was probably Captain Matthew Turner, who fished the waters off the Amur River and Kamchatka coast in 1863 (Cobb 1916:25).

Cod are bottomfish that move from deep waters to shallower waters in the late winter, spring, or early summer, when they spawn. Some cod also feed in shallow waters into later spring and summer. Estimates of spawning time vary by region and investigator; for example, in the western Bering Sea, Musienko (1970) states that cod spawn from January to March, while Vinnikov (1996:188) writes that the "peak of cod spawning occurs in the southwestern Bering Sea in late March–early April, and in the northwestern Bering Sea in April." Estimates for spawning in the eastern Bering Sea range from January to March (Bakkala 1993; Mito, Nishimura, and Yanagimoto 1999; Teshima 1985), to January to April (Witherell 2000), to January to May (Niggol 1982). For the Gulf of Alaska, estimates range from January to March (Niggol 1982), to January to April (DiCosmo and Kimball 2001), and February to July (Hirshberger and Smith 1983). Of course, all of these geographic regions are very large, and local conditions undoubtedly affect the timing of spawning in specific areas.

Pacific cod in the northerly part of their range are slower growing with greater longevity, and achieve larger size than those in the south (Alaska Department of Fish and Game 2008a). Pacific cod are schooling fish, and their distribution varies seasonally. For example, Turner (1886:89) observed that cod were most abundant during July and September in some places in the Aleutians, and during February and March elsewhere. Larger cod were caught in deeper water (Cobb 1916:51; Turner 1886:90). Water temperature affects seasonal patterns of cod abundance; in the Bering Sea, the offshore migration in fall is to avoid the cooling of inshore water when sea ice forms (Shimada and Kimura 1994). When the coastal shelf waters warm in the spring, the fish move inshore. In Puget Sound, the migration is reversed; Pacific cod move to deep waters in the summer to avoid the shallow warm water nearshore.

Recent genetic work has shown that the Asian and North American stocks of Pacific cod are genetically distinct, probably since 18,000 years ago, when they became separated at the height of the last ice age (Canino et al. 2006). The American cod samples show a genetic isolation-by-distance pattern, and because cod tend to return to the same spawning grounds each year, this indicates most mating occurs between adjacent groups. These different subpopulations may have different behaviors and respond to fishing pressure independently. Ongoing genetic research aims to map out such subpopulations in greater detail in order to treat them as distinct units for management.

Cod prey on clams, worms, crabs, shrimp, and juvenile fish; cod are eaten by halibut, sharks, seabirds, pinnipeds, and other marine mammals (Alaska Fisheries Science Center 2008). Today, fishers take cod with bottom longlines (45 percent), trawls (37 percent), and pots (16 percent) (Seafood Choices Alliance 2008). Pacific cod are usually harvested in "multispecies complexes," that is, several groundfish species are caught at the same time (NOAA 2008). In U.S. waters, Pacific cod are managed under two Fishery Management Plans: one for the Bering Sea/Aleutian Islands region and the other for the Gulf of Alaska. These plans control the fishery through permits and limited entry, catch quotas, gear restrictions, closed waters, seasons, bycatch limits and rates, etc. Trawlers dominate Japanese and Russian Pacific cod fisheries, although longliners also fish in Russian waters. Even though Pacific cod is found from California to Alaska and from Japan to Russia, the vast majority of the world's catch comes from Alaska waters. Small trawl fisheries operate in British Columbia and Washington State.

In 2006, another gadid, walleye pollock (*Theragra chalcogramma*), made up 71 percent of the groundfish catch off Alaska at 1.56 million metric tons (Alaska Fisheries Science Center 2008).* Pacific cod were next in abundance, accounting for 11 percent of the total 2006 groundfish catch (239,427 metric tons). For comparison, the amount of salmon taken per year from Alaska waters between 2000 and 2004 averaged 336,566 metric tons (Alaska Department of Fish and Game 2008b). Today, Pacific cod regularly rank second in catch and product value in the Alaska groundfish fishery, with most coming from the Bering Sea, Aleutian Islands, and western Gulf of Alaska region.

Pacific cod became the most important trawl caught fish in British Columbia waters by the early 1960s (Shimada and Kimura 1994). Hecate Strait is a prime location for Pacific cod fishing in British Columbia, with a vast expanse of sand and gravel substrates below shallow shelf waters (100–150 m). Sinclair and Crawford (2005) examined a variety of environmental factors to determine which were correlated with the largest age classes of Pacific cod between 1962 and 1995. Pacific cod abundance in Hecate Strait does not appear to correlate with climate changes such as regime shifts, El Niños, or La Niñas (Sinclair, 2006, pers. comm.). Sinclair and Crawford (2005) found that seawater temperature and the availability of prey were not strongly related to cod abundance either. Using barometrically adjusted sea level heights as a proxy for transport by currents, Sinclair and Crawford (2005) concluded that Pacific cod abundance in this region was most strongly affected by water currents propelled by southeast winter winds during and after spawning. If nearshore waters are calm, the cod eggs and larvae will develop and settle to the bottom as juvenile fish. In stormy weather, the currents will disperse the eggs and larvae during this critical time, resulting in lower numbers of maturing cod. These investigators estimated that during times of low productivity, the maximum fishery in Hecate Strait would be half of what it would be during times of high productivity (Sinclair and Crawford 2005:138).

Pacific Cod in Southeast Alaska

NOAA fisheries research biologist Grant Thompson (2008, pers. comm.) estimates that the waters of southeast Alaska contribute less than 5 percent of the Pacific cod taken from the Gulf of Alaska. This directed Pacific cod fishery is open year-round, but most fishing occurs in the winter. Alaska Department of Fish and Game groundfish biologist Cleo Brylinsky (2008, pers. comm.), headquartered in Sitka, explained that the quota for southeast Alaska of

---

* Pollock was not a major commercial fishery until the 1970s and early 1980s when their abundance increased, possibly as a result of the 1976–1977 regime shift (Hare and Mantua 2000).

1.25 million pounds (567 metric tons) is rarely approached because of the low price of cod. Trawls are banned in federal groundfish fisheries in southeast Alaska, so the fish are taken with longlines or pots. In 2004, about 123 metric tons of Pacific cod were landed in southeast Alaska (Kruse 2005:14). Pacific cod is also frequently targeted as bait for use in crab pots or for longlining for halibut and black cod (sablefish). Having worked as a processor in both the Bering Sea region and southeast Alaska, John Hickman (2008, pers. comm.) said that although the Pacific cod of western Alaska get larger than those in the southeast, the fish in the southeast are good in quality, although low in relative numbers. He said a "big day" for processing Pacific cod at Sitka Sound Seafoods was 20,000 pounds, while in a Bering Sea plant a big day would be 450,000 pounds. He considers the Sitka plant's processing of Pacific cod during the winter as "more as a service to the fleet" than a money-making venture. The relatively small scale of the commercial fishery for Pacific cod in southeast Alaska helps explain why the bulk of U.S.-funded research on this species is focused on the commercially important fisheries of the Bering Sea and western Alaska. To summarize, in southeast Alaska, Pacific cod have not been taken in the extremely large commercial quantities characteristic of the Bering Sea, Aleutian Islands, and western Gulf of Alaska. The southeast Alaska fishery for Pacific cod is also smaller than that of Hecate Strait–Queen Charlotte Sound and the north coast of Washington State to Vancouver Island (Bakkala et al. 1984:113).

Tlingit Use of Pacific Cod in Southeast Alaska

In an earlier paper, ethnographic information on Pacific cod use was compiled from a variety of sources (Bowers and Moss 2001:170–172). Cobb (1916:20) described the "few small" inshore banks where cod was taken in southeast Alaska:

> These banks, which vary from 5 to 7 fathoms in depth, are mainly in Chatham Straits, Lynn Canal, and Icy Straits. The fish are found on the banks in the summer, disappearing into the deeper water into the fall. The fish caught are comparatively small, examples more than 24 inches in length being rare.

One of the banks Cobb (1916:25) specifically identified was "Hoocheno Bank" in Chatham Strait, probably near Angoon (Figure 10.1). This is the place where the first Alaska vessel went for cod in 1879. Captain Haley of Wrangell did not fish for cod here himself, but

> purchased his fare from the natives who claimed the exclusive right to engage in the fishery. These fishermen used bark lines, with wooden iron-pointed hooks, and, as they considered a catch of 30 or 40 fish a good day's work, Capt. Haley had to wait quite a while before he could accumulate a cargo. (Cobb 1916:27)

This suggests that the Tlingit highly valued cod, and that the fishing banks were owned. The Yakutat reportedly fished for cod at depths of eight to twenty fathoms (de Laguna 1972:391). Emmons reported an earlier type of fishing gear than that described by Cobb (above); he wrote that cod (among other fish):

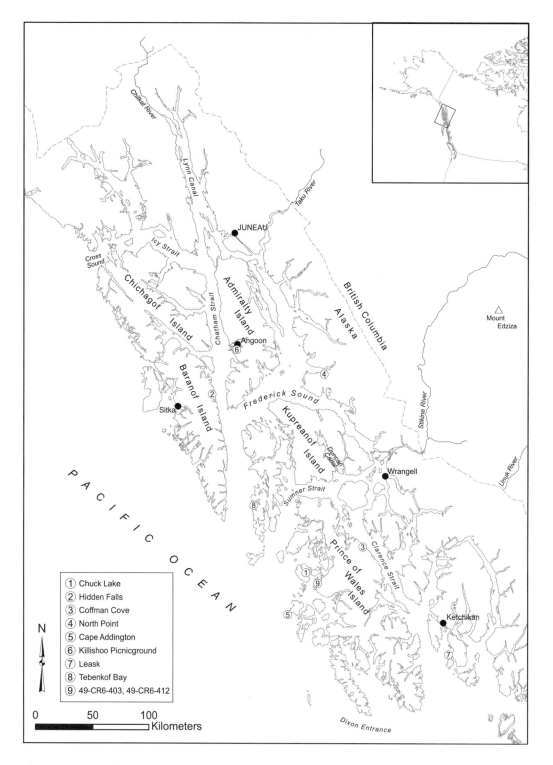

Figure 10.1. Map of southeast Alaska showing sites with abundant Pacific cod remains.

were taken with a simple compound V-shaped hook, made with a wooden shank and a barb of sharpened bone or wood. The line was of twisted spruce root, and the sinkers…were unworked stones, generally oblong or of a convenient shape for the attachment of the line. (Emmons 1991:121; see also de Laguna 1960:116)

In addition to lines made of spruce root, long lines of kelp were attached to these hooks (de Laguna in Emmons 1991:121; Stewart 1977).

Regarding the seasonality of Pacific cod fishing in southeast Alaska, Cobb (1916:20) mentioned that cod were found on inshore banks in the summer. We do not know what time of year Captain Haley was in Chatham Strait purchasing cod from the Angoon Tlingit. Most ethnographic information points to March as the key time for cod fishing. Oberg (1973:73) wrote that by February, "stored food had become impalatable" and fresh fish were highly desired. Oberg (1973:66) stated that the Tlingit took Pacific cod with hook and line during March, when schools of fish move to shallow waters. In an earlier paper, we assumed that it was during late winter and early spring, when schooling Pacific cod move to shallower waters (31–65 m), that they would have been most accessible to Aboriginal fishers (Bowers and Moss 2001).

Emmons (1991:148–149) stated that Pacific cod and some fish other than salmon were eaten fresh, and that they were not cured and "formed an unimportant part of the diet as long as fresh or dried salmon or halibut were to be had." This contradicts the experience of Tlingit elder Henry Katasse, who remembered fishing and processing cod in the early spring. Mr. Katasse considered Pacific cod as "the most delectable smoke-dried fish to have, by way of a change" (Newton and Moss 2005:14). Mr. Katasse described how cod were prepared: First the cod heads were cut off and the fish were split open lengthwise. Then they were laid out between hemlock branches in a shallow pit over a period of several days to remove any worms. Then the cod could be cooked for immediate consumption or smoked for storage. Smoked cod were soaked, roasted, or boiled. One of de Laguna's (1972:402) informants also described cod smoked in the spring: "The cod was just like wood when they dried it. You couldn't eat it. But when it was soaked in salt water for two, three days, it was just like fresh." Billy Jackson from Yakutat also reported that cod (along with halibut) were smoked on Big and Little Kayak islands offshore (Goldschmidt and Haas 1998:46, 127).

Pacific Cod in Southeast Alaska Archaeological Assemblages

Three gadid species occur in southeast Alaska: Pacific cod, walleye pollock (*Theragra chalcogramma*), and Pacific tomcod (*Microgadus proximus*). Pacific cod can reach 114 cm, walleye pollock, 91 cm, and tomcod, 30 cm (Eschmeyer, Herald, and Hammann 1983:97–98). Based on the size of the archaeological specimens, Pacific tomcod can be ruled out in the assemblages I've studied. In general, the cranial, pectoral, and pelvic elements of walleye pollock and Pacific cod are distinguishable. Even though their vertebrae cannot be distinguished, in my analyses, I have tended to identify gadid vertebrae as Pacific cod because of the strong Tlingit tradition of Pacific cod use in southeast Alaska (Newton and Moss 1984; Bowers and Moss 2001). Other analysts have taken different approaches.

Table 10.1. Fish assemblages from southeast Alaska with more than 100 NISP identified to family

Site #	Reference	Age	Mesh	Anoplopomatid	Chimaerid	Clupeid	Cottid	Embiotocid	Gadid	Hexagrammid	Hexanchid	Lamnid	Ophiodontid	Osmerid	Pholid	Pleuronectid	Rajid	Salmonid	Scorpaenid	Squalid	Stichaeid	Zoarchid	Total
49-CRG-188	Moss 2004	2000–300 cal BP	1/4"			27	11	11	649	14			39			910		789	102	1	8		2561
49-CRG-188	Moss 2004	2000–300 cal BP	1/8"			41	3		17				1			21		22	6			1	112
49-CRG-196	Greiser et al. 1994	3000–250 BP	1/4"															184					184
49-CRG-237	Ackerman et al. 1989	8200 BP	???	5		7	55		356	140			7			6		6	37				614
49-CRG-403	Greiser et al. 1994	800–500 BP	1/4"				3		724							9		70	1				811
49-CRG-412	Greiser et al. 1994	1800–700 BP	1/4"		1	1			120									1	1				124
49-CRG-443	Hanson & Broderick 1995; Pierce 2008	1700 BP	1/4"			91										15							106
49-CRG-459	Hanson 1996	1700 BP	1/8"			433																	433
	Hanson 1996	unknown	2 mm				2		13								1	1403					1419
49-DIX-046	Moss 2008a	3000–1700 cal BP	1/4"						2				2			6			90					100
49-DIX-053	Moss 2007a, 2008a	1600–200 cal BP	1/4"			10	5		4	13						1		1	33		21		93
49-KET-229	Minor et al.1986	2400–1800 BP	1/4"				6		59				8			6		2	45		18		144
	Minor et al.1986		1/8"		15	102			307			1				53		115	29	26			599
49-PET-067	Moss 2008b	5500–700 cal BP	1/4"			66	288		528	1			3			53		9463	229	12			10645
	Moss 2008b	5500–700 cal BP	.132"			88	54		3							4		242					391
49-PET-556	Moss 2007c	3000–2100 cal BP	1/4"			82	219		1308	1			117			85		128	448	47			2435
	Moss 2007c	3000–2100 cal BP	.132"			152	23		33				1			3		7	12	2			233
49-SIT-119	Moss 1989a	3000–1300 BP	1/4"	22		95		41	4789	3			3			152		3891	683				9679
	Moss 1989a	4600–3000 BP	1/4"			16			19							7		160	7				209
49-SIT-124	Moss 1989b, 2007b	1600–900 cal BP	1/8"			8	1		130	7	4		2			5	10	163	35		1		356
	Moss 1989b	1600–900 cal BP	2 mm			183	8		12	8				1	2			1016	15	18			1273
49-SIT-130	Moss 1989b	1000–700 cal BP	2 mm													3		1039					1042
49-SIT-132	Moss 1989b	900–700 cal BP	2 mm			147	4			1						1		202					356
49-SIT-171	Moss 1989b	1000–900 cal BP	2 mm			15			1									368					384
49-SIT-244	Moss 1989b	900–250 cal BP	2 mm			953	6		2						3	1		755	3		4		1731
49-SIT-299	Moss 1989b	1400–650 cal BP	2 mm			13	2		9	1						5	4	1074	1				1109
49-SIT-304	Moss 1989b	1200 cal BP	2 mm			916												3					919
49-SUM-025	Bowers & Moss 2001	2800–2000 cal BP	1/4"			6	4		2094	10						10		135	155	1			2415
49-XPA-029	Maschner 1992	1500–1200 cal BP	1/8"			2297			248	5			4			2		945	52				3553
		300–100 cal BP	1/8"			1223			101	4			4					5998	29				7359
		4400–4000 cal BP	1/8"			953			812									416	6				2187
49-XPA-039	Maschner 1992	3000–2600 cal BP	1/8"			579			1154									26	13				1772
		2000–1500 cal BP	1/8"			4033			454							30		141	40				4698
		1200–800 cal BP	1/8"			2162			592				8			5		273	21				3061
49-XPA-106	Maschner 1992	750–550 cal BP	1/8"			18			19									475	3				515
49-XPA-112	Maschner 1992	750–550 cal BP	1/8"			8			2									536	2				548
49-YAK-020	Davis 1996	<1000 BP	???															144					144

In reviewing all excavation reports from the region, I found twenty-six sites that contained a minimum of 100 NISP of fish identified to at least the family level. Table 10.1 presents these data along with the published and unpublished sources from which they were drawn. Species and genus identifications have been collapsed to family identifications to ensure comparability across taxa (Driver 1992). The following species or genera make up most of the NISP collapsed into the families they represent: *Clupea harengus pallasi* (clupeid), *Gadus macrocephalus* (gadid), *Hippoglossus stenolepis* (Pleuronectid), *Oncorhynchus* spp. (salmonid), *Sebastes* spp. (scorpaenid), and *Squalus acanthias* (squalid). Note that screen size has a significant impact on assemblage composition; where screen sizes are equal to or finer than one-eighth inch, herring bones are often present in greater numbers than they are in samples recovered with quarter-inch screens (Moss, Butler, and Elder, this volume).

Another factor to consider when evaluating fish assemblage composition is how bone density can structure taxonomic abundances and element representation (Butler and Chatters 1994; Smith 2008; Smith et al., this volume). Some portions of Pacific cod bones are denser than salmon and halibut bones, so Pacific cod may be overrepresented compared to other taxa. Because the details of element representation in the assemblages from the twenty-six sites are not available, I did not follow Smith's protocol to examine the effect of bone density for this set of sites. Smith (2008) has done so, however, for two assemblages included here: North Point (49-SUM-025) and Cape Addington Rockshelter (49-CRG-188). At neither of these sites did density-mediated attrition affect cod, salmon, or halibut representation, but this cannot be ruled out for the other assemblages. Therefore, the observations below should be understood within the context of differential bone density.

Of the sites listed, twelve contain abundant cod remains. These sites range from the 8,200-year-old shell midden at Chuck Lake (49-CRG-237), where cod make up 58 percent of the fish identified to family, to the 300–100 cal BP component at Elena Bay Village (49-XPA-029). Sites with abundant cod include two located on the outer islands of the Alexander Archipelago, including Cape Addington (49-CRG-188) on Noyes Island and Chuck Lake (49-CRG-237) on Heceta Island. More protected locations on the west side of Prince of Wales Island are represented by 49-CRG-403 and 412. Hidden Falls (49-SIT-119) and Killisnoo Picnicground (49-SIT-124) are located along Chatham Strait while Elena Bay and Step Island villages (49-XPA-029, -039) in Tebenkof Bay are located on the west side of Kuiu Island, along the entrance to Chatham Strait from the south. Recall that Chatham Strait is one of the waterways known for abundant cod from ethnographic and historical sources cited in the previous section. Henry Katasse's account cited above is drawn from his experiences with his grandparents in Saginaw Bay (Newton and Moss 2005:14), located on the north end of Kuiu Island near the confluence of Chatham Strait and Frederick Sound. The North Point site (49-SUM-025) is located on the mainland, not far from Frederick Sound. Sites located on the east side of Prince of Wales Island in the Clarence Strait area (and along its entrance from the south) include the Coffman Cove sites (49-PET-067, -556) and the Leask Site (49-KET-229) on Annette Island. Alaska Department of Fish and Game biologist Deidre Holum (2008, pers. comm.) has stated that the Pacific cod in the Coffman Cove area today move north into Clarence Strait from the south, en route to Sumner Strait and Duncan Canal on Kupreanof Island where they feed (see Chapter 13, this volume).

Of the fourteen sites listed in Table 10.1 at which cod are not abundant, most assemblages are dominated by salmon. From Shallow Water Town (49-YAK-020) near Yakutat, the salmon probably were taken from the nearby Lost River. Nine sites are located within fairly protected estuaries where salmon are abundant, including six sites located on southwest Admiralty Island (49-SIT-130, -132, -171, -244, -299), 49-CRG-459 near Craig, 49-CRG-196, also on Prince of Wales Island, and 49-XPA-106 and -112 in somewhat protected locations in Tebenkof Bay on Kuiu Island. Two outer coast sites, Kit'n'Kaboodle Cave (49-DIX-46) on Dall Island and Elderberry Cave (49-DIX-53) on Lowrie Island, appear to be campsites where people took advantage of rockfish in the immediate site vicinity. Two assemblages, 49-CRG-443 from Craig and 49-SIT-304 from Garnes Point (east of Angoon on Admiralty Island), are dominated by herring in the absence of most other taxa. This may be a consequence of limited sampling, but it may also indicate specialization on herring in two locations where herring are known to have been remarkably abundant (Moss, Butler, and Elder, this volume). These spatial trends show, not unexpectedly, that much of the variation in fish assemblage composition can probably be explained by proximity to fish habitats. Unfortunately, the seasonal movements and availability of many fish are not as well documented as archaeologists might hope.

Regarding temporal trends, the twenty-six assemblages in Table 10.1 do not adequately represent all time periods in southeast Alaska precontact history. No Late Pleistocene sites are represented, and only one Early Holocene site, Chuck Lake, has produced a sizable assemblage of fish. Yet Chuck Lake does demonstrate that Pacific cod has been important to people in southeast Alaska for more than 8,000 years (Ackerman 1992, 1996). The next oldest assemblage is Coffman Cove (49-PET-067), where cod are abundant in the 5000–4000 cal BP levels. The Component II subassemblage at Hidden Falls (49-SIT-119) is nearly as old (ca. 4600–3000 BP), but while cod are present, they are not abundant. In the Component III subassemblage at Hidden Falls (ca. 3000–1300 BP), however, cod are the most abundant constituent. Cod are also abundant at the 4400–4000 BP component at Step Island Village (49-XPA-039) in Tebenkof Bay. All other assemblages in Table 10.1 postdate 3000 cal BP and these appear to reflect the environmental variation explained above, as opposed to temporal trends. Pacific cod dominate the faunal assemblage from the Russian-American Company site at Sitka, occupied between AD 1830 and 1840 (Petruzelli and Hanson 1998). Unfortunately, this historic assemblage could not be included in Table 10.1 because the primary data from it have not been published. Although the twenty-six assemblages do not adequately represent all time periods, they do suggest long-term, significant use of Pacific cod throughout southeast Alaska precontact history.

Concluding Assessment

Pacific cod are common in archaeological assemblages of southeast Alaska. From the twenty-six sites in Table 10.1, cod bones are abundant at twelve assemblages and present in an additional seven assemblages. Although zooarchaeologists across the North Pacific recognize how common Pacific cod bones are in the assemblages they have studied, the economic importance of Pacific cod has not figured into broader understandings about

Northwest Coast resource use practices and culture change. Part of the anthropological neglect of this species may be due to the emphasis on salmon and the lesser role played today by Pacific cod in the commercial fisheries of southeast Alaska. As discussed earlier in this paper, the economic importance of Pacific cod is not widely recognized, in contrast to the attention Atlantic cod has garnered in recent years as the "fish that changed the world." The slight difference in texture between Pacific and Atlantic cod does not adequately explain the lack of analytical attention focused on Pacific cod.

Due to lack of research, many questions remain:

1. What roles did Pacific cod play in the fishing economies of southeast Alaska and the larger North Pacific coast? Were cod of equivalent importance to all social groups living across southeast Alaska?

2. Were codfish seasonal staples or not? When were Pacific cod available during the annual cycle? In southeast Alaska, were March and April the primary cod-fishing seasons, or were cod routinely caught during summer and at other times?

3. How reliable and predictable were cod stocks from year to year? As Betts and colleagues (this volume) show in the Sanak Island case, are cod better characterized as "the fish that stops" because of their cyclic availability? How did cod populations respond to shifts in climate during the Holocene? This latter question is also addressed by Betts and colleagues.

4. Were Pacific cod processed for long-term storage as were salmon? The limited ethnographic information about Tlingit practices shows that cod were dried and smoked during the twentieth century, but can archaeological evidence of storage be identified? Partnow and Kopperl (this volume) systematically address this question.

5. Did cod fishing affect the abundance and distribution of this species locally or across the region? Did the scale of fishing affect cod population sizes in southeast Alaska? How did fishing for cod affect other organisms in the marine food web?

6. Do ethnographic sources adequately represent the cultural importance of cod to the Tlingit? Cod do not appear to have been a crest animal and are generally not represented in Tlingit artworks.* Does this imply that Pacific cod was of lesser cultural importance than salmon, shark, halibut, and herring, all of which are crest animals? Alternatively, do archaeological data better reflect the economic (if not social) importance of cod?

More data on the ecology and spatial distribution of Pacific cod in southeast Alaska are desperately needed. More complete reports of archaeological fish assemblages, including

---

* Although most Tlingit crest animals are mammals and birds, some fish do serve as crests. Salmon is a crest of fourteen Raven moiety lineages, shark is a crest for eight Eagle moiety lineages, herring is a crest for one Raven lineage, and halibut is a crest for one Eagle lineage (Jonaitis 1986). For the Nuu-chah-nulth, McMillan and St. Claire (2005:9–10, 12) mention and illustrate two codfish as part of a painted design on a house that once stood at Ts'ishaa.

element representation, need to be published to enable analysts to discern the effects of bone density on assemblage composition (Butler and Chatters 1994; Smith 2008; Smith et al., this volume). Clearly this species was of economic significance to the Tlingit and their ancestors. Northwest Coast archaeological sites are archives of environmental and cultural data that, ideally, should help us understand the long-term history of key fish species. Deeper historical ecological analyses should contribute to improved fisheries management and help restore marine ecosystems. Yet archaeological phenomena are best suited to studying long-term trends and larger-scale processes that are not equivalent in scale or detail to records that fisheries biologists and managers have compiled over the last few decades. Nevertheless, archaeological studies have clearly shown the importance of Pacific cod to First Peoples of the Northwest Coast for thousands of years. Through interdisciplinary communication and exchange, we should be able to fill in the many gaps in the long-term history of Pacific cod, and this should help us better plan for the future.

Acknowledgments

Deidra Holum, Grant Thompson, Alan Sinclair, Cleo Brylinksky, and John Hickman generously shared their insights and experience. Ross Smith provided a copy of his important master's thesis. In this paper, I have drawn from some of the data compiled by Virginia Butler, Tait Elder, and myself in association with Tom Thornton's Herring Synthesis Project, supported by the North Pacific Research Board. Other colleagues who have shared my enthusiasm for cod over the years include Pete Bowers and Paul Souders. Aubrey Cannon co-organized the 2008 SAA symposium where part of this chapter was presented, and his insightful comments led to substantial revision. Virginia Butler served as our perceptive and incisive discussant at the symposium, and her impressive body of work on fish remains sets a standard for all of us working on the coast. Finally, I appreciate the continuing support of Jon and Erik Erlandson, both of whom have been served many meals of baked Pacific cod.

References Cited

Ackerman, Robert E. "Earliest Stone Industries on the North Pacific Coast of North America." *Arctic Anthropology* 29 (1992): 18–27.

———. "Early Maritime Culture Complexes of the Northern Northwest Coast." In *Early Human Occupation in British Columbia*, R. L. Carlson and L. Dalla Bona, eds., pp. 123–132. Vancouver: UBC Press, 1996.

Ackerman, R. E., K. C. Reid, and J. D. Gallison. "Heceta Island: An Early Maritime Adaptation." In *Development of Hunting-Fishing-Gathering Maritime Societies along the West Coast of North America*, A. Blukis Onat, ed., pp. 1–29. Proceedings of the Circum-Pacific Prehistory Conference, Seattle. Pullman: Washington State University, 1989.

Alaska Department of Fish and Game. Pacific Cod Species Profile: Southeast Alaska and Yakutat. http://www.cf.adfg.state.ak.us/region1/finfish/grndfish/pcod/pcodinfo.php (accessed February 6, 2008), 2008a.

———. Salmon Fisheries in Alaska. http://www.cf.adfg.state.ak.us/geninfo/finfish/salmon/salm-home.php, (accessed February 6, 2008), 2008b.

Alaska Fisheries Science Center. Pacific Cod. http://www.afsc.noaa.gov/species/Pacific_cod.php (accessed January 31, 2008).

Bakkala, R. G. Structure and Historical Changes in the Groundfish Complex of the Eastern Bering Sea. NOAA Technical Report NMFS 114, 1993.

Bakkala, R., S. Westrheim, S. Mishima, C. Zhang, and E. Brown. "Distribution of Pacific Cod (*Gadus macrocephalus*) in the North Pacific Ocean." *International North Pacific Fish Commission Bulletin* 42 (1984): 111–115.

Barrett, J. H., A. M. Locker, and C. M. Roberts. "The Origin of Intensive Marine Fishing in Medieval Europe: The English Evidence." *Proceedings of the Royal Society* B271 (2004a): 2417–2421.

———. "'Dark Age Economics' Revisited: The English Fish Bone Evidence AD 600–1600." *Antiquity* 78 (2004b): 618–636.

Barsh, Russel Lawrence. "*Netukulimk* Past and Present: Mikmaw Ethics and the Atlantic Fishery." *Journal of Canadian Studies* 37 (2002): 15–42.

Belcher, William R. "Prehistoric Fish Exploitation in East Penobscot Bay, Maine: The Knox Site and Sea-Level Rise." *Archaeology of Eastern North America* 17 (1989): 175–191.

Bourque, Bruce J., Beverly Johnson, and Robert S. Steneck. "Possible Prehistoric Fishing Effects on Coastal Marine Food Webs in the Gulf of Maine." In *Human Impacts on Ancient Marine Ecosystems: A Global Perspective*, T. C. Rick and J. M. Erlandson, eds., pp. 165–185. Berkeley: University of California Press, 2008.

Bowers, Peter M., and Madonna L. Moss. "The North Point Wet Site and the Subsistence Importance of Pacific Cod on the Northern Northwest Coast." In *People and Wildlife in Northern North America: Essays in Honor of R. Dale Guthrie*, S. Craig Gerlach and Maribeth S. Murray, eds., pp. 159–177. BAR British Archaeological Report International Series 944. Oxford: Archaeopress, 2001.

Butler, Virginia L., and Sarah K. Campbell. "Northwest Coast and Plateau Animals." In *Environment, Origins, and Population,* Handbook of North American Indians, vol. 3, D. H. Ubelaker, ed., pp. 263–273. Washington, DC: Smithsonian Institution, 2006.

Butler, Virginia L., and James C. Chatters. "The Role of Bone Density in Structuring Prehistoric Salmon Bone Assemblages." *Journal of Archaeological Science* 21 (1994): 413–424.

Canino, Michael, Kathryn Cunningham, Ingrid Spies, and Lorenz Hauser. Population Structure of Pacific Cod (*Gadus macrocephalus*) Over Broad and Local Geographic Scales. ftp://ftp.afsc.noaa.gov/posters/pCanino01_genetic-population-cod.pdf. Washington Sea Grant, 2006.

Cannon, Aubrey. "Contingency and Agency in the Growth of Northwest Coast Maritime Economies." *Arctic Anthropology* 35 (1998): 57–67.

Carlson, Catherine C. "Maritime Catchment Areas: An Analysis of Prehistoric Fishing Strategies in the Boothbay Region of Maine." Master's thesis, University of Maine, Orono, 1986.

Carlson, Roy L. "Coastal British Columbia in the Light of North Pacific Maritime Adaptations." *Arctic Anthropology* 35 (1998): 23–35.

Cobb, John N. *Pacific Cod Fisheries.* Appendix IV to the Report of the U.S. Commissioner of Fisheries for 1915. U.S. Bureau of Fisheries Document 830, 1916.

Davis, Stanley D. "The Archaeology of the Yakutat Foreland: A Socioecological View." PhD diss., Texas A&M University. Ann Arbor, MI: University Microfilms, 1996.

DiCosimo, J., and N. Kimball. "Groundfish of the Gulf of Alaska: Species Profile." Anchorage: North Pacific Fishery Management Council, 2001.

Driver, Jonathan C. "Identification, Classification, and Zooarchaeology." *Circaea* 9 (1992) 35–47.

Dugmore, A. J., M. J. Church, K. Mairs, T. H. McGovern, S. Perdikaris, and O. Vesteinsson. "Abandoned Farms, Volcanic Impacts, and Woodland Management: Revisiting Pjorsardalur, the Pompeii of Iceland." *Arctic Anthropology* 44 (2007): 1–11.

Eldridge, Morley, and Steven Acheson. "The Antiquity of Fish Weirs on the Southern Coast: A Response to Moss, Erlandson, and Stuckenrath." *Canadian Journal of Archaeology* 16 (1992): 112–116.

Emmons, George T. *The Tlingit Indians*. Frederica de Laguna, ed. Seattle: University of Washington Press; Vancouver: Douglas and McIntyre; New York: American Museum of Natural History, 1991.

Eschmeyer, William N., Earl S. Herald, and Howard Hammann. *A Field Guide to Pacific Coast Fishes of North America, from the Gulf of Alaska to Baja California*. Boston: Houghton Mifflin, 1983.

Goldschmidt, Walter R., and Theodore H. Haas. *Haa Aani, Our Land: Tlingit and Haida Land Rights and Use*. Thomas F. Thornton, ed. Seattle: University of Washington Press; Juneau: Sealaska Heritage Foundation, 1998.

Greiser, T. W., D. E. Putnam, S. Moorhead, and G. A. Walter. Cultural Resources Specialist Report, Control Lake Environmental Impact Statement, Prince of Wales Island, Alaska. Report submitted under Forest Service Contract 53–0109–3-00369, Tongass National Forest, Ketchikan, AK. Missoula, MT: Historical Research Associates, 1994.

Hanson, Diane K. Ballpark Midden Site (49-CRG-459), Craig, AK: Report of Faunal Remains Collected during Excavations in September 1996." Anchorage: Littoral Zooarchaeology, 1996.

Hanson, Diane K., and Frank Broderick. A Report on Test Excavations at the Craig Administrative Site 49-CRG-443, Prince of Wales Island, Southeast Alaska. MS on file, Craig Ranger District, Craig, AK, 1995.

Hardy, Gregg. Message posted to University of California, Davis, Seafood Archives. http://listproc .ucdavis.edu/archives/seafood/log0703/0016.html (accessed July 10, 2007).

Hare, Steven R., and Nathan J. Mantua. "Empirical Evidence for North Pacific Regime Shifts in 1977 and 1989." *Progress in Oceanography* 47 (2000): 103–145.

Hirshberger, W. A., and G. B. Smith. Spawning of Twelve Groundfish Species in the Alaska and Pacific Coast Regions, 1975–1981. U.S. Dept. of Commerce, NOAA Tech. Memo. NMFS F/ NWC: 44, 1983.

Hodgetts, Lisa, M.A.P. Renouf, Maribeth S. Murray, Darlene McCuaig-Balkwill, and Lesley Howse. "Changing Subsistence Practices at the Dorset Paleoeskimo Site of Phillip's Garden, Newfoundland." *Arctic Anthropology* 40 (2003): 106–120.

Jonaitis, Aldona. *Art of the Northern Tlingit*. Seattle: University of Washington Press, 1986.

Kopperl, Robert. "Cultural Complexity and Resource Intensification on Kodiak Island, Alaska." PhD diss., University of Washington, 2003.

Kruse, Gordon. "Fisheries Overview." In *Workshop Report: Southeast Alaska Synthesis of Marine Biology and Oceanography*, Ginny L. Eckert, ed., pp. 12–16. http://doc.nprb.org/web/04_prjs/ f0406_jan-jun05_workshop_report.pdf. North Pacific Research Board, 2005.

Kurlansky, Mark. *Cod: A Biography of the Fish that Changed the World*. New York: Penguin Books, 1998.

de Laguna, Frederica. *The Story of a Tlingit Community: A Problem in the Relationship between Archeological, Ethnological and Historical Methods*. Washington, DC: Smithsonian Institution Bureau of American Ethnology Bulletin 172, 1960.

———. *Under Mount Saint Elias: The History and Culture of the Yakutat Tlingit*. Washington, DC: Smithsonian Contributions to Anthropology, vol. 7, 1972.

Maschner, H.D.G. "The Origins of Hunter-Gatherer Sedentism and Political Complexity: A Case Study from the Northern Northwest Coast." PhD diss., University of California, Santa Barbara. Ann Arbor, MI: University Microfilms, 1992.

Matson, R. G. "The Evolution of Northwest Coast Subsistence." In *Long-Term Subsistence Change in Prehistoric North America*, D. R. Croes, ed., pp. 367–428. Research in Economic Anthropology Supplement 6. Greenwich, CT: JAI Press, 1992.

Matson, R. G., and Gary Coupland. *The Prehistory of the Northwest Coast*. San Diego: Academic Press, 1995.

McGraw, Jon. Message Posted to University of California, Davis, Seafood Archives. http://listproc .ucdavis.edu/archives/seafood/log0703/0016.html (accessed July 10, 2007).

McMillan, Alan D., and Denis St. Claire. *Ts'ishaa: Archaeology and Ethnography of a Nuu-Chah-Nulth Origin Site in Barkley Sound*. Burnaby, BC: Archaeology Press, Simon Fraser University, 2005.

Minor, Rick, D. C. Barner, and R. L. Greenspan. *The Leask Site: A Late Prehistoric Campsite on Annette Island, Southeast Alaska*. Eugene, OR: Heritage Research Associates Report 50, 1986.

Mito, K., A. Nishimura, and T. Yanagimoto. "Ecology of Groundfishes in the Eastern Bering Sea, with Emphasis on Food Habits." In *Dynamics of the Bering Sea*, T. R. Loughlin and K. Ohtani, eds., pp. 537–580. Fairbanks: University of Alaska Sea Grant, AK-SG-99-03, 1999.

Monks, Gregory G. "Prey as Bait: The Deep Bay Example." *Canadian Journal of Archaeology* 11 (1987): 119–142.

Moss, Madonna L. "Analysis of the Vertebrate Assemblage." In *The Hidden Falls Site, Baranof Island, Alaska*, S. D. Davis, ed., pp. 93–129. Anchorage: Aurora, Alaska Anthropological Association Monograph Series, 1989a.

———. "Archaeology and Cultural Ecology of the Prehistoric Angoon Tlingit." PhD diss., University of California, Santa Barbara. Ann Arbor, MI: University Microfilms, 1989b.

———. *Archaeological Investigation of Cape Addington Rockshelter: Human Occupation of the Rugged Seacoast on the Outer Prince of Wales Archipelago, Alaska*. Eugene: University of Oregon Anthropological Paper No. 63, 2004.

———. "Haida and Tlingit Use of Seabirds from the Forrester Islands, Southeast Alaska." *Journal of Ethnobiology* 27 (2007a): 28–45.

———. "The Killisnoo Picnicground Midden (49-SIT-124) Revisited: Assessing Archaeological Recovery of Vertebrate Faunal Remains from Northwest Coast Shell Middens." *Journal of Northwest Anthropology* 41 (2007b): 1–17.

———. Coffman Cove Ferry Terminal Site (49-PET-556): Shell Midden Compositional Analyses and Analysis of the Vertebrate and Invertebrate Remains. Submitted to Northern Land Use Research, Fairbanks, AK, December 2007c.

———. "Outer Coast Maritime Adaptations in Southern Southeast Alaska: Tlingit or Haida?" *Arctic Anthropology* 45 (2008a): 41–60.

———. Coffman Cove Village (49-PET-067): Analysis of the Vertebrate and Invertebrate Remains. Submitted to Northern Land Use Research, Fairbanks, AK, March 2008b.

Moss, Madonna L., and Jon M. Erlandson. "A Comparative Chronology of Northwest Coast Fishing Features." In *Hidden Dimensions: The Cultural Significance of Wetland Archaeology*, K. Bernick, ed., pp. 180–198. Vancouver: UBC Press, 1998.

Moss, Madonna L., Jon. M. Erlandson, and Robert Stuckenrath. "Wood Stake Weirs and Salmon Fishing on the Northwest Coast: Evidence from Southeast Alaska." *Canadian Journal of Archaeology* 14 (1990): 143–158.

Musienko, L. N. "Reproduction and Development of Bering Sea Fishes." In *Soviet Fisheries Investigations in the Northeastern Pacific*, Part V, pp. 161–224. Israel Program for Scientific Translation (1972), 1970.

Newton, Richard G., and Madonna L. Moss. *The Subsistence Lifeway of the Tlingit People: Excerpts of Oral Interviews*. Juneau: USDA Forest Service, Alaska Region Report No. 179, 1984.

———. *Haa Atxaayi Haa Kusteeyix Sitee, Our Food Is Our Tlingit Way of Life: Excerpts of Oral Interviews.* Juneau: USDA Forest Service, Alaska Region, R10-MR-30, 2005.

Niggol, K. Data on Fish Species from Bering Sea and Gulf of Alaska. U.S. Dept. of Commerce, NOAA Tech. Memo NMFS F/NWC-29, 1982.

NOAA (National Oceanic and Atmospheric Administration) National Marine Fisheries Service. FishWatch—U.S. Seafood Facts. http://www.nmfs.noaa.gov/fishwatch/species/pac_cod.htm (accessed February 6, 2008).

Norfisk. About the Specie: North Atlantic Cod. http://www.norfisk.is/pcod_aboutspec.html (accessed February 29, 2008), 2007.

Oberg, Kalervo. *The Social Economy of the Tlingit Indians.* Seattle: University of Washington Press, 1973.

Orchard, Trevor. *An Application of the Linear Regression Technique for Determining Length and Weight of Six Fish Taxa: The Role of Selected Fish Species in Aleut Paleodiet.* BAR British Archaeological Reports, International Series 1172. Oxford: Archaeopress, 2003.

Perdikaris, Sophia, George Hambrecht, Seth Brewington, and Thomas McGovern. "Across the Fish Event Horizon: A Comparative Approach." In *The Role of Fish in Ancient Time*, Heidemarie Hüster Plogmann, ed., pp. 51–62. Proceedings of the 13th Meeting of the ICAZ Fish Remains Working Group, Basel. Rahden/Westf: Verlag Marie Leidorf GMbH, 2007.

Petruzelli, Renee, and Diane K. Hanson. "Fauna from Mid-Nineteenth Century Structures at Castle Hill State Park, Sitka, Alaska." Paper presented at the 8th International Congress of the International Council for Archaeozoology, Victoria, BC, 1998.

Pierce, Shona. Faunal Analysis of CRG-443, Craig Administrative Site Midden. USDA Forest Service Alaska Region Heritage Resources Project Report R2007100551046. MS on file, Craig Ranger District, Craig, AK, 2008.

Rojo, Alfonso L. "Faunal Analysis of Fish Remains from Cellar's Cove, Nova Scotia." *Archaeology of Eastern North America* 18 (1990): 89–108.

Seafood Choices Alliance. Cod, Pacific, *Gadus macrocephalus.* http://www.seafoodchoices.com/smartchoices/species_cod.php (accessed July 13, 2008).

Shimada, Allen M., and Daniel K. Kimura. "Seasonal Movements of Pacific Cod, *Gadus macrocephalus*, in the Eastern Bering Sea and Adjacent Waters Based on Tag-Recapture Data." *Fishery Bulletin* 92 (1994): 800–816.

Sinclair, A. F., and W. R. Crawford. "Incorporating an Environmental Stock-Recruitment Relationship in the Assessment of Pacific Cod (*Gadus macrocephalus*)." *Fisheries Oceanography* 14 (2005): 138–150.

Smith, Ross E. "Structural Bone Density of Pacific Cod (*Gadus macrocephalus*) and Halibut (*Hippoglossus stenolepis*): Taphonomic and Archaeological Implications." Master's thesis, Portland State University, 2008.

Souders, Paul. "Ellikarrmiut Economy: Animal Resource Use at Nash Harbor (49-NI-003), Nunivak Island, Alaska." Master's paper, University of Oregon, 1997.

Stevenson, Ann. "Wet-Site Contributions to Developmental Models of Fraser River Fishing Technology." In *Hidden Dimensions: The Cultural Significance of Wetland Archaeology*, K. Bernick, ed., pp. 220–238. Vancouver: UBC Press, 1998.

Stewart, Hilary. *Indian Fishing: Early Methods on the Northwest Coast.* Seattle: University of Washington Press, 1977.

Teshima, K. "Maturation of Pacific Cod in the Eastern Bering Sea." *Bulletin of the Japanese Society of Scientific Fisheries* 51 (1985): 29–31.

Turner, Lucien M. *Results of Investigations Made Chiefly in the Yukon District and the Aleutian Islands; Conducted under the Auspices of the Signal Service, U.S. Army, Extending from May 1874 to August 1881*. Washington, DC: Government Printing Office, 1886.

Vinnikov, A. V. "Pacific Cod (*Gadus macrocephalus*) of the Western Bering Sea." In *Ecology of the Bering Sea: A Review of Russian Literature*, O. A. Mathisen and K. O. Coyle, eds., pp. 183–202. Fairbanks: University of Alaska Sea Grant College Program Report No. 96–01, 1996.

Westrheim, S. J. "On the Pacific Cod (*Gadus macrocephalus*) in British Columbia Waters, and a Comparison with Elsewhere, and Atlantic Cod (*G. morhua*)." Canadian Technical Report Fisheries and Aquatic Sciences 2092, 1996.

Wissler, Clark. "Material Cultures of the North American Indians." *American Anthropologist* 16 (1914): 447–505.

Witherell, David. "Groundfish of the Bering Sea and Aleutian Islands Area: Species Profiles." Anchorage: North Pacific Fishery Management Council, 2000.

CHAPTER 11

Zooarchaeology of the "Fish That Stops"

Using Archaeofaunas to Construct Long-Term Time Series of Atlantic and Pacific Cod Populations

Matthew W. Betts, *Archaeology and History Division, Canadian Museum of Civilization*
Herbert D. G. Maschner, *Department of Anthropology, Idaho State University*
Donald S. Clark, *Fisheries and Oceans Canada, St. Andrews Biological Station, New Brunswick*

Overexploitation of the marine ecosystem and its impact on commercial sustainability are watershed issues in modern science. Responding to Pauly's (1995) concept of the "shifting baseline," where benchmark parameters for a healthy population can degrade through time, fisheries researchers have recently attempted to incorporate a long-term perspective into the analysis of fish stocks (Swain et al. 2003; Swain, Sinclair, and Hanson 2007; Willis and Birks 2006; Worm et al. 2006, see papers in Starkey, Holm, and Barnard 2008). The goal of such research is often to (1) determine a "benchmark" or "natural" ecosystem or population state (e.g., Jackson 2001), and (2) introduce management policies that will establish and maintain this "preindustrial" equilibrium condition (see Lyman 2006 for discussion). Such research is typically based on the longest time-series records available, which for oceanic fisheries data currently extends about fifty to sixty years into the past (e.g., Barot et al. 2004; Clark and Perley 2006; Thompson et al. 2007; Thompson, Dorn, and Nichol 2006).

These "modern" data are obviously heavily biased—nearly all are derived from fish populations that have been under extreme exploitation pressure for more than a century (and often more) and are responding to a half-decade-long episode of unprecedented climate change. Recognizing these biases, and responding to Pauly's (2005) "shifting baselines syndrome," researchers have turned to historical documents such as ships' and merchants' logs to extend our understanding of fish populations hundreds of years into the past (e.g., McKenzie 2008; Rosenberg et al. 2005). Unfortunately, these studies often suggest that baselines shifted even further back than historical records can track; some of the earliest records from North America indicate that fish stocks were already heavily altered by exploitation pressure hundreds of years ago (e.g., Rosenberg et al. 2005:86). Therefore, though critical for understanding the history of fish populations, historical documents are unlikely to provide the types of "unbiased" population and ecosystem data fisheries managers require. Further, the types of detailed population measures (e.g., size or age of individuals) necessary for comparison with modern data are often absent from these records.

Realizing the limitations of modern datasets and historical records, researchers are beginning to turn to paleoecological data to create preindustrial baselines of fish populations (e.g., Finney et al. 2002; Jackson 2001; Jackson et al. 2001; Maschner et al. 2008). Zooarchaeologists have been tracking fluctuations in the taxonomic abundance and average size of fish for some time (Amorosi et al. 1996; Amorosi, McGovern, and Perdikaris 1994; Butler 2000; Butler and Delacorte 2004; Kenchington and Kenchington 1993; Leach and Davidson 2001; Leach, Davidson, and Horwood 1997; Rojo 1986, 1987, 1990; Spiess and Lewis 2001; Wheeler and Jones 1989), and are beginning to adapt this research to the service of conservation biology (e.g., Carder, Reitz, and Crock 2007; Jackson et al. 2001; Maschner et al. 2008; papers in this volume). These works reveal the potential for archaeofaunas to provide the types of deep time-series population data so crucial to an ecosystem approach to fisheries management.

Many types of proxy measures currently used by fisheries researchers can be reconstructed from archaeologically derived fish bones: size, length at age, fecundity (based on size relationships), DNA profiles, and isotopic signatures are all potential candidates for reconstruction from well-preserved fish bone. In this paper we reconstruct size-based indicators (see definition below) of shifts in fish populations from the archaeological record and discuss methodological issues surrounding the integration of these datasets with modern fisheries records. To highlight the effectiveness of a multiregional approach, we present two very different case studies from opposite sides of the North American continent—Pacific cod (*Gadus macrocephalus*) populations from the western Gulf of Alaska (GOA) and Atlantic cod (*Gadus morhua*) populations from the Gulf of Maine (GOM). Current records show that cod in both regions vary in size on an interdecadal cycle, in response to both environmental shifts and exploitation pressure. Our intent is to track long-term centennial- and millennial-scale shifts in these two very different populations and to assess changes in their structure within the context of contrasting climate cycles and exploitation histories. In the process, we demonstrate the importance of paleofisheries research for understanding modern fisheries data.

Size-Based Indicators and Sustainability

Size is an important metric in fisheries management research. Fish grow throughout their life span and most life history traits in fish stocks are correlated with size (Reiss 1989).

Size also constrains energy assimilation; as a result, changes in fish size may be correlated with environmental change (Shin et al. 2005:392) and have been linked to changes in trophic structure (Jennings et al. 2002). Short- and long-term changes in fish size are strongly linked to exploitation pressure. Fishing is size selective and modern trawl and net fishing techniques tend to remove larger fish from populations. In fact, heavy exploitation has been demonstrated to force evolutionary changes reflected in smaller size at age and age/size at maturation (for cod, consult Andersen et al. 2007; Swain, Sinclair, and Hanson 2007; see also Bianchi et al. 2000; Rochet 1998; Trippel 1995).

Size-based indicators, or SBIs, are "statistics summarizing the size distribution of fish assemblages and populations" (Shin et al. 2005:384). They have increasingly been a focus of modern fisheries research, and metrics such as mean length, mean length at age, and maximum length have been used to assess modern fish populations (Babcock et al. 1999; Bellail et al. 2003; Gedamke and Hoenig 2006; Swain, Sinclair, and Hanson 2007). Interest in mean length is a relatively recent phenomenon (Shin et al. 2005: Table 1), and empirical studies reporting temporal and spatial trends are not numerous. Proponents of the metric suggest, however, that it provides a robust quantitative indication of changes in populations, and especially of the combined changes in large and small individuals (e.g., Shin et al. 2005). We utilize mean length (fork length) as a population parameter primarily because it is easily reconstructed from faunal remains and it can be linked easily to fecundity levels (e.g., Maschner et al. 2008). While age can be reconstructed from fish remains (providing the opportunity to construct length at age metrics), it is often difficult to associate a length and an age to the same element—a problem that would greatly reduce sample sizes.

The response of an average metric, in this case mean fork length, to changes in population structure is complicated and must involve a careful consideration of many possible factors. One useful way to think of the problem is to visualize the impact of different size and age cohorts on mean population length. Mean length may decrease because of a drop in the abundance of large fish, the consequence of exploitation pressure by size-maximizing gear types (e.g., Swain, Sinclair, and Hanson 2007). Environmental effects, such as a decrease in oceanic productivity, may also negatively affect large fish (Shin et al. 2005:392). Alternatively, a reduction in mean length might be caused by an increase in recruitment, where more small fish enter the system due to increased fecundity (linked to survivorship of large fish). Decreased fecundity, due to a loss of large fish, interruption of spawning, or constraints on food, may actually result in an increase in the mean length of a fish population, as fewer juveniles are recruited into the population. The point of this discussion is that interpreting changes in size-based indicators in fish populations is challenging, and must be conducted by thoroughly weighing available data against length frequencies to adequately assess causality. With this paper we intend to both introduce these problems and provide some possible solutions.

Spatial and Temporal Aspects of the Archaeofaunal Samples

The Gulf of Alaska Pacific cod samples were recovered from Sanak Island, the largest island in a small archipelago southwest of the Alaska Peninsula. The island is located at

Table 11.1. Radiocarbon dates from Sanak Island

AHRS site number 49-XFP	Material dated	Lab number	Radiocarbon date	Calibrated date range (2 sigma)	Calibrated mean date
111	Wood charcoal	CAMS-127700	4025±35 BP	2630 BC–2460 BC	2545 BC
111	Wood charcoal	CMAS-127699	4040±35 BP	2670 BC–24670 BC	2570 BC
054	Wood charcoal	BETA-194363	3410±50 BP	1880 BC–1680 BC	1750 BC
036	Wood charcoal	CMAS-127688	2645±35 BP	900 BC–770 BC	840 BC
061	Wood charcoal	CAMS-110666	2480±35 BP	770 BC–410 BC	595 BC
058	Wood charcoal	CAMS-110660	2070±35 BP	180 BC–AD 20	80 BC
056 Upper	Wood charcoal	CAMS-110658	1540±45 BP	AD 420–AD 620	AD 520
056 Lower	Wood charcoal	CAMS-110659	1005±50 BP	AD 890–AD 1170	AD 1030
110	Wood charcoal	CAMS-110686	395±35 BP	AD 1440–AD 1640	AD 1540
110	Wood charcoal	CAMS-110681	385±40 BP	AD 1440–AD 1640	AD 1540

Note: The stratigraphy for site 036 is complex, and recent radiocarbon assays may revise the dating of the site.

the edge of the continental shelf, separated from the mainland peninsula by about 50 km of open water. Sanak has about 92 km of shoreline, most of which is rocky intertidal. Archaeological evidence suggests that Sanak has been inhabited for nearly 6,000 years, with significant shell midden deposits developing after ca. 4,500 years ago.

The assemblages described herein were recovered from eight discrete archaeological contexts spanning the period ca. 2550 cal BC to cal AD 1540 (see Table 11.1). All sites are located adjacent to rocky intertidal zones, and all are associated with large shell midden deposits. The faunal samples themselves are derived from single stratigraphic units (in a shell matrix), and all dates were provided by AMS (accelerator mass spectrometer) determinations on wood charcoal. All dates utilized in the figures in this paper represent the mean of single radiocarbon assays, or pooled means from multiple assays, and we note that none of the age ranges for the samples overlap (thus avoiding the possibility of chronological redundancy; see Table 11.1).

For the Gulf of Maine, samples are from the Turner Farm site, located on North Haven Island in outer Penobscot Bay. North Haven is the second largest island in the Fox Island group, and the site itself is positioned on a terrace system above a gravel beach. Bourque (1995:13) describes Turner Farm as "one of the largest and deepest Fox Island sites," whose "deposits preserve what is probably the longest and most complex cultural sequence on the Gulf of Maine coast."

Based on radiocarbon assays, stratigraphy, and artifact types, Bourque divided the history of the site into several occupations, which he numbered sequentially from oldest to latest. While stratigraphic control of the excavation was excellent, absolute dating of the reported faunal assemblages is complicated by a lack of stratigraphic separation during the analysis of the faunal material (see description in Spiess and Lewis 2001), a sampling and analysis issue common in zooarchaeology. Therefore, the assemblages described in this paper are not presented with absolute dates, but instead have been aggregated into four "occupations" that each represent a broad chronological range and multiple stratigraphic units. The "occupations" span the period 2400 cal BC to approximately cal AD 350 (Bourque, Johnson, and Steneck 2008). The radiocarbon sequence is documented exhaustively by Bourque (1995) and others (Jackson et al. 2001; Spiess and Lewis 2001), and hence we do not repeat those data here. The radiocarbon dating from Turner Farm is currently being reevaluated by the excavators (e.g., Bourque, Johnson, and Steneck 2008:168). Instead of confounding the issue, we use mean dates recently published by Bourque, Johnson, and Steneck (2008:169), which are based on radiocarbon dates, stratigraphic associations, and artifact assemblages: Occupation 1 (ca. 2400 cal BC), Occupation 2 (ca. cal 2150 BC), Occupation 3 (ca. cal 1600 BC), and Occupation 4 (ca. cal AD 350). Each occupation can be envisioned as a discrete chronological (and therefore analytical) entity, though we note the duration of each occupation is not the same.

In this paper, and in contrast to recently published work (e.g., Bourque, Johnson, and Steneck 2008; Jackson et al. 2001), we omit the "plow zone" assemblage (Occupation 5) from the length-frequency analysis. We do this primarily because of the potential stratigraphic mixing of fish assemblages that may have occurred as a result of repeated tilling (e.g., Bourque, Johnson, and Steneck 2008:168).

Size-Based Indicators and Comparability with Modern Samples

There are several potential obstacles to integrating archaeological data with modern fishing records. These relate to the way in which the archaeological sample was created, preserved, and recovered, and its subsequent comparability to modern records with their own unique sampling histories. The issues are not dissimilar to comparing modern records from different studies—as Shin et al. (2005:390) have noted, fisheries researchers interested in comparing SBIs must control for differences caused by the "gear used, time, and location." Not surprisingly, these same factors are crucial when comparing archaeofaunal-derived proxy measures and those derived from modern fish populations. Below we discuss differences in gear type, sampling location, seasonality, and a related zooarchaeological consideration, recovery methods.

Gear Type

All fishing gear, whether modern or ancient, is designed to take a specific size range of individuals from a population. When comparing length distributions or developing size/length sequences, it is imperative that the samples being compared were obtained with similar gear types or, at the minimum, gear types that select for a similar range of sizes. In the following case studies, we chose modern samples taken with hook-based gear types that best approximate the jigging technology used by prehistoric groups.

For the Gulf of Alaska we used a modern sample taken with longline gear with baited circle hooks (Thompson, Dorn, and Nichol 2006). Ivory and bone fishhook barbs recovered from levels directly linked to the Sanak Island faunal samples indicate that prehistoric Aleut hooks would have been comparable in size to a range of modern hooks approximately 3.5 to 6 cm in length, a size consistent with most modern longline gear.

There is some indication in the fisheries literature that longline gear may subtly select for a smaller size range than handlines (i.e., jigs; Halliday 2002), a possibility we address directly in our analyses below.

This difference in gear type does not affect the GOM sample. Here we utilized a modern sample taken with handlines that used baited "J" hooks. Bone fishhooks recovered from Turner Farm range in size between 2.5 and 6.0 cm in length (Bourque 1995: Figures 5.26 and 7.13), a size range similar to "J" hooks used by modern handliners.

Sampling Location and Seasonality

Cod in both the GOA and the GOM are seasonal migrators and spawners (Robichaud and Rose 2004; Shimada and Kimura 1994). In Atlantic cod, size and age cohorts are known to be spatially heterogeneous, and therefore sampling location may affect the mean size of the population (Beacham 1982; Clark and Green 1991; Halliday 2002:72; Sinclair 1992). We controlled for seasonality in the GOA by choosing only archaeological samples that were derived from fall and winter shell midden deposits associated with semisubterranean house ruins. We determined the seasonality of the shell midden deposits by (1) their association with semisubterranean houses, a house type described ethnographically as primarily a fall and winter structure (Ponomarev and Glotov 1988; Veniaminov 1984); (2) the presence of migratory birds, which only stage in the area in the fall months; and (3) salmon element representation (i.e., lack of cranial elements; see Hoffman, Czederpiltz, and Partlow 2000). We compared these assemblages with modern fisheries data derived from fall and winter (September to December) cod surveys conducted in 2005 (Thompson, Dorn, and Nichol 2006). All faunal assemblages were derived from sites on the eastern coast of Sanak Island, and all were associated with sheltered bays with rocky intertidal coastlines. The modern longline samples were derived from nearshore waters (25–140 fathoms) that mimic the shelf ecosystem that surrounds Sanak Island (Thompson, Dorn, and Nichol 2006:170), where we assume much of the cod in the Sanak middens originated.

Seasonality is more difficult to assess for the Gulf of Maine samples. The Turner Farm assemblages likely represent year-round occupations, but analysis of cod vertebrae annuli (Spiess and Lewis 2001:88) indicate they were universally harvested during the summer and early fall, from June to October. Our comparative modern sample from the Gulf of Maine is primarily derived from excursions during the normal fixed-gear fishing season from June to December (for discussion of seasonal fish harvesting in the GOM, see Mayo and O'Brien 2007:5). It was collected from North Atlantic Fisheries Organization (NAFO) subregion 4Xr, a fisheries management region located 100 km northwest of the mouth of Penobscot Bay. This is a nearshore location at the head of a bay (the northwestern tip of the Nova Scotia peninsula, near the Bay of Fundy), similar to the marine environment that surrounds Turner Farm.

Recovery

Zooarchaeologists have known for some time that recovery methods are size selective and can significantly skew taxonomic representation of fish (James 1997). Due to the potential to select for larger fish, coarser recovery methods will potentially also skew length frequencies reconstructed from archaeofaunal assemblages (Nagaoka 2005).

All of the Sanak Island samples were screened through 8 mm mesh. We tested the efficacy of this mesh size by dual screening bulk-sampled deposits through both 8 mm and 3 mm mesh and comparing the recovered cod remains. While the quantity of spines, ribs, and rays varied substantially between the two mesh sizes, there was no difference in the recovery of cod trunk and caudal vertebrae or the majority of cranial elements (see methods below). This was not unexpected; cod element, or "specimen," sizes are so large in the Gulf of Alaska that 8 mm mesh appears to recover most of their preserved elements.

In regards to recovery, the Turner Farm samples pose a particular challenge, because screening protocols were not consistently utilized in the collection of fauna from any of the excavated strata. This likely biased the recovery of smaller fish taxa (Spiess and Lewis 2001:7), although its effects on larger taxa, such as cod, are unknown. In his discussion of screening methods, Cannon (1999) notes the importance of assessing the impact of specimen sizes on the recovery of animal remains. Specifically, he suggests that a "hand-picked" zooarchaeological sample can be adequate if specimen sizes are large. In his comparison of fish remains from across arctic sites, Whitridge (2001), following Cannon (1999), indicated that careful hand picking can result in adequate recovery of many northern fish taxa, since these are often very large. That the Turner Farm fauna was meticulously recovered by hand sampling has been noted (Bourque 1995; Spiess and Lewis 2001:7), and indeed we draw attention to the taxonomic diversity of the Turner Farm sample, which clearly demonstrates the success of hand picking in the recovery of large quantities of fish much smaller than Atlantic cod, such as winter flounder (*Pseudopleuronectes americanus*) and sculpin (Cottidae). In fact, Bourque (1995:21–31) suggests this method may have actually increased the recovery of fish bones, which were fragile and susceptible to crushing in shell-laden screens. Given the average specimen size of Atlantic cod bones, we consider the recovery methods used at Turner Farm to have relatively little impact on the size distribution of cod collected from its middens.

Analytical Methods

Reconstructing the length of live fish from their bones is a relatively simple process involving allometric regression, as long as the regression formulae exist for the taxon under study. Accurate regression formulae for commercially important fish, such as salmonids, have been used by zooarchaeologists for decades (Wheeler and Jones 1989). It is therefore a scientific irony that regression formulae for cod, perhaps the most historically important commercial fish, have only recently been created (see Orchard 2001, 2003; Rojo 1986, 1987, 1990, 2002).

For the Sanak Island sample, cod were identified using ichthyofaunal reference collections at the Idaho Museum of Natural History and the Canadian Museum of Civilization. In cases of potential ambiguity between the identification of walleye pollock (*Theragra chalcogramma*) and Pacific cod, we chose not to include that questionable element in our study. Our length reconstructions are based on the following measurements: (1) the maximum length of the ascending process of premaxillae and (2) the anterior-posterior length of the centra of abdominal (trunk) vertebrae. We chose these elements because they were abundant and consistently had the best preserved anatomical landmarks with corresponding regression formulae, which we obtained from Orchard (2001, 2003). Sample sizes for the Sanak assemblages ranged from a low of 10 to a high of 507 elements (see Maschner et al. 2008 for a complete presentation of the dataset).

The GOM measurement data are derived from Spiess and Lewis's (2001:89) account of the Turner Farm fauna, and are based on the maximum diameter of the centra of trunk (abdominal) vertebrae. We reanalyzed these data using regression equations created by Rojo (1987) from Atlantic cod harvested in the northern GOM. Sample sizes for each of the Turner Farm occupations ranged between 33 and 117 elements (see Spiess and Lewis 2001: Table 3–55).

The modern mean lengths utilized in this study are based on length frequency data generated during modern fisheries surveys taken in 2005 (GOA), and between the years 2000 and 2007 (GOM). Sample sizes ranged from 3,308 individuals in the GOA longline sample, to 112 in the GOM handline sample. In each case, lengths were recorded as frequencies in size "bins," and we utilized procedures described in Gedamke and Hoenig (2006:485) to convert these data to mean lengths.

Results

We present the mean length (with associated standard errors) for each of the Gulf of Alaska and Gulf of Maine assemblages in Figures 11.1 and 11.2, respectively. The two figures are very different, suggesting significant contrasts in the population histories of Atlantic and Pacific cod; by scrutinizing their individual structures we can highlight these differences. The Sanak Island sequence is notable in three aspects. First, precontact mean length varies in a cyclical fashion of increasing and then decreasing size, in apparent multicentennial cycles. It is uncertain if this cyclical oscillation is simply an artifact of the relatively uneven chronological resolution or a result of real differences in mean length. If the latter, this can only be elucidated by comparison to an external context of reference (see below). The second pattern observable in Figure 11.1, again notable by its contrast to the Turner Farm sequence, is that mean lengths appear to vary minimally. The largest sequential change in length is only 7.6 cm, and the maximum range of variability over the entire 4,500-year sequence is only 9 cm. Lastly, and perhaps crucially, the modern mean length of Pacific cod in the Sanak Island sequence appears to fall within the range of precontact variability. In fact, T-tests reveal that the modern mean is not significantly different from five of the precontact assemblages ($p < 0.05$; see Maschner et al. 2008 for details).

In contrast to the GOA sequence, the GOM sequence suggests a significant increase in cod size over the Late Holocene, during Occupations 1–4, but a drastic decline in mean length in the modern era. Unlike the GOA sequence, T-tests indicate that all precontact means are significantly different from the modern mean ($p < 0.01$). Perhaps most noticeable is the gradual and significant increase in mean size over the first 5,000 years

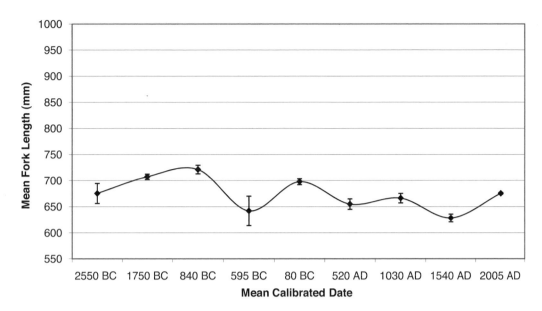

Figure 11.1. Reconstructed mean Pacific cod fork lengths from the Gulf of Alaska (Sanak Island). Dates are presented as calibrated means; for ranges please consult Table 11.1.

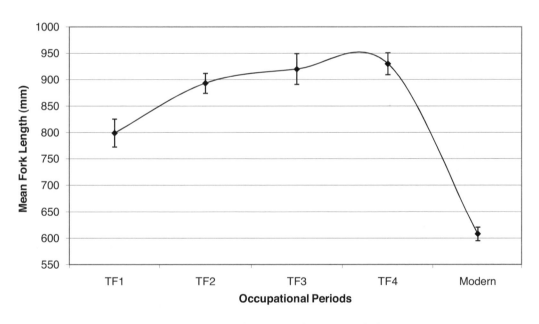

Figure 11.2. Reconstructed mean Atlantic cod fork lengths from the Gulf of Maine (Turner Farm).

of the sequence (Occupations 1–4). The duration, direction, and magnitude of this trend are intriguing, especially when one views it in light of the relatively stable interdecadal records of the modern era, and especially when viewed from the perspective of hypothesized increasing human population levels (and therefore human exploitation) over time (Bourque 1995; see discussion in Jackson et al. 2001).

Bourque, Johnson, and Steneck (2008) suggest that carbon isotopes from Turner Farm cod bones change little over this prehistoric part of the sequence, indicating that cod did not change their trophic position significantly, despite their increasing size through time. Yet when comparing this sequence to carbon isotopes from modern cod, Bourque, Johnson, and Steneck (2008:177) noted a significant change (from a $\delta^{13}C$ value of –13 percent to –17 percent). Bourque, Johnson, and Steneck attribute this to a major ecosystem change (from a kelp/sea grass–dominated ecosystem to a phytoplankton-based ecosystem), which we suggest could impact both the size and numbers of cod.

Histories of Exploitation

Perhaps the most obvious contrast in the two graphs is the difference in the position of the end points of each series, which we expect marks the reaction of cod to modern commercial exploitation. The graphs imply a significant contrast in the exploitation histories of these two stocks, and it will therefore be informative to explore the differences in the historical length, intensity, and type of gear emphasized between these two industries. In the discussion that follows we assume that heavily exploited populations should experience a reduction in mean size, as larger fish are removed from the population and as sustained exploitation pressure forces a reduction in length at age and age at maturation (e.g., Barot et al. 2004).

In the Gulf of Maine, cod has been intensively exploited in commercial quantities since the early seventeenth century, when merchants in Bristol first began sponsoring fishing excursions to the area. By 1624, the town of Gloucester, Massachusetts, supported "no less than 50 vessels" (Babson 1860, in Lear 1998), and just a hundred years later, Massachusetts alone supported 5,000 fishermen and about 800 vessels (Lear 1998:52). The traditional center of this fishery was initially inshore, focused on the Gulf of Maine and the nearshore Scotian Shelf, but over time, adjacent Georges Bank became equally important.

Handlining, or jigging, using a line with a single baited hook, was the only technique used to catch Atlantic cod for nearly 250 years. This method, where small groups of individuals handlined cod from small vessels, was essentially the same as fishing practices used by Indigenous inhabitants of the GOM, though vastly amplified in scale and intensity (see Barsh 2002). In the early 1850s larger vessels began using "bultow," or longline gear, composed of a long fishing line with several hundred baited hooks, spaced at intervals towed behind a moving vessel. This technology greatly increased catch rates, and by 1880, 294 million pounds of cod were landed, the largest catch in the history of the New England fishery (Lear 1998:60).

A far more efficient gear type, the otter trawl, was first experimented with in the GOM in the late 1800s, but fishermen were constrained in their use of the technology—they were unable to efficiently preserve the large amounts of fish caught using the technique.

This problem was solved with the introduction of shipboard freezing systems in the 1920s, and by 1930 otter trawling accounted for 59 percent of the total GOM catch. By 1935, 69 percent of all groundfish caught in the GOM was by otter trawl, and by 1994, it accounted for approximately 85 percent of the total GOM catch (Claesson, 2008, pers. comm.; Mayo and O'Brien 2006:5; Wallace 2007). Today approximately 65 percent of U.S. landings are made with otter trawls (Mayo and O'Brien 2006:5).

Before the mid-nineteenth century, few historical data from the GOM indicate how Atlantic cod may have reacted to increasing commercial exploitation pressure. An eighteenth-century length-frequency sample from the Terrance Bay wreck, a ship that sank off the central Scotian coast, indicates that the mean length of cod in the 1750s was 89.1 cm, well within the range of prehistoric variation for Turner Farm (Kenchington and Kenchington 1993:339). Although size differences have been observed between cod stocks in the GOM and adjacent waters (e.g., Beacham 1982; Clark and Green 1991), data suggest that the mean length of cod for the stock represented by the Terrance Bay wreck (NAFO fishery region 4VsW) have been historically smaller than 4X cod (e.g., compare Beacham [1982] with data presented below). Thus, the very large size of the cod in the Terrance Bay wreck implies that overexploitation, at least as measured as a reduction in mean length, was not evident in the 1750s.

After 1850, historical documents record the impact that industrial fishing operations had on local cod. These data, taken from ships' logs and merchant accounts, suggest that by 1852, during the early days of the longline fishery, GOM fishermen were already noting a decline in the numbers of fish caught (Rosenberg et al. 2005:86). Between 1859 and 1882, a clear reduction in catch per unit effort has been identified from these records (Rosenberg et al. 2005:87). Not surprisingly, Canadian fishermen at this time moved to ban the use of longlines, which they considered responsible for the declines in cod (Lear 1998:60). Yet, while cod stocks were clearly responding to some form of exploitation pressure, landings increased throughout the late nineteenth century, presumably a result of the increased fishing effort due to the use of longlines.

Without adequate faunal remains from the early historical period, it is difficult to assess the point at which mean cod lengths began to react to the pressures of industrial fishing. If the Terrance Bay data are applicable, the major shift must have occurred between the mid-eighteenth and mid-twentieth centuries, a period that witnessed the widespread adoption of both longlining and trawling. Given the historical evidence, it seems clear that longlining must have had a major impact on cod stocks (see data in Rosenberg et al. 2005). Jackson et al. (2001:631), however, speculate that it was the development of otter trawls in the 1920s that precipitated much of the decline in both the numbers and size of Atlantic cod. Furthermore, Bourque, Johnson, and Steneck (2008; see also Jackson et al. 2001) indicate that cod populations may have already declined significantly in both size and number as early as the seventeenth century. These contact-era changes are difficult to interpret because they are based on very small cod bone assemblages from mixed plow zone deposits (Spiess and Lewis 2001: Tables 3–53, 3–55). Nevertheless, these data may imply that even a traditional handline fishery had a significant impact on local cod populations (Bourque, Johnson, and Steneck 2008:180), although we stress that the age of this context (ca. cal AD 1600) makes it difficult to determine if this sample represents a response of cod stocks to a traditional Aboriginal fishery or a commercial European one.

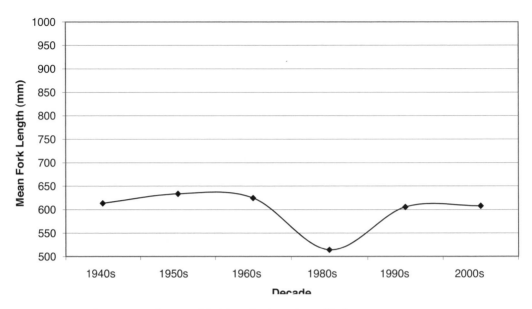

Figure 11.3. Modern mean Atlantic cod fork lengths from the Gulf of Maine (NAFO Region 4Xr).

It is unfortunate that we have so little data that allow us to gauge the exact response of the fishery to the introduction of bottom trawling. What we do know is that by 1940, when trawling was the dominant gear type in use, the average length of handlined cod in the study area had fallen to 61.3 cm, a size that has characterized the stock for much of the last sixty years (Figure 11.3), with the obvious exception of the 1980s stock crash.

Our data suggest that modern Atlantic cod lengths are on average 27.7 cm shorter than they were for much of the last 5,000 years. The relationship between this reduction in size and Atlantic cod abundance is clear. Biomass estimates for Atlantic cod calculated from historical records suggest that the current biomass of cod in the GOM and Scotian Shelf regions is only 4 percent of that in 1852 (Rosenberg et al. 2005:88). We note that by 1852, cod had already been intensively exploited in the GOM for over 250 years.

The potential impacts that these gear types may have had on the Atlantic fishery may be best explored by contrasting this history with the world's other great cod fishery in the North Pacific. The first major difference between the GOM and GOA fisheries is that the GOA industry has a much shorter history. Commercial cod fishing began in the eastern Aleutians in the 1870s, when Scandinavian fishermen, seeking more productive fishing grounds, settled in the Shumagin and Sanak islands. Word of the productive waters grew, and soon schooners from ports in the United States and Canada sailed to the region to fish. This was predominately a shore-based handline fishery, though larger schooners also employed longlines. By 1915, at least seventeen shore stations were operating in the region, taking over one million fish each year (Shields 2001:20). Although specific data on catch rates are not available, after 1915 cod catches began to decline, and by 1930 shore stations began to close. The situation worsened; between 1942 and 1975 the commercial fishery was all but defunct.

The well-documented 1975 oceanic regime shift (Benson and Trites 2002; Yang 2004) corresponds with the return of large numbers of Pacific cod and the cod fishing industry. After 1975, the industry primarily employed longlines and cod pots, while bottom trawling has had a relatively limited penetration into the industry. Trawling never reached the intensity that it did in the Atlantic, and by its peak in 2002, bottom trawling accounted for only 50 percent of the total groundfish catch in the GOA (Enticknap 2002:2). Recent bans have further reduced the trawl catch to even lower levels, in stark contrast to the GOM cod fishery.

That this exploitation history is associated with a modern size profile within the range of prehistoric variability is significant. Put simply, we can find no evidence for exploitation effects on cod in the Gulf of Alaska based on SBIs. While this seems implausible in an era of global fisheries crises (e.g., Jackson et al. 2001; Worm et al. 2006), this should perhaps not be unexpected: We know from modern management records that Gulf of Alaska cod populations are both healthy and not being overexploited (Thompson et al. 2007; Thompson, Dorn, and Nichol 2006).

The comparatively large decline in the size of Atlantic cod in the modern era was caused by sustained fishing pressure and the removal of large fish from the ecosystem. The reduction in size may reflect evolutionary changes caused by fishing pressure, such as a reduction in length at age and age at maturation. These latter processes are still ongoing in the Gulf of Maine (Barot et al. 2004; Swain et al. 2003; Swain, Sinclair, and Hanson 2007). Our analysis suggests this was a long historical process, involving the cumulative weight of centuries of exploitation pressure and increasing fishing effort. Certain threshold decisions involving the use of specific catch-maximizing gear types, most notably the longline and otter trawl, may have had significant impacts on the fishery. Given the vast differences in the GOM and GOA sequences, we believe that the contrasting histories can serve as case studies for long-term management scenarios.

Correlations with Climate

While the bulk of the analysis to this point has focused on the use of the archaeological record to understand the impact of fishing pressure on fish size, a primary goal of the paper is to expose the potential for tracking climate impacts on fish stocks using a deep-time perspective. Recent fisheries literature has tracked the impacts of climate on heavily exploited fish populations (e.g., Drinkwater 2005). A major obstacle to these types of analyses is that they are in essence measuring the response of already-stressed fish stocks to short-term climate fluctuations, another consequence of the "shifting baselines syndrome" (Pauly 1995). The zooarchaeological record provides the means to assess climate impacts on fish stocks under preindustrial exploitation pressure.

For the Gulf of Alaska, we created a model (Table 11.2) of changing climate based on proxy records of prehistoric atmospheric and sea surface temperature, storminess, and productivity (Calkin, Wiles, and Barclay 2001; D'Arrigo et al. 2005; Finney et al. 2002; Heusser, Heusser, and Peteet 1985; Hu et al. 2001; Jordan and Krumhardt 2003; Loso et al. 2006; Mann 2001; Mann and Hamilton 1995; Mason and Jordan 1993; Misarti 2007). From a marine ecosystem perspective, these perturbations may be referred to as macro-level regime shifts (Benson and Trites 2002; Polovina 2005). We have not attempted to develop a climate model for the Turner Farm sequence—the lack of temporal resolution in the dating of the Turner Farm samples makes this type of analysis unproductive. However, there are other means to track environmental impacts on these samples, as we demonstrate below.

Table 11.2. Paleoclimate model for Gulf of Alaska and Sanak Archipelago

Period	Date range	Approximate climate conditions
Early Holocene	7000–4200 cal BC	Cool and dry
Middle Holocene	4200–2550 cal BC	Warm and wet
Neoglacial	2550–1200 cal BC	Cool and wet
	1200–100 cal BC	Cool and wet (increased storminess)
Pre-Medieval Warm period	100 BC–cal AD 300	Warm
	cal AD 300–900	Cold
Medieval Warm period	cal AD 900–1250	Warm and dry
Little Ice Age	cal AD 1250–1850	Cold and wet
Modern era	cal AD 1850–present day	Very warm

Note: Lightly shaded boxes represent periods of cooler and stormier conditions, which are generally more productive, while the darkly shaded boxes represent warmer and generally less productive conditions.

Figure 11.4 displays the GOA climate model projected against our Sanak Island Pacific cod length sequence. A visual inspection clearly indicates significant correspondence between climate and cod size, especially after the Neoglacial period. In general, the trends appear to suggest an increase in mean cod length with warm regimes, and a decrease in mean cod length with cold and stormy regimes. We draw attention to the correspondence between size and the Medieval Warm period (cal AD 900–1250) and the Little Ice Age (cal AD 1250–1850), when cod increase and then decrease in size, respectively. We also note the slight increase in cod length at the end of the sequence, during the modern era. Given the analysis above, we propose that this latter increase may in fact reflect a stock response to global warming, rather than a response to exploitation pressure (Maschner et al. 2008). If so, then the response is similar to the increase in mean size that occurred during the Medieval Warm period and other warm regimes in the past.

That cold and productive regimes are associated with a drop in mean fish size may initially seem counterintuitive. Yet ecologically, this is exactly what we expect if climate shifts are affecting cod populations. During cooler and more productive regimes (Chavez et al. 2003), fecundity should increase, and so should cod abundance due to increased recruitment; with the increase in the number of young individuals in the population, mean size should decrease (Shin et al. 2005:392). This correspondence implies that mean cod length is tracking ecologically forced changes in cod population structure in the GOA. Furthermore, it supports the proposition that modern global warming may be a complicating factor in developing management policies for Pacific cod (Benson and Trites 2002; Maschner et al. 2008; Yang 2004).

We test this hypothesis by comparing prehistoric cod lengths to their relative abundance. Here our assumption is that cod abundance will also fluctuate with oceanic productivity, and therefore climate, and that these shifts in population abundances will be

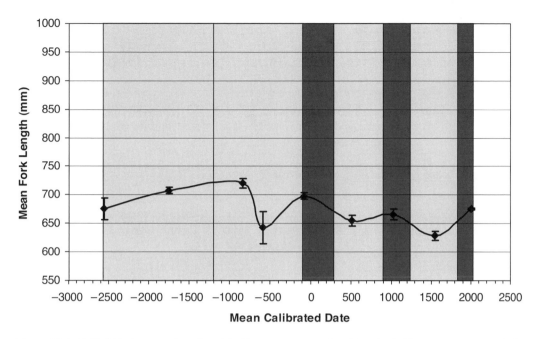

Figure 11.4. Gulf of Alaska mean lengths with climate reconstruction. The shaded boxes represent periods of significant changes in air temperature, sea surface temperature, storminess, and ocean circulation that drive ocean productivity. The lightly shaded boxes represent periods of cooler and stormier conditions, which are generally more productive, while the darkly shaded boxes represent warmer and generally less productive conditions. Dates are presented as calibrated means; for ranges please consult Table 11.1.

discernable from midden faunas. In the analysis that follows, we utilize an abundance index to track fluctuations in the relative abundance of cod. We prefer abundance indices because not only do they track changes in relative frequency, but when the comparative taxa are generally smaller in body size, they incorporate size-based caloric relationships, which can be used to speculate about changes in return rates (e.g., Betts and Friesen 2006; Broughton 1994, 1997, 1999; Butler 2000; Cannon 2000; Grayson 2001; Hildebrandt and Jones 1992; Janetski 1997). We calculated an abundance index for Sanak cod using Scorpaeniformes order fish in the denominator, using the following formula: AI = NISP Pacific cod/NISP Pacific cod + NISP Scorpaeniformes. We utilize Scorpaeniformes for the following reasons: (1) when aggregated, the taxa in Scorpaeniformes are the second most abundant behind cod in the middens, and (2) because they are groundfish, which tend to be caught as bycatch when fishing for cod.

In the calculation of such relative abundance indices, one must always consider the problem of "closed arrays" (Grayson 1984; Lyman 2008), or the possibility that the bivariate proportional measure we are using is actually tracking shifts in the frequencies of the comparison taxa (in this case, fish in the order Scorpaeniformes), rather than shifts in the taxon of interest (Pacific cod). We believe we avoid these problems because the Scorpaeniformes taxon used here is actually composed of multiple fish taxa, including *Sebastes* sp. (rockfish), Hexagrammidae (greenling), and Cottidae (sculpin), and therefore the AI captures variability in multiple groundfish species. More importantly, the prey model suggests that when the abundance of cod, a high-ranked fish, declines, larger numbers of the

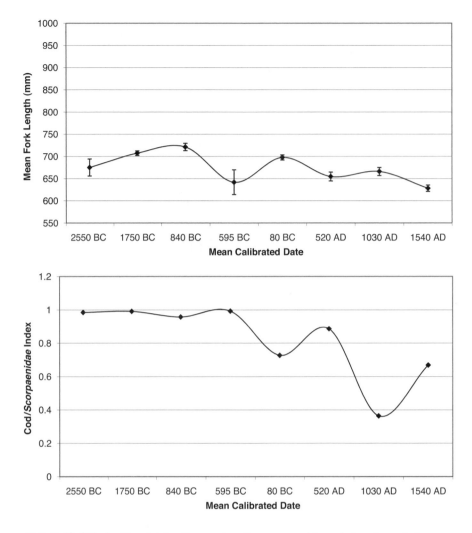

Figure 11.5. Gulf of Alaska (Sanak Island) mean lengths compared to cod abundance index.

lower-ranked (bycatch) fish in Scorpaeniformes would be harvested to mitigate any short-falls (assuming a goal of constant returns). Therefore any decline or increase in Scorpaeniformes should, theoretically, be linked to a concomitant shift in cod abundance, and not fishing effort. Because of how the index is calculated, these relationships would amplify the magnitude of changes in the abundance index, but not its direction.

When we compare shifting cod lengths to their associated cod–Scorpaeniformes AI, we see that abundance trends are inversely correlated with size trends (Figure 11.5). That is, when mean size increases, abundance decreases, and when mean size decreases, abundance increases. This is exactly the correlation we expect; when productivity increases during cold regimes, cod populations increase, but mean length will decline because of the large numbers of juveniles in the population. Cod fecundity increases dramatically with size (Karp 1982), so while fish may live longer and grow larger under good conditions, the increased fecundity of the large fish leads to proportionally larger increases in juveniles. These smaller, younger fish will tend to dominate the harvest.

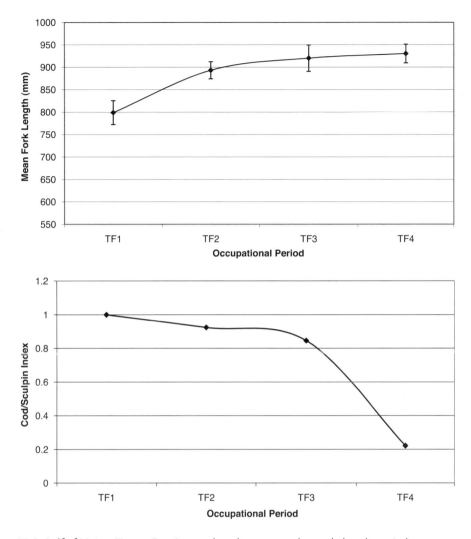

Figure 11.6. Gulf of Maine (Turner Farm) mean lengths compared to cod abundance index.

This climate-based effect on mean size is quite different from what we might expect from sustained overfishing. In the case of climate, the reduction in mean size is caused by increased fecundity and more juveniles in the population. In the overfishing scenario, the reduction is caused by sustained removal of large fish from the population, with concomitant evolutionary changes in the size at age and age at maturation.

How might these relationships inform us about climate impacts on the Turner Farm sequence? A comparison of a similar abundance index for the Turner Farm samples (again utilizing Scorpaeniformes in the denominator) suggests a similar relationship as that noted in the GOA (Figure 11.6). As mean size increases, the abundance index declines. To us, this suggests that Atlantic cod responded to climate regime shifts in the same manner as Pacific cod, and specifically that our size reconstruction here is tracking natural changes in cod populations rather than increased exploitation pressure. We note that the impact of precommercial human predation on Atlantic cod stocks in the GOM is currently debated (e.g., Barsh 2002; Bourque, Johnson, and Steneck 2008; Jackson et al. 2001). The data we have presented suggest that climate may also have had significant impacts on the abundance of prehistoric cod populations.

Conclusions

The population histories presented above represent a 5,000-plus-year relationship between humans and cod in the Northern Hemisphere. Management decisions based on little more than half a century of data will not provide the temporal depth necessary to adequately assess the modern state of fish populations. The shifting baseline syndrome (Pauly 1995) can only be adequately addressed by developing a preindustrial baseline that extends beyond the origins of the commercial fishery. For many areas of the world, this can only be achieved by probing the archaeological record for appropriate proxy data.

Our analysis indicates that precontact-period cod populations in the Gulf of Maine and the Gulf of Alaska were variable and exhibit clear, and previously unknown, multi-centennial cycles that appear to correspond to climate shifts (see Finney et al. 2002 for a similar conclusion about salmon in the Gulf of Alaska). More specifically, we believe our sequences indicate that Pacific cod populations in the Gulf of Alaska are resilient and have maintained a relatively constant and largely self-correcting length (and therefore fecundity and age) structure over some 4,500 years, despite significant commercial exploitation in the modern era. This is not the case for the Gulf of Maine cod stocks, which appear to have experienced both a progressive shift toward increasing mean length and decreasing abundance (which may be linked to climate) prior to the historic era, followed by a profound impact by modern overexploitation (although see Bourque, Johnson, and Steneck 2008). Despite these differences, in both sequences cod abundance and mean cod size fluctuated, sometimes considerably, before commercial exploitation. These climate-linked variations in cod populations under subsistence-level exploitation pressure should be taken into account when analyzing modern stock trends and developing management policies (see Drinkwater [2005] for an analysis of the impact of recent climate change on cod).

The perturbations in the numbers and size of cod are intriguing given the historical evidence for the periodic disappearance and sudden reappearance of cod in the GOA during the historic period. As discussed above, the significant decline in cod abundance in the GOA in the 1930s and 1940s, followed by their sudden return in 1975, appears to be related to an "oceanic regime shift" that occurred during the same period (Benson and Trites 2002; Stephens et al. 2001). That such oceanic shifts may have abruptly influenced the numbers of cod in the North Pacific may be evident in the Aleut language. The Aleut word for cod, *atxidaq*, literally translates as "the fish that stops." From the Gulf of Maine, the Mi'kmaq and Maliseet words for Atlantic cod, *peju* and *nuhkomeq*, mean "the fish we wait for" and "the fish there isn't enough of," respectively. This linguistic evidence also implies some uncertainty in cod populations. These Indigenous names expose the implicit traditional ecological knowledge about fish populations imbedded in Aboriginal languages, and further reinforce our identification of variability in precommercial cod stocks.

The contrasts in the histories of cod exploitation in the Gulf of Maine and the Gulf of Alaska are striking. While we can find no evidence for exploitation impacts on fish lengths in the GOA, the drastic decline in cod length at the end of the GOM sequence is almost certainly a response to overexploitation (Clark and Perley 2006; Jackson 2001; Jackson et al. 2001). Clearly, the cumulative historical weight of 400 years of increasing fishing efforts resulted in the severely stunted size profile that currently characterizes Atlantic cod in the GOM. Modern fisheries research suggests that this may reflect evolutionary changes caused by fishing pressure, such as a reduction in length at age and age at maturation. These processes are still ongoing in the Gulf of Maine (see Barot et al. 2004; Swain et al. 2003; Swain, Sinclair, and Hanson 2007).

Nevertheless, the contrast in the use of certain gear types clearly raises questions; Jackson et al. (2001:631) speculate that it was the rapid development of an otter trawl fishery after the 1920s, and the increase in fishing effort that entails, that precipitated much of the declines in both the numbers and size of Atlantic cod. This is an interesting hypothesis considering the relative lack of penetration of otter trawling in the GOA and its correspondence with a cod length structure within the range of precommercial variation. While bottom trawling was likely a significant factor in the changes in the GOM cod stock, we believe our analysis exposes the detrimental effect of the *cumulative* weight of fishing pressure—the sustained increases in fishing effort over decades and centuries—on cod stocks. Viewed from this perspective, the adoption of bottom trawling in the Gulf of Maine was the straw that broke the camel's back, the final blow in a gradual assault of more efficient gear types and increasing fishing effort that can be traced historically to at least the early 1850s with the adoption of commercial longline gear (Rosenberg et al. 2005), and perhaps as early as the seventeenth century (e.g., Bourque, Johnson, and Steneck 2008; Jackson et al. 2001) with the start of the commercial European handline fishery. In short, over a period of ca. 350 years the commercial cod fishery in the Gulf of Maine finally managed to exert so much fishing effort and inflict so much mortality that cod stocks were devastated, as clearly reflected in the stunted modern size profile unlike anything seen in the previous 4,500 years of human-cod interactions. Yet we also note that our data suggest cod stocks responded dramatically to changing climate, and this was likely a complicating factor in the recent decline in cod.

Our study had two primary goals: (1) to provide preindustrial benchmarks of cod populations and (2) to integrate these baselines in meaningful ways with modern fisheries management data. The ultimate aim of our bicoastal study was to provide solutions to the shifting baseline problem that affects modern fisheries management (e.g., Pauly 1995) by integrating archaeofaunal data with modern population measures, in a comparative and historically contextualized approach. By integrating zooarchaeological, historical, and modern data we possess the ability to document the absolute magnitude of the changes that can occur in a heavily exploited fish stock. Ultimately, we believe our comparison highlights the fact that management successes are possible, and indeed that significant fishing pressure can be sustained by cod stocks without fundamentally altering their length structure, as long as fishing effort is closely monitored and fishing mortality restricted to moderate levels. In the Gulf of Alaska, this appears to have occurred through a combination of a fortunate historical coincidence (i.e., the late penetration of commercial fishing) and the effectiveness of recent management policies.

Acknowledgments

We thank Stefan Claesson for providing statistics on historic catch rates in the GOM. We benefited from conversations with Bruce Bourque, Robert Steneck, and Arthur Spiess about the original Turner Farm length frequency analysis. We also thank the many students who helped us sort through tens of thousands of fish bones from our Sanak Island samples. Finally, we acknowledge Madonna Moss and Aubrey Cannon for inviting us to participate in a wonderful session on Pacific paleofisheries at the 2008 SAA conference in Vancouver.

References Cited

Amorosi, Thomas, Thomas McGovern, and Sophia Perdikaris. "Bioarchaeology and Cod Fisheries: A New Source of Evidence." *ICES Maritime Science Symposium* 198 (1994): 31–48.

Amorosi, Thomas, James Woollett, Sophia Perdikaris, and Thomas McGovern. "Regional Zooarchaeology and Global Change: Problems and Potentials." *World Archaeology* 28 (1996): 126–157.

Andersen, K. H., K. D. Farnsworth, U. H. Thygesen, and J. E. Beyer. "The Evolutionary Pressure from Fishing on Size at Maturation of Baltic Cod." *Ecological Modeling* 204 (2007): 246–252.

Babcock, Russell C., Shane Kelly, Nick T. Shears, Jarrod W. Walker, and Trevor J. Willis. "Changes in Community Structure and Temperate Marine Reserves." *Marine Ecology Progress Series* 189 (1999): 125–134.

Babson, John J. *History of the Town of Gloucester, Cape Anne, Including the Town of Rockport.* Gloucester: Proctor Brothers, 1860.

Barot, Sébastien, Mikko Heino, Loretta O'Brien, and Ulf Dieckmann. "Reaction Norms for Age and Size at Maturation: Study of the Long Term Trend (1970–1998) for Georges Bank and Gulf of Maine Cod Stocks." *Ecological Applications* 14 (2004): 1257–1271.

Barsh, Russel L. "*Netukulimk* Past and Present: Mikmaw Ethics and the Atlantic Fishery." *Journal of Canadian Studies* 37 (2002): 15–42.

Beacham, Terry D. "Variability in Median Size and Age at Sexual Maturity of Atlantic Cod (*Gadus morhua*) on the Scotian Shelf in the Northwest Atlantic Ocean." *Fishery Bulletin* 81 (1982): 303–312.

Bellail, R., J. Bertrand, O. Le Pape, J. C. Mahe, J. Morin, J. Poulard, M-J. Rochet, I. Schlaich, A. Souplet, and V. Trinkel. "A Multispecies Dynamic Indicator-Based Approach to the Assessment of the Impact of Fishing on Communities." *ICES Document*, CM 2003/V:02, 2003.

Benson, Ashleen J., and Andrew W. Trites. "Ecological Effects of Regime Shifts in the Bering Sea and Eastern North Pacific." *Fish and Fisheries* 3 (2002): 95–113.

Betts, Matthew W., and T. Max Friesen. "Declining Foraging Returns from an Inexhaustible Resource? Abundance Indices and Beluga Whaling in the Western Canadian Arctic." *Journal of Anthropological Archaeology* 25 (2006): 59–81.

Bianchi, G., H. Gislason, K. Graham, L. Hill, X. Jin, K. Koranteng, S. Manickchand-Heileman, I. Paya´, K. Sainsbury, F. Sanchez, and K. Zwanenburg. "Impact of Fishing on Size Composition and Diversity of Demersal Fish Communities." *ICES Journal of Marine Science* 57 (2000): 558–571.

Bourque, Bruce J. *Diversity and Complexity in Prehistoric Maritime Societies: A Gulf of Maine Perspective.* New York: Plenum, 1995.

Bourque, Bruce J., Beverly Johnson, and Robert S. Steneck. "Possible Prehistoric Fishing Effects on Coastal Marine Food Webs in the Gulf of Maine." In *Human Impacts on Ancient Marine Ecosystems: A Global Perspective*, T. C. Rick and J. M. Erlandson, eds., pp. 165–185. Berkeley: University of California Press, 2008.

Broughton, Jack M. "Declines in Mammalian Foraging Efficiency during the Late Holocene, San Francisco Bay, California." *Journal of Anthropological Archaeology* 13 (1994): 371–401.

———. "Widening Diet Breadth, Declining Foraging Efficiency, and Prehistoric Harvest Pressure: Ichthyofaunal Evidence from the Emeryville Shellmound, California." *Antiquity* 71 (1997): 845–862.

———. *Resource Depression and Intensification during the Late Holocene, San Francisco Bay: Evidence from the Emeryville Shellmound Vertebrate Fauna.* Berkeley: University of California Anthropological Records 32, 1999.

Butler, Virginia L. "Resource Depression on the Northwest Coast of North America." *Antiquity* 74 (2000): 649–661.

Butler, Virginia L., and Michael G. Delacorte. "Doing Zooarchaeology as If It Mattered: Use of Faunal Data to Address Current Issues in Fish Conservation Biology in Owens Valley, California." In *Archaeology and Conservation Biology*, R. Lee Lyman and Kenneth P. Cannon, eds., pp. 25–44. Salt Lake City: University of Utah Press, 2004.

Calkin, Parker E., Gregory C. Wiles, and David J. Barclay. "Holocene Coastal Glaciation of Alaska." *Quaternary Science Reviews* 20 (2001): 449–461.

Cannon, Michael D. "A Mathematical Model of the Effects of Screen Size on Zooarchaeological Relative Abundance Measures." *Journal of Archaeological Science* 26 (1999): 205–214.

———. "Large Mammal Relative Abundance in Pithouse and Pueblo Period Archaeofaunas from Southwestern New Mexico: Resource Depression among the Mimbres-Mogollon?" *Journal of Anthropological Archaeology* 19 (2000): 317–347.

Carder, Nanny, Elizabeth J. Reitz, and John G. Crock. "Fish Communities and Populations during the Post-Saladoid Period (AD 600/800–1500), Anguilla, Lesser Antilles." *Journal of Archaeological Science* 34 (2007): 588–599.

Chavez, Francisco P., John Ryan, Salvador E. Lluch-Cota, and Miguel Ñiquen C. "From Anchovies to Sardines and Back: Multidecadal Change in the Pacific Ocean." *Science* 10 (2003): 217–221.

Clark, Donald S., and John M. Green. "Seasonal Variation in Temperature Preference of Juvenile Atlantic Cod (*Gadus morhua*), with Evidence Supporting an Energetic Basis for Their Diel Vertical Migration." *Canadian Journal of Zoology* 69 (1991): 1302–1307.

Clark, Donald S., and Peter Perley. Assessment of Cod in Division 4X in 2006. Department of Fisheries and Oceans. Canadian Science Advisory Secretariat Research Document 2006/087, 2006.

D'Arrigo, Rosanne, Erika Mashig, David Frank, Rob Wilson, and Gordon Jacoby. "Temperature Variability over the Past Millennium Inferred from Northwestern Alaska Tree Rings." *Climate Dynamics* 24 (2005): 227–236.

Drinkwater, Kenneth, F. "The Response of Atlantic Cod (*Gadus morhua*) to Future Climate Change." *ICES Journal of Marine Science* 62 (2005): 1327–1337.

Enticknap, Ben. "Trawling the North Pacific: Understanding the Effects of Bottom Trawl Fisheries on Alaska's Living Sea Floor." Alaska Marine Conservation Council Research Report, 2002.

Finney, Bruce P., Irene Gregory-Eaves, Marianne S. V. Douglas, and John P. Smol. "Fisheries Productivity in the Northeastern Pacific Ocean over the Past 2,200 Years." *Nature* 416 (2002): 729–733.

Gedamke Todd, and John M. Hoenig. "Estimating Mortality from Mean Length Data in Nonequilibrium Situations, with Application to the Assessment of Goosefish." *Transactions of the American Fisheries Society* 135 (2006): 476–487.

Grayson, Donald K. *Quantitative Zooarchaeology.* Orlando: Academic Press, 1984.

———. "Explaining the Development of Dietary Dominance by a Single Ungulate Taxon at Grotte XVI, Dordogne, France." *Journal of Archaeological Science* 28 (2001): 115–125.

Halliday, R. G. "A Comparison of Size Selection of Atlantic Cod (*Gadus morhua*) and Haddock (*Melanogrammus aeglefinus*) by Bottom Longlines and Otter Trawls." *Fisheries Research* 57 (2002): 63–73.

Heusser, C. J., L. E. Heusser, and D. M. Peteet. "Late-Quaternary Climate Change on the American North Pacific Coast." *Nature* 315 (1985): 485–487.

Hildebrandt, William R., and Terry L. Jones. "Evolution of Marine Mammal Hunting: A View from the California and Oregon Coasts." *Journal of Anthropological Archaeology* 11 (1992): 360–401.

Hoffman, Brian W., Jessica M. C. Czederpiltz, and Megan A. Partlow. "Heads or Tails: The Zooar-chaeology of Aleut Salmon Storage on Unimak Island, Alaska." *Journal of Archaeological Science* 27 (2000): 699–708.

Hu, Feng Sheng, Emi Ito, Thomas A. Brown, B. Brandon Curry, and Daniel R. Engstrom. "Pronounced Climatic Variations in Alaska during the Last Two Millennia." *Proceedings of the National Academy of Sciences* 98 (2001): 10552–10556.

Jackson, Jeremy B. "What Was Natural in the Coastal Oceans?" *Proceedings of the National Academy of Science* 98 (2001): 5411–5418.

Jackson, Jeremy B., Michael X. Kirby, Wolfgang H. Berger, Karen A. Bjorndal, Louis W. Botsford, Bruce J. Bourque, Roger H. Bradbury, Richard Cooke, Jon Erlandson, James A. Estes, Terence P. Hughes, Susan Kidwell, Carina B. Lange, Hunter S. Lenihan, John M. Pandolfi, Charles H. Peterson, Robert S. Steneck, Mia J. Tegner, and Robert R. Warner. "Historical Overfishing and the Recent Collapse of Coastal Ecosystems." *Science* 293 (2001): 629–638.

James, Steven R. "Methodological Issues Concerning Screen Size Recovery Rates and Their Effects on Archaeofaunal Interpretations." *Journal of Archaeological Science* 24 (1997): 385–397.

Janetski, Joel C. "Fremont Hunting and Resource Intensification in the Eastern Great Basin." *Journal of Archaeological Science* 24 (1997): 1075–1088.

Jennings, S., S. P. R., Greenstreet, L. Hill, G. J. Piet, J. K. Pinnegar, and K. J. Warr. "Long-term Trends in the Trophic Structure of the North Sea Fish Community: Evidence from Stable Isotope Analysis, Size-spectra and Community Metrics." *Marine Biology* 141 (2002): 1085–1097.

Jordan, James W., and Andrea P. Krumhardt. "Postglacial Climate and Vegetation on the Western Alaska Peninsula." *Alaska Journal of Anthropology* 1 (2003): 16–33.

Karp, William A. "Biology and Management of Pacific Cod (*Gadus macrocephalus*) in Port Townsend, Washington." PhD diss., University of Washington, 1982.

Kenchington, Trevor J., and E. L. Kenchington. "An Eighteenth Century Commercial Length-Frequency Sample of Atlantic Cod, *Gadus morhua*, Based on Archaeological Data." *Fisheries Research* 18 (1993): 335–347.

Leach, Foss, and Janet Davidson. "The Use of Size-Frequency Diagrams to Characterize Prehistoric Fish Catches and to Assess Human Impact on Inshore Fisheries." *International Journal of Osteoarchaeology* 11 (2001): 150–162.

Leach, Foss, Janet Davidson, and L. M. Horwood. "The Estimation of Live Fish Size from Archaeological Cranial Bones of the New Zealand Blue Cod *Parapercis colias*." *International Journal of Osteoarchaeology* 7 (1997): 481–496.

Lear, William H. "History of Fisheries in the Northwest Atlantic: The 500 Year Perspective." *Journal of Northwest Atlantic Fishery Science* 23 (1998): 41–73.

Loso, Michael G., Robert S. Anderson, Suzanne P. Anderson, and Paula J. Reimer. "A 1500-Year Record of Temperature and Glacial Response Inferred from Varved Iceberg Lake, Southcentral Alaska." *Quaternary Research* 66 (2006): 12–24.

Lyman, R. Lee. "Paleozoology in the Service of Conservation Biology." *Evolutionary Anthropology* 15 (2006): 11–19.

———. *Quantitative Paleozoology*. New York: Cambridge University Press, 2008.

Mann, Daniel H. "Climate During the Past Millennium." *Weather* 56 (2001): 91–102.

Mann, Daniel H., and Thomas D. Hamilton. "Late Pleistocene and Holocene Paleoenvironments of the North Pacific Coast." *Quaternary Science Reviews* 14 (1995): 449–471.

Maschner, Herbert D. G., Matthew W. Betts, Katherine Reedy-Maschner, and Andrew W. Trites. "A 4500 Year Time Series of Pacific Cod (*Gadus macrocephalus*): Archaeology, Regime Shifts, and Sustainable Fisheries." *Fisheries Bulletin* 106 (2008): 386–394.

Mason, Owen K., and James W. Jordan. "Heightened North Pacific Storminess during Synchronous Late Holocene Erosion of Northwest Alaska Beach Ridges." *Quaternary Research* 40 (1993): 55–69.

Mayo, Ralph, and Loretta O'Brien. Atlantic Cod: Status of Fishery Resources off the Northeastern US. Northeast Fisheries Science Center, National Oceanic and Atmospheric Administration, 2007.

McKenzie, Matthew G. "Baiting Our Memories: The Impact of Offshore Technology Change on the Species around Cape Cod, 1860–1895." In *Oceans Past: Management Insights from the History of Marine Animal Populations,* David J. Starkey, Poul Holm, and Michaela Barnard, eds., pp. 77–89. London: Earthscan, 2008.

Misarti, N. "Six Thousand Years of Change in the Northeast Pacific: An Interdisciplinary View of Maritime Ecosystems." PhD diss., University of Alaska Fairbanks, 2007.

Nagaoka, Lisa. "Differential Recovery of Pacific Island Fish Remains." *Journal of Archaeological Science* 32 (2005): 941–955.

Orchard, Trevor. "The Role of Selected Fish Species in Aleut Palaeodiet." Master's thesis, University of Victoria, 2001.

———. *An Application of the Linear Regression Technique for Determining Length and Weight of Six Fish Taxa: The Role of Selected Fish Species in Aleut Palaeodiet.* BAR British Archaeological Reports, International Series 1172. Oxford: Archaeopress, 2003.

Pauly, Daniel. "Anecdotes on the Shifting Baseline Syndrome of Fisheries." *Trends in Ecology and Evolution* 10 (1995): 430.

Polovina, Jeffrey J. "Climate Variation, Regime Shifts, and Implications for Sustainable Fisheries." *Bulletin of Marine Science* 76 (2005): 233–244.

Ponomarev, S. T., and Stepan G. Glotov. "A Report by the Cossack Savin T. Ponomarev and the Promyshlennik Stepan G. Glotov, Concerning their Discovery of New Islands in the Aleutian Chain, September 12, 1762." In *Russian Penetration of the North Pacific Ocean: Three Centuries of Russian Eastward Expansion* (Vol. 2, 1700–1797), Basil Dmytryshyn, E. A. P. Crownhart-Vaughan, and Thomas Vaughan, eds., pp. 19–27. Portland: Oregon Historical Society Press, 1988.

Reiss, Michael J. *The Allometry of Growth and Reproduction.* Cambridge: Cambridge University Press, 1989.

Robichaud, David, and George A. Rose. "Migratory Behavior and Range in Atlantic Cod: Inference from a Century of Tagging." *Fish and Fisheries* 5 (2004): 185–214.

Rochet, Marie-Joëlle. "Short-term Effects of Fishing on Life History Traits of Fishes." *ICES Journal of Marine Science* 55 (1998): 371–391.

Rojo, Alfonso. "Live Length and Weight of Cod (*Gadus morhua*) Estimated from Various Skeletal Elements." *North American Archaeologist* 7 (1986): 329–351.

———. "Excavated Fish Vertebrae as Predictors in Bioarchaeological Research." *North American Archaeologist* 8 (1987): 209–226.

———. "Faunal Analysis of Fish Remains from Cellar's Cove, Nova Scotia." *Archaeology of Eastern North America* 18 (1990): 89–108.

———. *Morphological and Biometric Study of the Bones of the Buccal Apparatus of Some Nova Scotia Fishes of Archaeological Interest.* NSMNH Curatorial Report 96. Halifax: Nova Scotia Museum, 2002.

Rosenberg, Andrew A., W. Jeffrey Bolster, Karen E. Alexander, William B. Leavenworth, Andrew B. Cooper, and Matthew G. McKenzie. "The History of Ocean Resources: Modeling Cod Biomass Using Historical Records." *Frontiers in Ecology and the Environment* 3 (2005): 84–90.

Shields, Captain Ed. *Salt of the Sea: The Pacific Coast Cod Fishery and the Last Days of Sail.* Lopez Island, WA: Pacific Heritage Press, 2001.

Shimada, A. M., and D. K. Kimura. "Seasonal Movements of Pacific Cod, *Gadus macrocephalus*, in the Eastern Bering Sea and Adjacent Waters Based on Tag-Recapture Data." *Fishery Bulletin* 92 (1994): 800–816.

Shin, Yunne-Jai, Marie-Joelle Rochet, Simon Jennings, John G. Field, and Henrik Gislason. "Using Size-based Indicators to Evaluate the Ecosystem Effects of Fishing." *ICES Journal of Marine Science* 62 (2005): 384–396.

Sinclair, Alan F. "Fish Distribution and Partial Recruitment: The Case of Eastern Coastal Shelf Cod." *Journal of Northwest Atlantic Fisheries Science* 13 (1992): 15–34.

Spiess, Arthur E., and Robert A. Lewis. *The Turner Farm Fauna: 5000 Years of Hunting and Fishing in Penobscot Bay, Maine.* Occasional Publications in Maine Archaeology 11. Augusta: Maine State Museum, 2001.

Starkey, David J., Poul Holm, and Michaela Barnard, eds. *Oceans Past: Management Insights from the History of Marine Animal Populations.* London: Earthscan, 2008.

Stephens, Cathy, Sydney Levitus, John Antonov, and Timothy P. Boyer. "On the Pacific Ocean Regime Shift." *Geophysical Research Letters* 28 (2001): 3721–3724.

Swain, Douglas P., Alan F. Sinclair, M. Castonguay, G. A. Chouinard, Kenneth F. Drinkwater, L. P. Fanning, and Donald S. Clark. "Density- versus Temperature-dependent Growth of Atlantic Cod (*Gadus morhua*) in the Gulf of St. Lawrence and on the Scotian Shelf." *Fisheries Research* 59 (2003): 327–341.

Swain, Douglas P., Alan F. Sinclair, and J. Mark Hanson. "Evolutionary Response to Size Selective Mortality in an Exploited Fish Population." *Proceedings of the Royal Society* 274 (2007): 1015–1022.

Thompson, Grant G., Martin W. Dorn, and Daniel G. Nichol. "Assessment of the Pacific Cod Stock in the Gulf of Alaska." In *Stock Assessment and Fishery Evaluation Report for the Groundfish Resources of the Gulf of Alaska.* Anchorage: North Pacific Fisheries Management Council, 2006.

Thompson, Grant G., James N. Ianelli, Martin W. Dorn, and Mark Wilkins. "Assessment of the Pacific Cod Stock in the Gulf of Alaska." In *Stock Assessment and Fishery Evaluation Report for the Groundfish Resources of the Gulf of Alaska.* Anchorage: North Pacific Fisheries Management Council, 2007.

Trippel, Edward A. "Age at Maturity as a Stress Indicator in Fisheries." *BioScience* 45 (1995): 759–771.

Veniaminov, Ivan. *Notes on the Islands of the Unalaska District.* Lydia T. Black and R. H. Goeghegan, trans. Richard A. Pierce, ed. Alaska History No. 27. Kingston, ON: Limestone Press, 1984.

Wallace, Scott. *Dragging Our Assets: Towards an Ecosystem Approach to Bottom Trawling in Canada.* Vancouver: David Suzuki Foundation, 2007.

Wheeler, Alwyne, and Andrew K. Jones. *Fishes.* New York: Cambridge University Press, 1989.

Whitridge, Peter. "Zen Fish: A Consideration of the Discordance between Artifactual and Zooarchaeological Indicators of Thule Inuit Fish Use." *Journal of Anthropological Archaeology* 20 (2001): 3–72.

Willis, Kathy J., and Harry J. Birks. "What Is Natural? The Need for a Long-Term Perspective in Biodiversity Conservation." *Science* 314 (2006): 1261–1265.

Worm, Boris, Edward B. Barbier, Nicola Beaumont, J. Emmett Duffy, Carl Folke, Benjamin S. Halpern, Jeremy B. C. Jackson, Heike K. Lotze, Fiorenza Micheli, Stephen R. Palumbi, Enric Sala, Kimberley A. Selkoe, John J. Stachowicz, and Reg Watson. "Impacts of Biodiversity Loss on Ocean Ecosystem Services." *Science* 314 (2006): 787–790.

Yang, Mei-Sun. "Diet Changes of Pacific Cod (*Gadus macrocephalus*) in Pavlov Bay Associated with Climate Changes in the Gulf of Alaska Between 1980 and 1995." *Fisheries Bulletin* 102 (2004): 400–405.

CHAPTER 12

Processing the Patterns

Elusive Archaeofaunal Signatures of Cod Storage on the North Pacific Coast

Megan A. Partlow, *Department of Anthropology, Central Washington University*
Robert E. Kopperl, *SWCA Environmental Consultants, Seattle*

Archaeological research in the eastern North Pacific, from the Aleutians to the Washington coast, has tended to focus on the evidence for salmon fishing and storage in the prehistoric record (Ames 1981, 1994; Butler and Chatters 1994; Cannon 1991, 2001; Coupland 1998; Fladmark 1975; Hayden 1997; Hoffman, Czederpiltz, and Partlow 2000; Matson 1983, 1992; Matson and Coupland 1995; Partlow 2000, 2006; Schalk 1977). There are several key reasons for this. First, salmon have been historically important to eastern North Pacific peoples (Davydov 1977; Emmons 1991; Veniaminov 1984). Secondly, salmon fit Testart's (1982:523) preconditions for large-scale intensive storage: they are spatially and temporally predictable and abundant, and with the right technology, large numbers of salmon could be harvested, processed, and stored for the winter. Finally, intensive storage of species such as salmon can help cushion against seasonal resource scarcity (i.e., the winter months; Rowley-Conwy 1982:533), can lead to decreased mobility by both inhibiting it and making it less necessary (Testart 1982:524), and may contribute to increased social complexity (Ames 1985; Soffer 1989; Steffian, Saltonstall, and Kopperl 2006:99; Wesson 1999).

By contrast, discussion of the importance of Pacific cod in prehistory has been relatively rare. Yet, historically, Pacific cod has been an important resource for many eastern North Pacific peoples, including the Aleut (Jochelson 1933; Lantis 1984; Turner 1886), the Alutiiq of Prince William Sound (Birket-Smith 1953) and the Kodiak Archipelago (Davydov 1977; Holmberg 1985), the Eyak and Tlingit (de Laguna 1972, 1990a, 1990b; Oberg 1973), the

Haida (Langdon 1979), the Nuu-chah-nulth (Arima and Dewhirst 1990), and the Quinault (Hajda 1990). According to ethnohistorical records, some Native peoples dried Pacific cod for storage in the nineteenth and twentieth centuries (Davydov 1977; Emmons 1991; Gideon 1989; de Laguna 1972; Newton and Moss 1984; Oberg 1973; Turner 1886). The extent to which cod were dried for storage prior to the arrival of Russians in the eighteenth century and the beginning of the commercial cod fishery in the mid-nineteenth century, however, is unclear. Cod make up a substantial percentage of the fish remains from many North Pacific sites (Hanson 1995; Kopperl 2003; Maschner 1997; Moss 1989a; Orchard 2003; Yarborough 2000), and several researchers have suggested that Pacific cod, like salmon, may have been dried and stored prehistorically (e.g., Bowers and Moss 2001; Cannon 1991; Croes 1995; Croes and Hackenberger 1988; Moss 2004; Saltonstall and Steffian 2006).

To what extent do Pacific cod fit Testart's preconditions for large-scale intensive storage? Pacific cod are available year-round, although they are more abundant and easier to locate and catch during the spring and summer and along specific banks. They could be harvested in large numbers in a short period of time by longline fishing out of small boats (Cobb 1927; Shields 2001). Storage may have been made possible by air-drying or smoking, as with salmon. Therefore, Pacific cod meet Testart's four preconditions for intensive food storage: they are abundant, seasonally available, easily harvestable with longlines from small boats, and storable.

If Pacific cod were dried and stored by Native peoples prehistorically, how do we identify this in the archaeological record? Unfortunately, obvious storage facilities containing abundant cod bones have not been found, although some codfish (Gadidae) were found in the subfloor pit assemblage from a small house excavated at Agayadan Village (49-UNI-067) on Unimak Island (Hoffman 2002:115). Less direct evidence might be features for drying and smoking (e.g., drying racks, smudge pits), tools for cod capture or processing (e.g., cod hooks), and associated cod bones. Based on Pacific cod bones and hundreds of deep-sea fishing hooks associated with drying racks and smudge fires at the Hoko River Wet/Dry site (45-CA-213), Croes and Hackenberger (1988:76–77; Croes 1992, 1995) argued that Pacific cod were caught and processed for storage ca. 3000–2500 BP on the Washington coast. Based on large numbers of Pacific cod bones and processing tools associated with a distinctive (15–42 cm thick) layer of wood charcoal suggestive of a smokehouse at the Horseshoe Cove Site (49-KOD-415), Hays (2007) and Steffian and others (2006:116; Saltonstall and Steffian 2006:56, 83) argue that Pacific cod were caught and processed for storage ca. 3700–3300 BP in the Kodiak Archipelago.

Although the smoking and drying facilities at these sites are suggestive, we cannot be certain that they were used for cod. Another approach is to evaluate the faunal record of cod skeletal part representation. Differential skeletal part representation has been used to argue for the presence of stored salmon in North Pacific sites and also stored cod in North Atlantic sites (e.g., Amundsen et al. 2005; Barrett 1997; Butler and Chatters 1994; Hoffman, Czederpiltz, and Partlow 2000; Partlow 2000, 2006; Perdikaris 1999).

This paper summarizes the ethnohistorical evidence for Pacific cod storage by eastern North Pacific peoples and analyzes the skeletal part evidence from fifteen cod assemblages recovered from thirteen eastern North Pacific sites, including part of the Horseshoe Cove site and the Hoko River Wet/Dry site. This skeletal part evidence is compared against expectations for cod storage and also with selected North Atlantic assemblages interpreted as probable processing and storage sites for Atlantic cod during the historic dried codfish trade.

Pacific Cod

Pacific cod (*Gadus macrocephalus*), also known as true cod or gray cod, is an economically important groundfish, or demersal fish, of the true cod family, Gadidae. Other important North Pacific cod fishes of Gadidae include walleye pollock (*Theragra chalcogramma*), Pacific tomcod (*Microgadus proximus*), and saffron cod (*Eleginus gracilis*). Pacific cod are found just above the sea bottom, usually at depths of less than 350 m (Mecklenburg, Mecklenburg, and Thorsteinson 2002:296), along the large continental shelf area from northern California to the Alaska Peninsula (Kasahara 1961:16). They appear to be most abundant in the western part of the Gulf of Alaska around the Shumagin Islands, Chirikof Island, and the Kodiak Archipelago (Mundy and Hollowed 2005:91).

Pacific cod continue to grow throughout their lives. In the Gulf of Alaska, they mature between five and eight years of age, by which time they average about 67 cm in length (DiCosimo and Kimball 2001:3). They can live to be as old as eighteen or nineteen years (DiCosimo and Kimball 2001:3; Witherell 2000:2) and reach total lengths up to 120 cm (Mecklenburg, Mecklenburg, and Thorsteinson 2002:296). Trawl-caught fish today average 70–75 cm in total length and 4.5 kg (Mecklenburg, Mecklenburg, and Thorsteinson 2002:296).

In general, the distribution of adult Pacific cod is strongly tied to water temperature. Adult cod prefer waters between 0°C and 10°C (Kasahara 1961:66). Consequently, in winter they usually concentrate in the warmer waters along the shelf edge at depths between 100 and 200 m, moving back to shallower waters that warm up in the spring and summer (DiCosimo and Kimball 2001:3; Kasahara 1961:66–67). Along southcentral Alaska and British Columbia, however, the water temperature does not vary as much from summer to fall; hence in some areas, cod are available year-round, although they are more abundant in spring and summer than in winter (Cobb 1927).

Pacific cod have been fished commercially in the Bering Sea and the Aleutians since the last half of the nineteenth century (Bakkala 1984:157; Cobb 1927; Kasahara 1961:71; Shields 2001). By contrast, as late as the early 1960s, the waters south of the Alaska Peninsula had never been fished intensively for Pacific cod (Kasahara 1961:73). Significant contemporary catches of Pacific cod have occurred near Buldir Island and the Island of Four Mountains in the Aleutians, south of Kodiak Island and the Alaska Peninsula west to Unimak Island in the Aleutians, in Hecate Strait between the Queen Charlotte Islands and mainland British Columbia, off the north coast of Washington around the southern end of Vancouver Island, and just off the north end of Vancouver Island (Bakkala et al. 1984: Figures 2–3).

Cod Fishing in the Ethnohistorical Record

The eastern North Pacific, a large region that stretches from the Aleutians in the northwest to Washington in the south, was home to many different Native peoples. The location of abundant cod populations and the ways in which different people used those resources have probably changed through time (Bowers and Moss 2001:159; Causey et al. 2005; Hunt and Stabeno 2005; Shields 2001; Turner 1886). For example, according to Turner (1886:90), Pacific cod were not seen around Attu Island in the Aleutians prior to 1873. To what degree the ethnohistorical record accurately reflects subsistence practices prior to contact with Euro-Americans and Russians is debatable (Hanson 1995; Yarborough 2000); nevertheless, the ethnohistorical record is worth examining for evidence of the importance of Pacific cod to North Pacific peoples.

Although salmon and halibut were generally ranked the highest in lists of important fish resources (Rousselot, Fitzhugh, and Crowell 1988:153), the ethnographic literature indicates that Pacific cod were also an important subsistence resource to many North Pacific peoples (Arima and Dewhirst 1990; Birket-Smith 1953; Davydov 1977; Hajda 1990; Holmberg 1985; Jochelson 1933; de Laguna 1972, 1990a, 1990b; Langdon 1979; Lantis 1984; Oberg 1973; Turner 1886). On a cautionary note, although Suttles (1990:25) lists the true cods as one of the more important saltwater fishes to peoples of the Northwest Coast, he acknowledges that the term *cod* was applied to "a variety of fishes not always distinguished in the ethnographic literature." Likewise, Hanson (1995:31) suggests that the "cod" caught historically by the Straits Salish may have been lingcod (*Ophiodon elongatus*) or rockfish (*Sebastes* spp.).

Native fishers traditionally caught Pacific cod by line-fishing from small boats throughout the North Pacific (Black and Liapunova 1988:55; Clark 1984:189–190; Haggarty et al. 1991:84; Holmberg 1985:47; de Laguna 1972:391, 1990a, 1990b; Rousselot, Fitzhugh, and Crowell 1988:154). Large numbers were probably caught from inshore banks using hand lines, a well-documented technique for the early twentieth-century commercial fisheries (Cobb 1927:389; Jochelson 1933:11; Shields 2001:20). Although cod are currently available year-round, spring and summer may have been the prime seasons for Native cod fishing because of safer weather conditions, cod abundance during spawning, larger size, and better taste (Birket-Smith 1953:39; Cobb 1927:389; Crowell and Laktonen 2001:176; Davydov 1977; Haggarty et al. 1991:84; Jochelson 1933:11; Oberg 1973:73; Veniaminov 1984:361). During the nineteenth century, the Aleut caught cod primarily during the summer, when they were abundant and easily caught within two miles from shore, whereas during the winter, they kayaked more than ten miles from shore in order to catch cod (Jochelson 1933:11). The season when Pacific cod were most abundant closer to shore may have varied from one area to another (see Moss, this volume). In the Aleutians, cod were most abundant in February and March in some places, while in others they did not show up until July or September (Turner 1886:89). On the Northwest Coast, deep-sea fishing for cod near offshore islands was one of the principal subsistence activities from March to July among the Chilkat Tlingit (Oberg 1973:66, 73). Deep-sea cod fishing was replaced by salmon fishing in July and August (Oberg 1973:74).

Cod Storage in the Ethnohistorical Record

According to ethnohistorical accounts, cod were dried by some Native peoples during both the colonial Russian period and the commercial fishing period in the mid-nineteenth century. To what extent this was a product of Russian and American expectations and economies or a continuation of previous subsistence practices is unclear. This section summarizes the ethnohistorical evidence for cod storage by North Pacific peoples.

For the Aleutians, there is conflicting ethnohistorical evidence regarding both cod storage for personal use and the importance of storage in general. According to one account, Aleuts were trading dried cod to the Russians by 1764 (Jochelson 1933:5). Similarly, in 1802, Sauer (1802:161) wrote of a group of Aleut drying cod for the winter, but noted it was only for trade with visiting Russians. According to Veniaminov (1984:276), who came to the Aleutians during the early to mid-nineteenth century, wet weather did not "permit the Aleuts to store much because the only possible method of fish preservation is by [air] drying." On the other hand, in his history of the Russian-American Company, Tikhmenev (1978:407) wrote that cod were dried on Unalaska Island during the mid-1800s, and Turner (1886:90) observed the Aleut drying large numbers of cod for the winter for their own use in the late nineteenth century. In their recent summary, McCartney and Veltre (2000:506, 512) argued that "[f]ood storage was a critical aspect of Aleut life" enabling them to survive the winter when resources were scarce. In his history of fishing in the Aleutians, Jacka (1999:219) includes "deep water" fish, presumably including cod, among the fish that the Aleuts "split and air-dried on racks," before storing them in sea lion stomachs to prevent them from growing moldy in the damp climate. According to Lantis (1984:176), however, the Aleut did not store much food prehistorically "except for the festivals," as the damp weather made drying difficult and because resources were available year-round.

Like the Aleutians, the Kodiak Archipelago is a prime cod-fishing area today, but it is uncertain the degree to which cod were dried and stored prehistorically. With the arrival of the Russians on Kodiak Island in the late eighteenth century, the Alutiiq were put to work catching and drying or salting thousands of fish for the Russians (Gibson 1978:372). How many of these dried fish were cod is unclear. Because of the endless rains, most fish were salted, while those that were simply dried were often not considered edible by the Russians (Gibson 1978:372). Gideon (1989:41) mentions dried cod among the Alutiiq during his visit to the Kodiak Archipelago ca. 1803–1809. On the other hand, Davydov (1977:232), who visited the region around the same time, wrote that cod were not stored by the Alutiiq but rather eaten fresh just until the salmon began running. The Alutiiq considered rotting fish heads delicacies and cod heads were so valued for this purpose that the cod were killed so as not to damage the head (Davydov 1977:172, 174).

Northern Northwest Coast accounts conflict regarding drying of cod as well. De Laguna (1972:401–402) noted that some people in Yakutat smoked spring-caught cod historically, whereas Emmons (1991:149) claimed that the Tlingit never dried cod, but always ate it fresh. On the other hand, in the late nineteenth century, Knapp and Childe (1896:27) describe cod on drying racks in their idea of a typical Tlingit village. At the end of the nineteenth century, Kamenskii (1985) wrote that the Tlingit knew how to dry all kinds of fish (possibly including Pacific cod) for the long winters. Finally, de Laguna (1990b:206) cautioned that no single subsistence economy characterized the Tlingit; rather, families and local groups in different regions might focus on different resources at any one time. No doubt, the same was true elsewhere in the eastern North Pacific.

Such contradictory accounts suggest that not everyone dried Pacific cod. Therefore, we should not expect to see evidence for dried cod in all assemblages from all time periods and regions. The challenge is to discern where and when Pacific cod may have been dried in the past. For this reason, we chose to look at assemblages from different regions and time periods in the North Pacific.

Expectations for Faunal Evidence of Cod Storage

Skeletal part representation is one line of evidence used to support arguments for or against fish storage. For Pacific salmon, an overabundance of vertebrae has been interpreted as the remains of stored salmon (e.g., Butler and Chatters 1994; Hoffman, Czederpiltz, and Partlow 2000; Partlow 2000, 2006), while an overabundance of heads has been interpreted as the remains of salmon processing at a fish camp (Hoffman, Czederpiltz, and Partlow 2000). For Atlantic cod, an overabundance of cleithra and tail vertebrae has been interpreted as the remains of stored cod, while an overabundance of heads, underabundance of cleithra, and the presence of more abdominal than tail vertebrae has been interpreted, for the most part, as the discarded remains of cod processed for the dried cod trade (e.g., Amundsen et al. 2005; Barrett 1997). Expectations for Pacific cod storage can be derived from ethnohistorically documented methods of butchery, economic anatomy, and less directly, from practices used during the historic commercial cod trade (which included salting the drying fish, a practice not known to have been used by Native peoples in the eastern North Pacific prior to Euro-American contact).

Based on the few ethnohistorical accounts, Pacific cod may have had their heads removed, but may also have had some of their vertebrae removed as well. For example, in one twentieth-century Tlingit method, cod heads were removed prior to drying (Bowers and Moss 2001). Mitchell and Donald (1988:320) describe cod air-dried or smoked for storage on the Northwest Coast as "filleted with the skin on and intact at belly or back, so that the whole was spread out to dry." According to Turner's (1886) mid-nineteenth-century Aleutian account, dried Pacific cod retained tails, but may have retained other vertebrae and even the head. Turner (1886:90) describes how the Aleut processed cod for storage as follows:

> The head is partly severed from the body at the throat, the gills are taken out, a slit along the belly and the entrails are removed, the backbone is cut out on each side and either removed as far as the tail, which is left to hold the two sides together to allow them to be hung over a pole, or else it is left in and dried with the body. When fish are abundant this is rarely done. The sides are then cut transversely through the flesh to the skin and the body then hung up by the tail to dry.

Based on economic anatomy, we expect cod heads to be treated differently from bodies. Because fish heads tend to have the most fat, they are the part of the fish least conducive to drying. For example, according to one Makah elder (cited in Croes 1992:348), dried halibut heads only lasted a few months because of their high oil content. The same may have been true for Pacific cod heads, which are not only the fattiest part of the fish but the choicest part (Cobb 1927; Davydov 1977). For these reasons, cod heads probably were not dried very often, but rather eaten separately either fresh or as a rotting delicacy.

In both the Pacific and Atlantic commercial dried cod fisheries of the nineteenth and twentieth centuries, cod heads and often the abdominal vertebrae were removed during processing (Barrett 1997:619–620; Cobb 1927; Shields 2001). The dried cod retained the tail and may have retained the cleithra as well. The cleithrum is a large and robust pectoral fin element that often traveled with the dried Atlantic cod in the fish trade to help hold the fillet together (Amundsen et al. 2005:135–136; Barrett 1997:619–620). It is important to note, however, that while salt often was used to dry Atlantic and Pacific cod during the historic commercial fish trade, it was not used by Native peoples prehistorically in the eastern North Pacific.

In summary, it is likely that because of their higher fat content and value as a delicacy (e.g., Cobb 1927; Davydov 1977), Pacific cod heads may have been cut off and treated differently than bodies during processing for drying and storage. If Pacific cod were processed for storage in a manner similar to that described historically for the Northwest Coast as well as for the dried Pacific cod commercial fish industry, we might expect to see a dichotomy in Pacific cod assemblages: (1) those dominated by bodies and/or pectoral elements (i.e., cleithra), with significantly fewer heads indicative of the remains of stored cod, and (2) those dominated by heads with significantly fewer bodies indicative of the remains of cod processing and/or the consumption of fresh or rotting cod heads. If, however, Pacific cod were processed for storage similar to the manner described by Turner (1886), then the remains of stored cod may be difficult to distinguish from the remains of cod eaten fresh on site; both types of remains would include cod heads and bodies. Likewise, in the absence of separate disposal areas on site, assemblages containing the remains of both stored cod bodies and cod heads eaten either fresh or as rotting delicacies may be indistinguishable from assemblages containing whole cod eaten fresh.

Methods

For this analysis, we compiled the skeletal parts data from radiocarbon-dated cod assemblages throughout the eastern North Pacific. In order to ensure reasonable samples of skeletal parts, we selected assemblages with 300 or more specimens (NISP) identified to element from Pacific cod or cod family (Gadidae) bones recovered with quarter-inch or smaller screens. At sites with multiple components, we separately selected components that met our criteria. The final sample was fifteen assemblages from thirteen sites (Figure 12.1, Table 12.1). The majority of the assemblages are from the Kodiak Archipelago, where we have conducted much of our research.

Screen size was considered in the sample selection because it can affect skeletal part representation (Barrett 1997; Grayson 1984; Moss 1989b, 2004; Nagoaka 2005). Three studies suggest that assemblages recovered using quarter-inch screen adequately represent skeletal parts from medium- to large-size Pacific cod. First, Moss (2004:163) found quarter-inch screen to be satisfactory (when compared to finer screen fractions) in recovering the large Pacific cod bones from the Cape Addington Rockshelter site. Two other

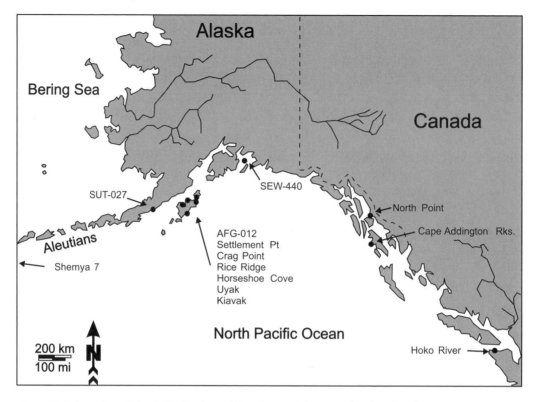

Figure 12.1. Location of North Pacific sites with cod assemblages analyzed in this chapter.

tests were made by the senior author for this study. In the first, a comparison of gadid skeletal parts (%MAU) by three screen sizes (half-inch, quarter-inch, eighth-inch) from the Settlement Point midden and house floor assemblages indicated no significant difference in cod skeletal parts based on screen size. All three screen sizes (half-inch, quarter-inch, eighth-inch) were cranially dominated. In the second test, a comparison of Pacific cod vertebra types (thoracic, precaudal, caudal) by screen size at the 49-AFG-012 site demonstrated no difference between the quarter-inch and eighth-inch samples.

A potential bias in skeletal part representation for gadids that is not present for other fishes such as salmon is the small size of characteristic tail elements. There is a strong likelihood of cod ultimate vertebrae being underrepresented due to their small size (Barrett 1997:625). Cod ultimate vertebrae are rarely identified in archaeological assemblages. To what degree missing tail elements in a cod assemblage are a product of screen-size bias or postdepositional destruction of the ultimate vertebrae rather than differential processing is unclear. Barrett (1997:625) believes a screen size as small as 1 mm is needed to recover Atlantic cod ultimate vertebrae. Because caudal vertebrae are smaller than thoracic and precaudal vertebrae, it is possible that low numbers of caudal vertebrae in a cod assemblage is also a function of screen size rather than cod processing. For this reason, low numbers of cod tails in an assemblage may simply be a function of the smaller size of the tail elements compared to the rest of the skeleton, rather than an indication that cod tails were removed from the site prior to deposition.

Table 12.1. North Pacific cod assemblages analyzed in this chapter

Site #	Site name (assemblage)	Region[a]	Sample used (taxon, NISP)	Screen size	Age (RYBP)[b] (# of dates)	Source
49-KOD-099	Kiavak (bulk samples)	Kodiak	Gadidae, 690	1/8"	331±32 (2 dates)	Clark 1974; Partlow 2000
49-AFG-012	None	Kodiak	Gadidae, 4469	1/8"	350±42 (2 dates)	Partlow 2000
49-SEW-440	None	PWS	Gadidae, 373	1/4"	380±60 (1 date)	Yarborough 2000
49-AFG-015	Settlement Point (midden)	Kodiak	Gadidae, 4022	1/8"	392±28 (4 dates)	Partlow 2000
49-AFG-015	Settlement Point (HF1)	Kodiak	Gadidae, 3110	1/8"	600±38 (2 dates)	Partlow 2000
49-KOD-145	Uyak (HF 7)	Kodiak	Pacific cod, 390	1/4"	1270±100 (1 date)	Kopperl 2003
49-SUT-027	None	Alaska Peninsula	Gadidae, 551	1/4"	1470±35 (2 dates)	Vanderhoek and Myron 2004; Schaaf and Bender, pers. comm. 4/28/2006
49-CRG-188	Cape Addington (Stratum V)	NWC	Pacific cod, 466	1/4"	1785±38 (3 dates)	Moss 2004
49-KOD-044	Crag Point (HF 1B)	Kodiak	Pacific cod, 529	1/4"	2194±47 (3 dates)	Kopperl 2003
49-ATU-061	Shemya 7 (Pit 1)	Aleutians	Gadidae, 7705	1/8"	2670±86 (2 dates)	Lefevre, West, and Corbett 2001; Partlow 1997
45-CA-213	Hoko River Wet	NWC	Gadidae, 592	hydraulic	2861±55 (3 dates)	Croes 1995; Croes, pers. comm. 11/28/2006
49-SUM-25	North Point Wet	NWC	Pacific cod, 2089	1/4"	2996±48 (2 dates)	Bowers and Moss 2001
49-KOD-415	Horseshoe Cove (SD2)	Kodiak	Gadidae, 1513	1/8"	3364±34 (4 dates)	Hays 2007, pers. comm. 7/5/2008; Saltonstall and Steffian 2006
49-KOD-363	Rice Ridge (HF Level B)	Kodiak	Pacific cod, 1689	1/4"	3930±80 (1 date)	Kopperl 2003
49-KOD-363	Rice Ridge (Level C)	Kodiak	Pacific cod, 566	1/4"	5070±40 (1 date)	Kopperl 2003

a. PWS = Prince William Sound; NWC = Northwest Coast.
b. If there was more than one date, then the radiocarbon dates were averaged using CALIB REV 5.0.1 (Stuiver and Reimer 1993).

The skeletal parts data can be expected to vary by cod treatment and site type. All else being equal (including density-mediated attrition; see Smith et al., this volume), if cod are not stored, then we expect all skeletal parts to be represented. If cod are stored, then fish camps would have discarded portions such as heads (presumably also consumed on site) and abdominal vertebrae, whereas multiseason winter villages could be expected to lack these elements. These can be considered initial expectations against which to compare the skeletal part representation in the assemblages.

Analyzing Skeletal Part Representation

Faunal analysts often quantify skeletal parts using minimal animal units (MAU; Binford 1984), calculated by dividing the minimum number of elements (MNE; Bunn 1982) by the frequency of that element in a particular animal. Often, MAUs are normed relative to the highest MAU in the assemblage to produce %MAUs. Recently, Grayson and Frey (2004) suggested using "normed NISPs" (NNISP) in place of MAUs as a simpler measure that can serve the same purpose. Normed NISPs are calculated by dividing the number of identified specimens (NISP) per element by the frequency of that element in a complete skeleton. For assemblages in which the bones are not highly fragmentary, MNE and NISP values will be very similar. Rather than compare MAU or NNISP values for each skeletal element, some researchers compare the frequencies of "butchering units" (Lyman 1979). This approach has been used by faunal analysts comparing skeletal parts for salmon and Atlantic cod (e.g., Amundsen et al. 2005; Partlow 2006). In these cases, %MAUs or %NNISPs are calculated for body regions that are treated differently during processing for drying (i.e., cranial, pectoral, thoracic vertebral, precaudal vertebral, caudal vertebral) based on the maximum MAU or NNISP value for each region.

Another method of measuring skeletal part representation is to simply count all the MNEs (rather than the highest one) across the different body regions without calculating MAUs or NNISPs and compare them with the percentages of MNEs expected for a whole fish (e.g., Norton, Kim, and Bae 1999; Wigen 2005). This method, while simpler, may suffer from identification biases. For example, often faunal analysts do not identify all the elements in a fish assemblage, particularly ribs, spines, and branchial arch bones, which makes it difficult to compare the assemblage to the expected MNEs for a whole fish. Additionally, some elements are more distinctive to taxon than others. By taking the highest MNE for a particular body region instead, potential bias due to interanalyst identification protocols is eliminated.

We chose Grayson and Frey's (2004) NNISP measure, allowing us to compare a variety of assemblages and include those reported as NISP counts. There is little concern here with problems of differential fragmentation in using NISP rather than MNE counts, because the cod remains from the North Pacific assemblages tend not to be highly fragmentary. In addition, we chose to analyze skeletal part distribution across three body regions: cranial, pectoral, and vertebral, with a further breakdown of vertebral into thoracic, precaudal, and caudal for assemblages that had those data. The pelvic and tail regions were excluded from our analysis because (1) the pelvic region in cod is represented by the basipterygium, which is easily broken and not often found in assemblages, and (2) the tail region is represented by the ultimate vertebrae, which is one of the smallest bones in a cod, as explained above. To norm the vertebrae NISPs, the number of each type of vertebrae was counted in Partlow's comparative Pacific cod specimens. The vertebrae NISPs were then divided by the number of times that vertebrae type appeared in a Pacific cod skeleton (thoracic = 5; precaudal = 15; caudal = 33).

The cod assemblages analyzed here are mixtures of Pacific cod (*Gadus macrocephalus*) and cod family (Gadidae) data. For some assemblages, cod remains were identified to the family level only, although the majority of the cod remains were probably Pacific cod.

For other assemblages, cod remains, including vertebrae, were identified as Pacific cod. For a third type of assemblage, although some elements were identified to species, many (particularly vertebrae) were not. For these latter assemblages, since vertebrae are critical in skeletal part representations, we combined cod skeletal parts data from species and family-level identifications and analyzed them at the family level. In the following tables and figures, AC = Atlantic cod, G = Gadidae, and PC = Pacific cod.

Results and Discussion

The skeletal parts data using three body regions from the fifteen North Pacific cod assemblages are displayed on the left side of Figure 12.2 (see also Table 12.2). The proportions for a complete cod are shown for comparison. On the right side of the graph are the skeletal part representations of selected North Atlantic cod assemblages. Three major observations can be made about the North Pacific assemblages. First, heads are well represented at the North Pacific sites, all of which have more than 34 percent of NNISP composed of cranial elements (33 percent is expected for a complete cod). Second, none of the North Pacific assemblages look like the inland Icelandic North Atlantic site of Hofstaðir, which is dominated by pectoral and vertebral elements but contains very few cranial elements. These have been interpreted as the remains of consumed stored cod, where the heads were cut off elsewhere during processing and the bodies with pectoral fins and some vertebrae were transported to the site (Amundsen et al. 2005). Third, the North Pacific assemblages broadly resemble the other North Atlantic sites, all of which are coastal. Several North Pacific assemblages are most similar to the Scottish site of Earl's Bu, which has been interpreted as a combination of the remains of stored cod and cod brought to the site whole (Barrett 1997).

Despite the variation, many North Pacific assemblages have an overabundance of heads and an underabundance of vertebrae compared to a complete cod. The Horseshoe Cove and Hoko River Wet site assemblages, both suggested in the literature as possible locations of cod processing for storage, are not significantly different from the majority of other North Pacific assemblages. The most unusual site here is the North Point Wet Site, which has more vertebrae than expected for a complete fish and an underabundance of pectoral elements.

In Figure 12.3 (see also Table 12.3), the vertebral region was further divided into three sections: thoracic, precaudal, and caudal, since it is possible that abdominal vertebrae (thoracic and precaudal) may have been cut out of a cod during processing for storage, while caudal vertebrae may have traveled with the dried cod. Since not all North Pacific assemblages provided these data, only six of them are considered and shown to the left of the complete cod.

There is considerable variation in these six North Pacific assemblages as well, but all have more cranial bones and fewer caudal vertebrae than a complete cod. The lack of caudal vertebrae is inconsistent with stored cod, based on the ethnographic and historic evidence noted earlier that stored cod retained tails. Alternatively, these six assemblages could match expectations of remnants of processing for storage if this activity took place

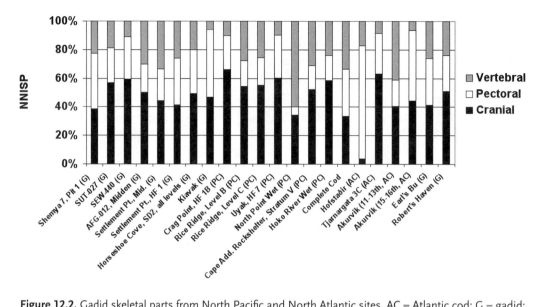

Figure 12.2. Gadid skeletal parts from North Pacific and North Atlantic sites. AC = Atlantic cod; G = gadid; PC = Pacific Cod. "Complete Cod" indicates skeletal parts expected from a whole fish. Data from Table 12.2.

Table 12.2. Cod assemblage skeletal part representation by three body regions (NNISP)

Assemblage[a]	Cranial	Pectoral	Vertebral	Source (NISP counts)
North Pacific				
Shemya 7, Pit 1 (G)	122.00	124.00	71.19	Partlow n.d.
49-SUT-027 (G)	14.00	6.00	4.67	Schaaf and Bender, pers. comm., 4/28/2006
49-SEW-440	15.00	7.50	2.80	Yarborough 2000:appendix A
49-AFG-012, Midden (G)	80.00	31.50	47.80	This paper
Settlement Pt., Midden (G)	57.00	28.50	42.80	This paper
Settlement Pt., HF 1 (G)	46.00	36.00	28.90	This paper
Horseshoe Cove, SD2 (G)	39.00	24.00	16.00	Hays, pers. comm., 7/5/2008
Kiavak (G)	29.00	29.50	3.56	This paper
Crag Point, HF 1B (PC)	28.00	10.00	4.25	This paper
Rice Ridge, Level B (PC)	44.00	14.50	22.38	This paper
Rice Ridge, Level C (PC)	17.00	6.00	7.80	This paper
Uyak, HF 7 (PC)	19.00	9.50	2.98	This paper
North Point Wet (PC)	20.00	3.50	35.00	Bowers and Moss 2001, table 4
Cape Add. Rockshelter, Stratum V (PC)	11.00	3.50	6.50	Moss 2004, tables 9–12
Hoko River Wet (PC)	10.00	3.00	4.10	Croes, pers. comm., 11/28/06
North Atlantic				
Hofstaðir (AC)	1.00	20.50	4.44	Amundsen et al. 2005, table 7
Tjarnargata 3C (AC)	715.00	322.50	94.60	Amundsen et al. 2005, table 7
Akurvik (11th–13th, AC)	46.00	21.00	46.70	Amundsen et al. 2005, table 7
Akurvik (15th–16th, AC)	124.00	137.50	17.50	Amundsen et al. 2005, table 7
Earl's Bu (G)	88.00	69.00	55.00	Barrett 1997, tables 1 and 3
Robert's Haven (G)	258.00	127.50	119.80	Barrett 1997, tables 1 and 3

a. (AC) = Atlantic cod; (G) = cod family; (PC) = Pacific cod

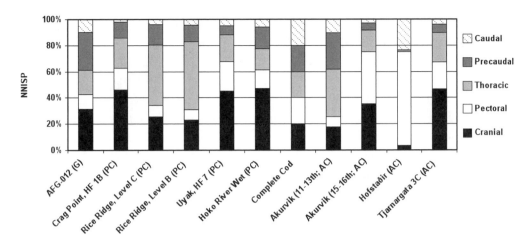

Figure 12.3. Gadid skeletal parts from North Pacific and North Atlantic sites (including vertebrae type). AC = Atlantic cod; G = gadid; PC = Pacific cod. "Complete Cod" indicates skeletal parts expected from a whole fish. Data from Table 12.3.

Table 12.3. Cod assemblage skeletal part representation by five body regions (NNISP)

Assemblage[a]	Cranial	Pectoral	Thoracic	Precaudal	Caudal	Source (NISP counts)
North Pacific						
AFG-012 (G)	80.0	28.5	46.6	73.5	24.7	This paper
Crag Point, HF 1B (PC)	28.0	10.0	14.2	7.3	1.3	This paper
Rice Ridge, Level B (PC)	44.0	14.5	98.4	24.0	8.5	This paper
Rice Ridge, Level C (PC)	17.0	6.0	30.8	10.8	2.5	This paper
Uyak, HF 7 (PC)	19.0	9.5	8.6	2.8	2.1	This paper
Hoko River Wet (PC)	10.0	3.0	3.4	3.6	1.2	Croes, pers. comm., 11/28/2006
North Atlantic						
Akurvik (11th–13th; AC)	46.0	21.0	94.5	72.0	27.3	Amundsen et al. 2005, table 7
Akurvik (15th–16th, AC)	124.0	137.4	58.5	18.7	11.0	Amundsen et al. 2005, table 7
Hofstaðir (AC)	1.0	20.5	0.3	0.1	6.8	Amundsen et al. 2005, table 7
Tjarnargata 3C (AC)	715.0	322.5	344.5	100.9	58.4	Amundsen et al. 2005, table 7

a. (AC) = Atlantic cod; (G) = cod family; (PC) = Pacific cod

at these sites. When the North Atlantic sites are viewed in this way, Hofstaðir is again the most distinctive. This site contains mostly pectoral elements and caudal vertebrae, consistent with the consumption of stored cod. In contrast, the coastal North Atlantic sites contain very few caudal vertebrae, consistent with processing for the trade and transport of cod tails away from the site.

To provide context for interpretation, one could look to skeletal part patterns found for other potentially stored fish in the North Pacific. For example, Norton and others (1999) interpreted the skeletal part representation of large fish at the coastal Korean site of Konam-Ri as the remains of fish processed for drying and storing. This assemblage was cranially dominated (65 percent MNE) but also had postcranial bones (35 percent MNE) (Norton, Kim, and Bae 1999:159). A similar skeletal pattern was found at an Aleutian site interpreted as a salmon fish camp by Hoffman and others (2000:705). This assemblage was also cranially dominated (fifteen MAU cranial versus five MAU postcranial) but had substantial postcranial remains as well, a pattern contrasting sharply with that found at the Settlement Point site, Agayadan Village, and Keatley Creek (Butler and Chatters 1994: Figure 5; Hoffman, Czederpiltz, and Partlow 2000: Figure 4; Partlow 2006). These three village sites, overwhelmingly dominated by salmon vertebrae and/or fin elements and yielding very few head bones, appear to contain primarily the remains of stored salmon.

How do we interpret the North Pacific cod assemblages? Since the North Pacific sites broadly match the coastal North Atlantic sites, does that mean all are remnants of processing for cod storage? Or are they combinations of cod processing, storage, and/or consumption? The latter interpretation was made for the North Atlantic site Tjarnargata 3C (Figure 12.3, far right), which was considered a *combination* of the remains of larger cod processed for the fish trade (some of the pectoral and vertebral elements leaving the site as part of the dried cod) and the remains of smaller cod consumed whole on site (Amundsen et al. 2005).

A complication in the hunt for evidence of stored Pacific cod is that winter villages in the Aleutian and Kodiak regions tended to be located near prime cod-fishing banks (Clark 1987; Haggarty et al. 1991). Thus, even with storage, presumably some fresh fish were consumed and, therefore, discarded in winter villages. The majority of the assemblages come from sites interpreted as multiseason villages or campsites. The exceptions are the three Northwest Coast assemblages and the Horseshoe Cove site from Kodiak, all of which have been interpreted in the literature as spring/summer fish camps. Yarborough (2000:235) suggests the 49-SEW-440 site in Prince William Sound may have been occupied either during spring or fall. Thus, cod assemblages from many North Pacific winter villages reflect cod processing, consumption, and storage. Unless we can find different disposal contexts within these sites for these activities, any faunal signature of cod processing or storage may be obscured by other activities involving cod. What other lines of faunal evidence can be used?

Another way to look at the cod data is to calculate the proportion that cleithra contribute to the total cod NISP of an assemblage, a method that some researchers have used in the North Atlantic (e.g., Amundsen et al. 2005:136). If cleithra traveled with dried cod, low percentages are expected for the remains of cod processing, and larger percentages are expected

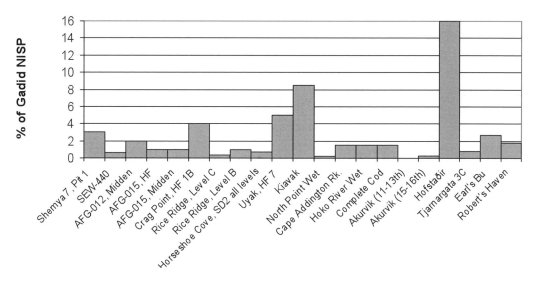

Figure 12.4. Cleithra as a proportion of gadid NISP. "Complete Cod" indicates percent expected from a whole fish. To calculate the percent cleithra in a complete cod, we did not count those elements frequently not identified in assemblages, such as ribs, spines, radials, hypurals, and branchial arch elements. The result was that cleithra make up approximately 1.5 percent of 130 cod bones frequently identified in assemblages.

Table 12.4. Archaeological Pacific cod live-length estimates

Region	Site	Length estimate (cm)	Source
Aleutians	Adak 009 (49-ADK-009)	44–112	Orchard 2003, table 6.16
Aleutians	Adak 011 (49-ADK-011	36–105	Orchard 2003, table 6.16
Aleutians	Buldir 008 (49-KIS-008)	54–107	Orchard 2003, table 6.16
Aleutians	Shemya 2 (49-ATU-021)	38–120	Orchard 2003, table 6.16
Aleutians	Shemya 7 (49-ATU-061)	35–125	Orchard 2003, table 6.16
Prince William Sound	Lowell Homestead (49-SEW-682)	47–68	Yarborough 2000, 185
Kodiak Archipelago	Settlement Point (49-AFG-015)a	46–80	This paper
Kodiak Archipelago	49-AFG-012[a]	57–68	This paper
Northwest Coast	Cape Addington Rk. (49-CRG-188)	Most >67	Moss 2004, 163

a. Live-length estimates based on dentary measurements using Orchard's (2003) dentary measurement #2 and corresponding regression formula.

for the remains of dried cod. This method has the advantage of simplicity and avoids the sensitivity to changes in one body region typical of the closed arrays in Figures 12.2 and 12.3.

Figure 12.4 shows the proportions of cleithra for the ten assemblages with these data. The majority have low percentages of cleithra (less than 4 percent). This matches expectations for cod processing sites in the North Atlantic (e.g., Amundsen et al. 2005) but also comes close to the proportion that cleithra compose of a complete Pacific cod. This low proportion of cleithra, however, does not match expectations for sites with stored cod, such as North Pacific winter villages. Assemblages with the highest percentages of cleithra include the Icelandic site of Hofstaðir and Kiavak village on Kodiak Island. These could be interpreted as resulting from consumption of dried cod with cleithra attached. The large

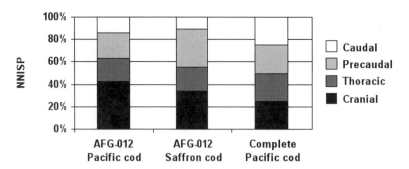

Figure 12.5. Gadid skeletal parts from 49-AFG-012. Data from Table 12.5.

Table 12.5. 49-AFG-012 skeletal part representation by large (Pacific) and small (Saffron) cod

Cod	Cranial[a]	Thoracic	Precaudal	Caudal
Pacific cod				
NISP	26.00	31.00	102.00	143.00
NNISP	13.00	6.20	6.80	4.33
Saffron cod				
NISP	67.00	205.00	1000.00	697.00
NNISP	67.00	41.00	66.60	21.12

a. The highest cranial NNISP for Pacific cod came from the premaxilla and for Saffron cod from the basioccipital.

number of cod head elements in the Kiavak assemblage, however, argues against the consumption of mostly dried cod bodies at this site.

Another consideration in interpretation of the skeletal part representation data is fish size. Small fish may be processed differently from large fish when drying for storage (e.g., Butler 1996; Greenspan 1998; Norton, Kim, and Bae 1999), so skeletal part signatures may vary by fish size. Indeed, the historic North Pacific and North Atlantic fish trade selected only larger cod (roughly 60–110 cm) for drying (Cobb 1927; Perdikaris 1999:390). Smaller cod were presumably consumed fresh. Live-length estimates from several North Pacific assemblages suggest most Pacific cod remains are from individuals large enough to be considered worth drying by North Atlantic standards (Table 12.4). Thus, it is possible that prehistorically, small cod were not dried, while large cod were. Alternatively, based on economic anatomy, small cod may have been dried whole, while larger cod had the fat-rich heads cut off and consumed on site while the bodies dried. In either case, the smaller cod remains should reflect whole cod deposited on site, while the larger cod remains should reflect differential treatment of heads and bodies.

One site where we can evaluate this is 49-AFG-012 on Afognak Island, for which Partlow recently identified selected gadid elements to species. Using Orchard's (2003) Pacific cod regression for premaxilla measurement #3, saffron cod (n = 54) are estimated to be 21–38 cm in fork length and Pacific cod (n = 21) 53–86 cm. The skeletal part frequencies for these small versus large cod at 49-AFG-012 are similar (Table 12.5, Figure 12.5). At least at this site, Pacific cod were not processed and disposed of differently than saffron cod, and thus we find no support for the storage hypothesis. Comparable data were not available for other North Pacific sites.

Conclusion

Neither the ethnohistorical nor the zooarchaeological records examined here provide *compelling* evidence for Pacific cod storage in the eastern North Pacific. Instead, both records provide conflicting evidence. Whereas some historic records report Native peoples drying cod (e.g., de Laguna 1972; Turner 1886), others report that cod were always eaten fresh and never dried (Davydov 1977; Emmons 1991). The skeletal parts represented at the majority of sites examined here have a slight overabundance of crania and slight underabundance of vertebrae compared to a complete cod. This is weakly consistent with expectations for storage processing, wherein there should be overrepresentation of heads and underrepresentation of vertebrae. But the fact that nearly all sites have this pattern regardless of time or location undermines the cod storage argument, unless we assume that this practice was nearly universal throughout the record. Furthermore, most of the sites are thought to be winter villages, which arguably should represent consumption of stored cod rather than processing, and thus be composed of an overabundance of vertebrae and underabundance of heads. The most unusual North Pacific site examined here, the North Point Wet Site, provides the best case for cod storage (see also Smith et al., this volume). This site stands out with its overrepresentation of vertebrae, nearly twice that expected for a complete cod or any other North Pacific site. On the other hand, it still has a moderate proportion of cod heads (34 percent NNISP, almost identical to that expected for a complete cod) and a very low proportion of cleithra, both of which could be viewed as inconsistent with the storage interpretation. Keeping in mind that a variety of factors can affect element frequencies, including transport, differential element destruction, and cultural preferences (Friesen 2001:316; Smith et al., this volume), and after comparing these assemblages with some of the Atlantic cod assemblages, we offer several observations.

First, perhaps Pacific cod were dried and stored for the winter prehistorically, but site disposal contexts obscure the faunal signature. Like some of the coastal North Atlantic assemblages, many of the Pacific cod assemblages may represent *combinations* of the remains of cod processed for storage and cod consumed fresh on site. Pacific cod assemblages at winter villages may be a mixture of large stored cod, cod heads consumed fresh as delicacies, and whole small cod consumed fresh year-round, while those interpreted as fish camps may be combinations of cod processing and cod consumed fresh on site. Archaeofaunal signatures of Pacific cod storage may be obscured by the co-location of processing, storage, and consumption at these sites.

Second, neither the Pacific cod nor many of the Atlantic cod assemblages present compelling evidence for cod storage, based on skeletal parts representation alone. Multiple lines of evidence are necessary and could include cutmarks and live-length estimations (e.g., Amundsen et al. 2005; Barrett 1997; Perdikaris 1999), cod found in storage features, and obvious evidence of drying and/or smoking structures (e.g., Croes 1995:233; Saltonstall and Steffian 2006:56, 83). In the Atlantic region, stable isotope analyses of cod bones are being used to differentiate nonlocal and local cod populations (Barrett et al. 2008; Bourque, Johnson, and Steneck 2008). Since dried Pacific cod probably did not travel very far from where they were caught prior to the commercial era, this technique probably would not aid in finding evidence of prehistorically stored Pacific cod.

Third, the role of differential element survivorship in cod skeletal part representation needs to be addressed. Beyond the fragility of cod basipterygia, cod elements may have significant differences in survivorship that could affect skeletal part distributions. Significant differences in head versus vertebrae bone densities and survivorship have been found for salmonids, with strong implications for the evaluation of salmon skeletal part data (Butler and Chatters 1994; Lubinski 1996). We need to assess to what degree missing cod elements may be due to an "accident of preservation, or an accident of sampling" (White 1956:402). Smith's (2008; Smith et al., this volume) study found basipterygia the least dense cod bone, but found little difference in bone density values between selected head elements and vertebrae. Based on these results, differential bone density would not explain the overrepresentation of heads found in most North Pacific assemblages. However, more density data are needed to bolster this assessment since the bones measured by Smith exclude those that made up the maximum NNISP values in our study (e.g., the basioccipital, parasphenoid, urohyal, and premaxilla). Interestingly, the dentary, which had the highest bone density value (Smith et al., this volume), gave us the highest NNISP value for Cape Addington Rockshelter. Another possible cause of differential survivorship is intentional destruction of some elements over others. For example, Smith et al. (this volume) argue that Pacific cod heads, with their high fat content, were more likely to be broken and boiled, hence less likely to be preserved in the archaeological record. But the majority of assemblages examined here have more heads than bodies, suggesting that in these cases cod heads were not preferentially broken and boiled compared to cod bodies.

Given the arguments for resource intensification and the record for salmon, the investigation of cod storage would benefit by inclusion of older assemblages. The debate continues over when and where people began storing salmon in earnest in the eastern North Pacific. Matson (1992:423) suggested the period from 3500–3000 BP marks the time on the Northwest Coast when the "salmon-storage economy came into being." Alternatively, based on their identification of a consistent pink salmon fishery (using ancient DNA) at the site of Namu, Cannon and Yang (2006; Cannon, Yang, and Speller, this volume) have argued that salmon storage occurred as early as 7000 BP on the Northwest Coast. In the Kodiak Archipelago, Steffian and others (2006:123) have argued that the period from 4000 to 3000 BP was a time of "intensified production and focused storage," in particular of salmon resources but also, perhaps, of Pacific cod. The majority of the cod assemblages examined here fall within the time period when increasing evidence of salmon storage is found in different regions of the eastern North Pacific. Cod assemblages from more ancient sites, such as the 8,200-year-old Chuck Lake site (49-CRG-237) in southeast Alaska (Ackerman, Reid, and Gallison 1989:13), could be particularly informative.

The question of whether, how much, and/or when different Native peoples stored Pacific cod in the eastern North Pacific is still unanswered. Nevertheless, the large numbers of Pacific cod remains at prehistoric sites throughout the region attest to the importance of this resource to North Pacific peoples. The availability of Pacific cod and other codfishes such as saffron cod and Pacific tomcod year-round in some areas may have been just as important as their abundance during the spring and early summer months. For example, Mishler (2001:38) writes of an incident when the salmon runs collapsed in 1935 and one Alutiiq man described surviving that winter by living off cod he jigged for from the shore. He said "the codfish saved us from starvation" (Mishler 2001:38).

Acknowledgments

An earlier version of this chapter was presented as a poster at the thirty-fourth annual meeting of the Alaska Anthropological Association in Fairbanks, Alaska, in 2007. We thank several people for generously sharing unpublished data, especially the Western Aleutians Archaeological and Paleobiological Project, Jeanne Schaaf and Susan Bender, Justin Hays, and Dale Croes. Aubrey Cannon, Madonna Moss, and especially Pat Lubinski supplied editorial advice and comments. This research made use of fish bone analyses at several sites undertaken by the authors in their dissertations, which were partially supported by the Afognak Native Corporation, Alaska Anthropological Association, National Science Foundation, U.S. Fish and Wildlife Service, and the University of Washington. This paper was inspired by a discussion of the importance of Pacific cod to North Pacific peoples from Bowers and Moss (2001).

References Cited

Ackerman, R. E., K. C. Reid, and J. D. Gallison. "Heceta Island: An Early Maritime Adaptation." In *Development of Hunting-Fishing-Gathering Maritime Societies along the West Coast of North America*, A. Blukis Onat, ed., pp. 1–29. Proceedings of the Circum-Pacific Prehistory Conference, Seattle. Pullman: Washington State University, 1989.

Ames, Kenneth M. "The Evolution of Social Ranking on the Northwest Coast of North America." *American Antiquity* 46 (1981): 789–805.

———. "Hierarchies, Stress, and Logistical Strategies among Hunter-Gatherers in Northwestern North America." In *Prehistoric Hunter-Gatherers: The Emergence of Cultural Complexity*, T. D. Price and J. A. Brown, eds., pp. 155–180. San Diego: Academic Press, 1985.

———. "The Northwest Coast: Complex Hunter-Gatherers, Ecology, and Social Evolution." *Annual Review of Anthropology* 23 (1994): 209–229.

Amundsen, Colin, Sophia Perdikaris, Thomas H. McGovern, Yekaterina Krivogorskaya, Matthew Brown, Konrad Smiarowski, Shaye Storm, Salena Modugno, Malgorzata Frik, and Monica Koczela. "Fishing Booths and Fishing Strategies in Medieval Iceland: An Archaeofauna from the Akurvik, North-West Iceland." *Environmental Archaeology* 10 (2005): 127–142.

Arima, Eugene, and John Dewhirst. "Nootkans of Vancouver Island." In *Northwest Coast*, Handbook of North American Indians, vol. 7, Wayne Suttles, ed., pp. 391–411. Washington, DC: Smithsonian Institution, 1990.

Bakkala, Richard G. "Pacific Cod of the Eastern Bering Sea." *International North Pacific Fisheries Commission, Bulletin* 42 (1984): 157–179.

Bakkala, R. S., Westrheim, S. Mishima, C. Zhang, and E. Brown. "Distribution of Pacific Cod (*Gadus macrocephalus*) in the North Pacific Ocean." *International North Pacific Fish Commission Bulletin* 42 (1984): 111–115.

Barrett, James H. "Fish Trade in Norse Orkney and Caithness: A Zooarchaeological Approach." *Antiquity* 71 (1997): 616–638.

Barrett, James, Cluny Johnstone, Jennifer Harland, Wim Van Neer, Anton Ervynck, Daniel Makowiecki, Dirk Heinrich, Anne Karin Hufthammer, Inge Bødker Enghoff, Colin Amundsen, Jørgen Schou Christiansen, Andrew K. G. Jones, Alison Locker, Sheila Hamilton-Dyer, Leif Jonsson, Lembi Lõugas, Callum Roberts, and Michael Richards. "Detecting the Medieval Cod Trade: A New Method and First Results." *Journal of Archaeological Science* 35 (2008): 850–861.

Binford, Lewis R. "Butchering, Sharing, and the Archaeological Record." *Journal of Anthropological Archaeology* 3 (1984): 235–257.

Birket-Smith, Kaj. *The Chugach Eskimo*. Copenhagen: Nationalmuseets Publikationsfond, 1953.

Black, Lydia T., and R. G. Liapunova. "Aleut: Islanders of the North Pacific." In *Crossroads of the Continents: Cultures of Siberia and Alaska*, William W. Fitzhugh and Aron Crowell, eds., pp. 52–57. Washington, DC: Smithsonian Institution, 1988.

Bourque, Bruce J., Beverly Johnson, and Robert S. Steneck. "Possible Prehistoric Fishing Effects on Coastal Marine Food Webs in the Gulf of Maine." In *Human Impacts on Ancient Marine Ecosystems: A Global Perspective*, T. C. Rick and J. M. Erlandson, eds., pp. 165–185. Berkeley: University of California Press, 2008.

Bowers, Peter M., and Madonna L. Moss. "The North Point Wet Site and the Subsistence Importance of Pacific Cod on the Northern Northwest Coast." In *People and Wildlife in Northern North America: Essays in Honor of R. Dale Guthrie*, S. Craig Gerlach and Maribeth S. Murray, eds., pp. 159–177. BAR British Archaeological Report International Series 944. Oxford: Archaeopress, 2001.

Bunn, Henry T. "Meat-eating and Human Evolution: Studies on the Diet and Subsistence Patterns of Plio-Pleistocene Hominids in East Africa." PhD diss., University of California, Berkeley, 1982.

Butler, Virginia L. "Tui Chub Taphonomy and the Importance of Marsh Resources in the Western Great Basin of North America." *American Antiquity* 61 (1996): 699–717.

Butler, Virginia L., and James C. Chatters. "The Role of Bone Density in Structuring Prehistoric Salmon Bone Assemblages." *Journal of Archaeological Science* 21 (1994): 413–424.

Cannon, Aubrey. *The Economic Prehistory of Namu: Patterns in Vertebrate Fauna*. Burnaby, BC: Archaeology Press, Simon Fraser University, 1991.

———. "Was Salmon Important in Northwest Coast Prehistory?" In *People and Wildlife in Northern North America: Essays in Honor of R. Dale Guthrie*, S. Craig Gerlach and Maribeth S. Murray, eds., pp. 178–187. BAR International Series 944. Oxford: Archaeopress, 2001.

Cannon, Aubrey, and Dongya Y. Yang. "Early Storage and Sedentism on the Pacific Northwest Coast: Ancient DNA Analysis of Salmon Remains from Namu, British Columbia." *American Antiquity* 71 (2006): 123–140.

Causey, Douglas, Debra G. Corbett, Christine Lefevre, Dixie L. West, Arkady B. Savinetsky, Nina K. Kiseleva, and Bulat F. Khassanov. "The Palaeoenvironment of Humans and Marine Birds of the Aleutian Islands: Three Millennia of Change." *Fisheries Oceanography* 14 (2005): 259–276.

Clark, Donald W. *Koniag Prehistory*. Tubinger Monographien Zur Urgeschichte, vol. 1. Verlag: W. Kohlhammer, 1974.

———. "Pacific Eskimo: Historical Ethnography." In *Arctic*, Handbook of North American Indians, vol. 5, David Dumas, ed., pp. 185–197. Washington, DC: Smithsonian Institution, 1984.

———. "On a Misty Day You Can See Back to 1805: Ethnohistory and Historical Archaeology on the Southeastern Side of Kodiak Island, Alaska." *Anthropological Papers of the University of Alaska* 21 (1987): 105–132.

Cobb, John N. *Pacific Cod Fisheries*. Bureau of Fisheries Document No. 1014. Washington, DC: Department of Commerce, Bureau of Fisheries, Government Printing Office, 1927.

Coupland, Gary. "Maritime Adaptation and Evolution of the Developed Northwest Coast Pattern on the Central Northwest Coast." *Arctic Anthropology* 35 (1998): 36–56.

Croes, Dale R. "Exploring Prehistoric Subsistence Change on the Northwest Coast." In *Long-Term Subsistence Change in Prehistoric North America*, Dale R. Croes, Rebecca A. Hawkins, and Barry L. Isaac, eds., pp. 337–366. Research in Economic Anthropology, Supplement 6. Greenwich, CT: JAI Press, 1992.

———. *The Hoko River Archaeological Site Complex*. Pullman: Washington State University Press, 1995.

Croes, Dale R., and Steven Hackenberger. "Hoko River Archaeological Complex: Modeling Prehistoric Northwest Coast Economic Evolution." In *Prehistoric Economies of the Pacific Northwest Coast*, B. L. Isaac, ed., pp. 19–86. Research in Economic Anthropology, Supplement 3. Greenwich, CT: JAI Press, 1988.

Crowell, Aron L., and April Laktonen. "*Sugucihpet*—Our Way of Living." In *Looking Both Ways: Heritage and Identity of the Alutiiq People*, Aron L. Crowell, Amy F. Steffian, and Gordon L. Pullar, eds., pp. 137–188. Fairbanks: University of Alaska Press, 2001.

Davydov, Gavrila Ivanovich. *Two Voyages to Russian America, 1802–1807*. Richard A. Pierce, ed. Colin Bearne, trans. Kingston, ON: Limestone Press, 1977.

DiCosimo, J., and N. Kimball. Groundfish of the Gulf of Alaska: Species Profile. Anchorage: North Pacific Fishery Management Council, 2001.

Emmons, George T. *The Tlingit Indians*. Frederica de Laguna, ed. Seattle: University of Washington Press; Vancouver: Douglas and McIntyre; New York: American Museum of Natural History, 1991.

Fladmark, Knut R. *A Paleoecological Model for Northwest Coast Prehistory*. Mercury Series Archaeological Survey of Canada Paper No. 43. Ottawa: National Museum of Man, 1975.

Friesen, T. Max. "A Zooarchaeological Signature for Meat Storage: Re-thinking the Drying Utility Index." *American Antiquity* 66 (2001): 315–331.

Gibson, James R. "European Dependence upon American Natives: The Case of Russian America." *Ethnohistory* 25 (1978): 359–385.

Gideon, Hieromonk. *The Round the World Voyage of Hieromonk Gideon 1803–1809*. Richard A. Pierce, ed. Lydia T. Black, trans. Alaska History No. 32. Alaska State Library Historical Monograph No. 9. Kingston, ON: Limestone Press, 1989.

Grayson, Donald K. *Quantitative Zooarchaeology*. Orlando: Academic Press, 1984.

Grayson, Donald K., and Carol J. Frey. "Measuring Skeletal Part Representation in Archaeological Faunas." *Journal of Taphonomy* 2 (2004): 27–42.

Greenspan, Ruth L. "Gear Selectivity Models, Mortality Profiles and the Interpretation of Archaeological Fish Remains: A Case Study from the Harney Basin, Oregon." *Journal of Archaeological Science* 25 (1998): 973–984.

Haggarty, James C., Christopher B. Wooley, Jon M. Erlandson, and Aron Crowell. *The 1990 Exxon Cultural Resource Program: Site Protection and Maritime Cultural Ecology in Prince William Sound and the Gulf of Alaska*. Anchorage: Exxon, 1991.

Hajda, Yvonne. "Southwestern Coast Salish." In *Northwest Coast*, Handbook of North American Indians, vol. 7, Wayne Suttles, ed., pp. 503–517. Washington, DC: Smithsonian Institution, 1990.

Hanson, Diane. "Subsistence During the Late Prehistoric Occupation of Pender Canal, British Columbia (DeRt-1)." *Canadian Journal of Archaeology* 19 (1995): 29–48.

Hayden, Brian. *The Pithouses of Keatley Creek*. Fort Worth: Harcourt Brace, 1997.

Hays, Justin M. "The Horseshoe Cove Site: An Example of Early Kachemak Subsistence Strategies from Faunal Remains in the Kodiak Archipelago." Master's thesis, University of Alaska, Anchorage, 2007.

Hoffman, Brian W. "The Organization of Complexity: A Study of Late Prehistoric Village Organization in the Eastern Aleutian Region." PhD diss., University of Wisconsin–Madison, 2002.

Hoffman, Brian W., Jessica M. C. Czederpiltz, and Megan A. Partlow. "Heads or Tails: The Zooarchaeology of Aleut Salmon Storage on Unimak Island, Alaska." *Journal of Archaeological Science* 27 (2000): 699–708.

Holmberg, Heinrich Johan. *Holmberg's Ethnographic Sketches*. Marvin W. Falk, ed. Fritz Jaensch, trans. Fairbanks: University of Alaska Press, 1985.

Hunt, George L., Jr., and Phyllis J. Stabeno. "Oceanography and Ecology of the Aleutian Archipelago: Spatial and Temporal Variation." *Fisheries Oceanography* 14 (2005): 292–306.

Jacka, Jerry. "Fishing." In *The History and Ethnohistory of the Aleutians East Borough*, Richard A. Pierce, Katherine L. Arndt, and Sarah McGowan, eds., pp. 213–242. Kingston, ON: Limestone Press, 1999.

Jochelson, Waldemar. *History, Ethnology and Anthropology of the Aleut*. Reprint. Salt Lake City: University of Utah Press, 1933.

Kamenskii, Fr. Anatolii. *Tlingit Indians of Alaska*. Sergei Kan, trans. Fairbanks: University of Alaska Press, 1985.

Kasahara, Hiroshi. *Fisheries Resources of the North Pacific Ocean,* Part 1. H. R. MacMillan Lectures in Fisheries. Vancouver: Institute of Fisheries, University of British Columbia, 1961.

Knapp, Frances, and Rheta Louise Childe. *The Thlinkets of Southeastern Alaska*. Chicago: Stone and Kimball, 1896.

Kopperl, Robert E. "Cultural Complexity and Resource Intensification on Kodiak Island, Alaska." PhD diss., University of Washington, 2003.

de Laguna, Frederica. *Under Mount Saint Elias: The History and Culture of the Yakutat Tlingit*. Washington, DC: Smithsonian Contributions to Anthropology, vol. 7, 1972.

———. "Eyak." In *Northwest Coast*, Handbook of North American Indians, vol. 7, Wayne Suttles, ed., pp. 89–196. Washington, DC: Smithsonian Institution, 1990a.

———. "Tlingit." In *Northwest Coast*, Handbook of North American Indians, vol. 7, Wayne Suttles, ed., pp. 203–228. Washington, DC: Smithsonian Institution, 1990b.

Langdon, Steve. "Comparative Tlingit and Haida Adaptation to the West Coast of the Prince of Wales Archipelago." *Ethnology* 18 (1979): 101–119.

Lantis, Margaret. "Aleut." In *Arctic*, Handbook of North American Indians, vol. 5, David Dumas, ed., pp. 161–184. Washington, DC: Smithsonian Institution, 1984.

Lefevre, Christine, Dixie West, and Debra G. Corbett. "Archaeological Surveys in the Near Islands: Attu Island and Shemya Island." In *Archaeology in the Aleut Zone of Alaska, Some Recent Research*, Don. E. Dumond, ed., pp. 235–250. University of Oregon Anthropological Papers No. 58. Eugene: Department of Anthropology and Museum of Natural History, 2001.

Lubinski, Patrick M. "Fish Heads, Fish Heads: An Experiment on Differential Bone Preservation in a Salmonid Fish." *Journal of Archaeological Science* 23 (1996): 175–181.

Lyman, R. Lee. "Available Meat from Faunal Remains: A Consideration of Techniques." *American Antiquity* 44 (1979): 536–546.

Maschner, Herbert D. G. "Settlement and Subsistence in the Later Prehistory of Tebenkof Bay, Kuiu Island, Southeast Alaska." *Arctic Anthropology* 34 (1997): 74–99.

Matson, R. G. "Intensification and the Development of Cultural Complexity: The Northwest versus the Northeast Coast." In *The Evolution of Maritime Cultures on the Northeast and the Northwest Coasts of America*, Ronald J. Nash, ed., pp. 125–148. Burnaby, BC: Department of Anthropology, Simon Fraser University, 1983.

———. "The Evolution of Northwest Coast Subsistence." In *Long-Term Subsistence Change in Prehistoric North America*, D. R. Croes, ed., pp. 367–428. Research in Economic Anthropology Supplement 6. Greenwich, CT: JAI Press, 1992.

Matson, R. G., and Gary Coupland. *The Prehistory of the Northwest Coast*. San Diego: Academic Press, 1995.

McCartney, Allen P., and Douglas W. Veltre. "Aleutian Island Prehistory: Living in Insular Extremes." *World Archaeology* 30 (2000): 503–515.

Mecklenburg, Catherine W., T. Anthony Mecklenburg, and Lyman K. Thorsteinson. *Fishes of Alaska*. Bethesda, MD: American Fisheries Society, 2002.

Mishler, Craig. *Black Ducks and Salmon Bellies: An Ethnography of Old Harbor and Ouzinkie, Alaska*. Anchorage: Alaska Department of Fish and Game, 2001.

Mitchell, Donald, and Leland Donald. "Archaeology and the Study of Northwest Coast Economies." In *Prehistoric Economies of the Pacific Northwest Coast*, Barry L. Isaac, ed., pp. 293–351. Research in Economic Anthropology, Supplement 3. Greenwich, CT: JAI Press, 1988.

Moss, Madonna L. "Analysis of the Vertebrate Assemblage." In *The Hidden Falls Site, Baranof Island, Alaska*, S. D. Davis, ed., pp. 93–129. Anchorage: Aurora, Alaska Anthropological Association Monograph Series, 1989a.

———. "Archaeology and Cultural Ecology of the Prehistoric Angoon Tlingit." PhD diss., University of California, Santa Barbara. Ann Arbor, MI: University Microfilms, 1989b.

———. *Archaeological Investigation of Cape Addington Rockshelter: Human Occupation of the Rugged Seacoast on the Outer Prince of Wales Archipelago, Alaska*. Eugene: University of Oregon Anthropological Paper No. 63, 2004.

Mundy, Phillip R., and Anne Hollowed. "Fish and Shellfish." In *The Gulf of Alaska: Biology and Oceanography*, Phillip R. Mundy, ed., pp. 81–97. Fairbanks: Alaska Sea Grant College Program, University of Alaska, 2005.

Nagaoka, Lisa. "Differential Recovery of Pacific Island Fish Remains." *Journal of Archaeological Science* 32 (2005): 941–955.

Newton, Richard G., and Madonna L. Moss. *The Subsistence Lifeway of the Tlingit People: Excerpts of Oral Interviews*. Juneau: USDA Forest Service, Alaska Region Report No. 179, 1984.

Norton, Christopher J., Byungmo Kim, and Kdong Bae. "Differential Processing of Fish during the Korean Neolithic: Konam-Ri." *Arctic Anthropology* 36 (1999): 151–165.

Oberg, Kalervo. *The Social Economy of the Tlingit Indians*. Seattle: University of Washington Press, 1973.

Orchard, Trevor J. *An Application of the Linear Regression Technique for Determining Length and Weight of Six Fish Taxa: The Role of Selected Fish Species in Aleut Palaeodiet*. BAR British Archaeological Reports, International Series 1172. Oxford: Archaeopress, 2003.

Partlow, Megan A. SH7 Pit 1 Fish Remains. MS on file, US Fish and Wildlife Service, Anchorage, November, 1997.

———. "Salmon Intensification and Changing Household Organization in the Kodiak Archipelago." PhD diss., University of Wisconsin–Madison, 2000.

———. "Sampling Fish Bones: A Consideration of the Importance of Screen Size and Disposal Context in the North Pacific." *Arctic Anthropology* 43 (2006): 67–79.

Perdikaris, Sophia. "From Chiefly Provisioning to Commercial Fishery: Long-term Economic Change in Arctic Norway." *World Archaeology* 30 (1999): 388–402.

Rousselot, Jean-Loup, William W. Fitzhugh, and Aron Crowell. "Maritime Economies of the North Pacific Rim." In *Crossroads of the Continents: Cultures of Siberia and Alaska*, William W. Fitzhugh and Aron Crowell, eds., pp. 151–172. Washington, DC: Smithsonian Institution, 1988.

Rowley-Conwy, Peter. "Comments on the Significance of Food Sharing among Hunter-Gatherers: Residence Patterns, Population Densities, and Social Inequalities." *Current Anthropology* 23 (1982): 533.

Saltonstall, Patrick G., and Amy F. Steffian. *The Archaeology of Horseshoe Cove*. Occasional Papers in Field Archaeology, No. 1. Anchorage: Bureau of Indian Affairs Alaska Region, 2006.

Sauer, Martin. *An Account of a Geographical and Astronomical Expedition to the Northern Parts of Russia*. London: T. Cadell, 1802.

Schalk, Randall. "The Structure of an Anadromous Fish Resource." In *For Theory Building in Archaeology*, L. Binford, ed., pp. 207–249. New York: Academic Press, 1977.

Shields, Captain Ed. *Salt of the Sea: The Pacific Coast Cod Fishery and the Last Days of Sail*. Lopez Island, WA: Pacific Heritage Press, 2001.

Smith, Ross E. "Structural Bone Density of Pacific Cod (*Gadus macrocephalus*) and Halibut (*Hippoglossus stenolepis*): Taphonomic and Archaeological Implications." Master's thesis, Portland State University, 2008.

Soffer, Olga. "Storage, Sedentism, and the Eurasian Paleolithic Record." *Antiquity* 63 (1989): 719–732.

Steffian, Amy F., Patrick G. Saltonstall, and Robert E. Kopperl. "Expanding the Kachemak: Surplus Production and the Development of Multi-season Storage in Alaska's Kodiak Archipelago." *Arctic Anthropology* 43 (2006): 93–129.

Stuiver, M., and P. J. Reimer. "Extended ¹⁴C Data Base and Revised Calib 3.0 ¹⁴C Age Calibration Program." *Radiocarbon* 35 (1993): 215–230.

Suttles, Wayne. "Environment." In *Northwest Coast*, Handbook of North American Indians, vol. 7, Wayne Suttles, ed., pp. 16–29. Washington, DC: Smithsonian Institution, 1990.

Testart, Alain. "The Significance of Food Storage among Hunter-Gatherers: Residence Patterns, Population Densities, and Social Inequalities." *Current Anthropology* 23 (1982): 523–530.

Tikhmenev, P. A. *A History of the Russian-American Company*. Richard A. Pierce and Alton S. Donnelly, eds., trans. Seattle: University of Washington Press, 1978.

Turner, L. M. *Contributions to the Natural History of Alaska*. Washington, DC: Government Printing Office, 1886.

Vanderhoek, Richard, and Rachel Myron. *Cultural Remains from a Catastrophic Landscape: An Archaeological Overview and Assessment of Aniakchak National Monument and Preserve*. Anchorage: National Park Service, Aniakchak National Monument and Preserve, 2004.

Veniaminov, Ivan. *Notes on the Islands of the Unalaska District*. Lydia T. Black and R. H. Goeghegan, trans. Richard A. Pierce, ed. Alaska History No. 27. Kingston, ON: Limestone Press, 1984.

Wesson, Cameron B. "Chiefly Power and Food Storage in Southeastern North America." *World Archaeology* 31 (1999): 145–164.

White, T. E. "The Study of Osteological Materials in the Plains." *American Antiquity* 21 (1956): 401–404.

Wigen, Rebecca J. "Vertebrate Fauna." In *The Hoko River Archaeological Site Complex: The Rockshelter (45CA21), 1,000–100 BP*, Dale. R. Croes, ed., pp. 71–114. Pullman: Washington State University Press, 2005.

Witherell, David. *Groundfish of the Bering Sea and Aleutian Islands Area: Species Profiles 2001*. Anchorage: North Pacific Fishery Management Council, 2000.

Yarborough, Linda Finn. "Prehistoric and Early Historic Patterns along the North Gulf of Alaska Coast." PhD diss., University of Wisconsin–Madison, 2000.

CHAPTER 13

Cod and Salmon
A Tale of Two Assemblages from Coffman Cove, Alaska

Madonna L. Moss, *Department of Anthropology, University of Oregon*

The variability in Indigenous use of Pacific cod and its economic importance vis-à-vis salmon are topics requiring further research across much of the North Pacific coast. Fish remains from two archaeological sites at Coffman Cove, located on Prince of Wales Island in southeast Alaska, provide an interesting contrast. At site 49-PET-556, Pacific cod is the most abundant fish taxon in an assemblage dating 3000–2000 cal BP, while at the nearby site 49-PET-067, salmon dominate the fish assemblage dating 5000–700 cal BP. After ruling out the effects of bone density and considering some alternatives, I argue that the residents of 49-PET-067 controlled the salmon resources in the vicinity of Coffman Cove, occupying the site during summer and fall, and likely during much of the year. The 49-PET-556 occupation, in contrast, appears to have been focused on Pacific cod fishing, probably during the spring. More direct evidence of site seasonality and more complete information about the behavior and distribution of Pacific cod in southeast Alaska would help substantiate an explanation of the difference between the two assemblages.

The Two Coffman Cove Sites

Coffman Cove is located on the northeast shore of Prince of Wales Island, falling within the territory of the Wrangell/Stikine Tlingit. According to Goldschmidt and Haas (1998:77), a former camp was located "near the head" of Coffman Cove and a fort was situated nearby. Site 49-PET-556 was discovered in 2005 during ferry terminal construction, hence the site is known as the

Coffman Cove Ferry Terminal Site. In September 2006, Northern Land Use Research (NLUR) archaeologists excavated six 1 m by 1 m units at the site. The faunal analyses involved study of bones recovered (1) during wet screening of fill through quarter-inch mesh and (2) in twenty-seven bulk samples (Moss 2007). Only the quarter-inch samples will be discussed in this paper and of these, 86 percent are fish. Of the 4,962 specimens, 2,525 were identified to a taxon beyond class. The site deposits were dated to ca. 3000–2000 cal BP, although no radiocarbon ages were obtained for the uppermost two levels of the deposit (Reger et al. 2007).

Site 49-PET-067 is located just over 600 m away from the Ferry Terminal Site. This site has been known since 1970 as the Coffman Cove Site, and has suffered continuous loss due to logging camp, road, residential, municipal, and other construction over the last fifty years. The site underwent archaeological testing by the Forest Service in the 1970s and the State Office of History and Archaeology in 1993 (Clark 1979, 1981; Reger 1995), as well as smaller-scale work (e.g., Rushmore, Colby, and Bell 1998). Although faunal remains were recovered during those investigations, they are not part of this analysis. The faunal remains reported here are those recovered by NLUR archaeologists in June and July 2006. This work involved excavations in three areas of the site: Parcels B-West, B-Northeast, and B-East. Altogether, twenty-one 1 m by 1 m units were excavated, and eight bulk samples were taken. The faunal analyses involved study of bones recovered (1) during wet screening of fill through quarter-inch mesh, (2) in 6.5 of the 8 bulk samples, and (3) in "grab" samples of matrix (Moss 2008). Only the quarter-inch mesh samples will be discussed here, and of these, 84 percent are fish. Of the 16,432 specimens, 11,066 were identified to a taxon beyond class. Discussion below will focus on samples from B-West, which comprised 99.6 percent of the quarter-inch samples. The site deposits range in age from 5500 to 700 cal BP. Even though there are no clear breaks in the history of site occupation, the following components have been defined, based on stratigraphic relationships and radiocarbon dates:

Component IV: 2000–700 cal BP (Level 1)
Component III: 3500–2000 cal BP (Level 2)
Component II: 4000–3500 cal BP (Levels 4 and 3)
Component I: 5000–4000 cal BP (Levels 6 and 5, including Features 4, 10, and 11)*

The Effects of Bone Density on Major Fish Taxa at Coffman Cove

Butler and Chatters (1994) and Smith (2008; Smith et al., this volume) have shown that differences in bone density can structure the taxonomic abundances and element representation in fish assemblages. Portions of many elements of Pacific cod are denser than salmon and halibut elements; therefore, Pacific cod may be overrepresented in fish assemblages vis-à-vis these other taxa. Butler and Chatters (1994) and Smith (2008) have developed protocols for identifying density-mediated destruction in fish assemblages. If such attrition can be ruled out, then archaeologists can proceed to consider cultural patterns of fishing,

---

* The oldest part of 49-PET-067 occurs in area B-Northeast and dates to 5500 cal BP. These materials are also considered Component I. Because the faunal remains under discussion here derive from B-west, I am using the oldest age estimate for this part of the site: 5000 cal BP.

fish processing, and disposal. Following Smith (2008), I examined the relationship between skeletal element representation and bone volume density for Pacific cod and salmon from the two Coffman Cove sites, where sample sizes permitted. Spearman's rank correlation was used following Drennan (1996:228–232). As specified by Smith (2008:80), only sample sizes greater than thirty, with more than half the element types represented, were used. The results are shown in Tables 13.1 and 13.2, where MAU is the minimum animal unit (MAU) required to account for all the elements and %MAU is element survivorship.

For the Ferry Terminal Site (49-PET-556), numbers were sufficient to examine whether Pacific cod bones were affected by density-mediated attrition for the site as a whole and for Component 2.* No significant correlation was identified between Pacific cod bone volume density and element survivorship. For the site overall, r_s = 0.513, and for Component 2, r_s = 0.505 (both 0.05 < p <0.1). In other words, bone density did not appear to have affected Pacific cod element representation at this site.

For the Coffman Cove Site (49-PET-067), numbers were sufficient to examine whether salmon bones were affected by density-mediated attrition for the site as a whole and for Components I and II. Here, Spearman's rank correlations are even lower. For the site overall, r_s = 0.297, for Component I, r_s = 0.198, and for Component II, r_s = 0.457. Having ruled out density-mediated attrition for the most abundant taxa in the two sites, we can consider cultural explanations.

A Tale of Two Assemblages at Coffman Cove: Cod and Salmon

Table 13.3 presents the identifications of vertebrate remains (beyond class) from the two Coffman Cove sites. In the Ferry Terminal site assemblage, of the fish identified to at least the family level, 54 percent is Pacific cod (Figure 13.1). Rockfish is next in abundance at 18 percent, followed by sculpin at 9 percent, and salmon and lingcod, both at ~5 percent. Herring and halibut contribute ~3 percent, spiny dogfish make up 2 percent, and greenling, Dover sole, and right-eyed flounders contribute less than 1 percent. Of the 128 salmon bones found in the assemblage, with the exception of one basipterygium, all were vertebrae. This suggests these are the remains of stored salmon, but the sample is small.

ADFG fisheries biologist Deidra Holum (2008, pers. comm.) explained that the Pacific cod in Clarence Strait (the main water channel east of Prince of Wales Island) likely move into the area from Hecate Strait. Today, the shallow waters of Snow and Kashevarof passages not only provide productive habitat but effectively funnel Pacific cod populations north to Sumner Strait and Duncan Canal. Duncan Canal is (and probably was) a special attraction for Pacific cod, with its abundant pink shrimp (*Pandalus*), one of their preferred foods. In Holum's (2008, pers. comm.) words, "one couldn't design more perfect fishing grounds" than these passages where the fish are forced up into depths accessible to those with hand-operated fishing gear. In her estimation, these shallows "concentrated

---

* Two components were identified at 49-PET-556, based on an abrupt smooth boundary between Levels 2B and 2C observed in the field. Analysis of twenty-seven bulk samples from the site, however, showed no significant difference in the composition of shell midden constituents between these two components (Moss 2007). In my study of the vertebrate fauna, I separated out two components following Reger et al. (2007). Because there are only three radiocarbon dates from specific proveniences, I have chosen to lump the two proposed components into one for the purpose of this chapter.

Table 13.1. Coffman Cove Ferry Terminal Site (49-PET-556) Pacific cod bone volume density (VD) and element representation (%MAU)

| Pacific cod element | VD rank | 49-PET-556 | | | |
| | | Overall | | Component 2 | |
		%MAU	Rank	%MAU	Rank
Dentary	1	78.3	3	84.6	3.5
Maxilla	2	73.9	4	84.6	3.5
Vomer	3	82.6	1.5	100.0	1
Articular	4	52.2	5	46.2	6
Cleithrum	5	0.0	10.5	0.0	10.5
Atlas Vertebra	6	0.0	10.5	0.0	10.5
Quadrate	7	47.8	6	61.5	5
Vertebrae	8	82.6	1.5	92.3	2
Ceratohyal	9	0.0	10.5	0.0	10.5
Opercle	10	4.3	8	7.7	8
Hyomandibular	11	8.7	7	15.4	7
Basipterygium	12	0.0	10.5	0.0	10.5
Summary statistics		MNE = 1308		MNE = 812	
		MAU = 23		MAU = 13	
		$r_s = 0.513$		$r_s = 0.505$	
		$0.05 < p < 0.1$		$0.05 < p < 0.1$	

Table 13.2. Coffman Cove Site (49-PET-067) salmon bone volume density (VD) and element representation (%MAU)

| Salmon element | VD rank | 49-PET-067 | | | | | |
| | | Overall | | Component I | | Component II | |
		%MAU	Rank	%MAU	Rank	%MAU	Rank
Vertebrae	1	100.0	1	100.0	1	100.0	1
Pectoral Fin Ray	2	0.7	10.5	0.0	10.5	1.1	9
Angular	3.5	2.8	7.5	4.8	7	2.2	8
Maxilla	3.5	8.3	5	19.0	3	8.7	5
Dentary	5	1.4	9	0.0	10.5	4.3	6
Hypural	6	9.0	4	4.8	7	10.9	4
Pterotic	7	4.1	6	14.3	4.5	3.3	7
Exoccipital	8.5	11.7	3	14.3	4.5	12.0	3
Basipterygium	8.5	51.7	2	38.1	2	54.3	2
Opercle	10.5	0.0	12	0.0	10.5	0.0	11
Coracoid	10.5	2.8	7.5	0.0	10.5	0.0	11
Ceratohyal	12	0.7	10.5	4.8	7	0.0	11
Summary statistics		NISP = 9463		NISP = 1370		NISP = 5986	
		MAU = 145		MAU = 21		MAU = 92	
		$r_s = 0.297$		$r_s = 0.198$		$r_s = 0.457$	
		$0.2 < p < 0.5$		$p > 0.5$		$0.1 < p < 0.2$	

Table 13.3. Vertebrate remains from Coffman Cove sites identified to beyond class

	49-PET-556	49-PET-067 - B-west				
Levels	all	1	2	3 and 4	5 and 6	Total
Components		IV	III	II	I	
Ages (cal BP)	3000–2100	2000-700	3500-2000	4000-3500	5000-4000	
Taxon		NISP	NISP	NISP	NISP	NISP
Pisces						
Clupea harengus pallasii	82		5	60	1	66
Cottid	218	3	67	202	14	286
Dasycottus setiger	1					
Gadus macrocephalus	1308	29	128	168	202	527
Hexagrammos sp.	1			1		1
Hippoglossus stenolepis	65		7	11	19	37
Lamna ditropis			1			1
Leptocottus armatus			1	1		2
Microstomus pacificus	1					
Oncorhynchus spp.	128	33	2061	5986	1370	9450
Ophiodon elongatus	117	1		2		3
Pleuronectid	19		4	9	3	16
Raja sp.			1			1
Sebastes melanops	1					
Sebastes spp.	447	16	95	108	10	229
Squalus acanthias	47	4	3	3	2	12
Aves						
Anatid	10		18	27	2	47
Ardea herodias	1			17		17
Brachyramphus marmoratus	1			1		1
Branta canadensis	1				2	2
Cyanocitta stelleri				1		1
Emberizidae				1		1
Fulmarus glacialis					9	9
Gavia cf. immer			3	1		4
cf. *Grus canadensis*	1					
Haliaeetus leucocephalus			1	1		2
Larus spp.					1	1
Melanitta spp.			8	20	2	30
Phalacrocorax spp.	1			3		3
Podiceps spp.				1		1
Turdidae				1		1
Marine Mammals						
Cetacea	1		1	2		3
Enhydra lutris	5			5	15	20
Eumetopias jubatus				1	1	2
Phoca vitulina	3	3	14	92	14	123
Phocoena dalli			5			5
Phocoena phocoena	6	1				1
Pinniped	6		11	7	3	21
Terrestrial Mammals						
Canis familiaris	19	1	10	33	11	55
Castor canadensis	1	3	6	17	2	28
Lontra canadensis			1		3	4
Martes americana			1	1		2
Mustela vison	4	1	2	1		4
Mustelid					1	1
Odocoileus hemionus sitkensis	29		2	11	16	29
Peromyscus keeni	1			5	2	7
Rodentia			1	1		2
Ursus americanus			1	4	3	8
Total	2525	95	2458	6805	1708	11066

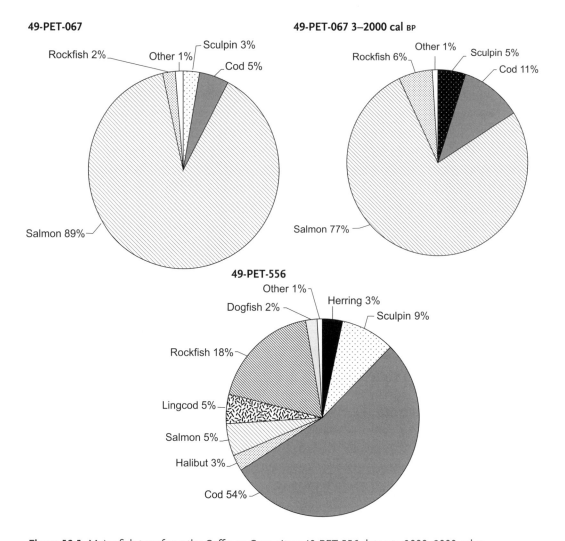

Figure 13.1. Major fish taxa from the Coffman Cove sites; 49-PET-556 dates to 3000–2000 cal BP.

migratory Pacific cod as they passed through and made them easy targets for the residents of Coffman Cove."

For 49-PET-067, four subassemblages can be considered. The Component I subassemblage, dating 5000–4000 cal BP, is moderate in size. Salmon make up 85 percent of the fish (identified to family), followed by Pacific cod at 12.5 percent. The Component II subassemblage is a robust sample, dating from 4000 to 3500 cal BP. Salmon comprise 91 percent of the fish, with no other fish taxon contributing more than 3.2 percent. The Component III subassemblage is moderate in size and dates to 3500–2000 cal BP. Salmon make up 87 percent of the fish, followed by Pacific cod at 5 percent, rockfish at 4 percent, and sculpin at 3 percent. The Component IV subassemblage is too small to make much of the numbers.

Of the fish identified to at least the family level in the entire 49-PET-067 assemblage, 89 percent are salmon, followed by Pacific cod at 5 percent, sculpin at ~3 percent, and rockfish at 2 percent (Figure 13.1). Herring, halibut, right-eyed flounders, spiny dogfish, and other minor taxa make up less than 1 percent each. The presence of salmon cranial elements shows that whole salmon were consumed and processed at 49-PET-067. This suggests that 49-PET-067 was occupied during salmon fishing season during the summer and fall. Based on the escapement surveys for five systems in the immediate vicinity of Coffman Cove, salmon are concentrated in the area during the last week of August until mid-September (Scott Forbes, 2008, pers. comm.). Assuming the timing of the salmon runs in these waters has not changed substantially over the past few thousand years, it would appear that this is one of the seasons when people occupied the Coffman Cove Site.

The major difference between the fish assemblages from the Ferry Terminal Site and the Coffman Cove Site is that cod predominate at the former, and salmon at the latter. The 49-PET-067 fish assemblage is significantly less diverse than that of 49-PET-556 where there is a more even spread of taxonomic abundances of cod, rockfish, and sculpin. A way to quantify this diversity is Simpson's measure of evenness (Krebs 1999:449), a measure that gives more weight to the common species in an assemblage than the rare ones. The index ranges from 0 to 1; if all species in an assemblage are present in the same proportion, then the index equals 1. Samples with an even distribution of taxonomic abundances will have a higher diversity than samples with the same number of taxa, but disproportionately high abundance of a few taxa. For the fish assemblage from 49-PET-556, Simpson's measure of evenness = 0.264, which is fairly low. For the 49-PET-067 fish assemblage, Simpson's measure of evenness is even lower (0.097). The extremely low value of the index for 49-PET-067 is because one taxon, salmon, dominates the assemblage.

Explanations for the Difference between the Coffman Cove and Ferry Terminal Faunal Assemblages

As is the case in many Northwest Coast shell middens, temporal resolution at both sites is relatively coarse (Stein, Deo, and Phillips 2003). Available radiocarbon dates force us to consider long periods of time. To better gauge change over time, we would need better temporal resolution and fewer overlapping radiocarbon dates (Moss 2008). Both 49-PET-556 and 49-PET-067 can be considered cumulative palimpsests as defined by Bailey (2007:204–205). Both sites are time-averaged accumulations that preclude detection of specific depositional episodes, but do allow identification of a significant difference in the faunal assemblages: the abundance of salmon at 49-PET-067 and of Pacific cod at 49-PET-556. I examine some potential explanations for the differences between the two assemblages in the sections that follow.

The 3000–2000 cal. BP Time Period

Even though the deposits at the Coffman Cove and Ferry Terminal sites are palimpsests, focusing specifically on the time period represented at both sites, 3000–2000 cal BP, is worthwhile because it controls for time. Perhaps an environmental or other factor during this time period favored fishing for Pacific cod, and perhaps this trend has been obscured by how temporal components have been defined at 49-PET-067.

The roughly 1,000-year-long occupation of 49-PET-556 overlaps with the latter part of the 1,500-year-long occupation of Component III at 49-PET-067. In an attempt to remove that part of the sample that dates from 3500–3000 cal BP at 49-PET-067, I compiled the numbers of fish remains found in Levels 2A and 2B from the five particularly faunal-rich units that contained at least one level in addition to Levels 2A and 2B: N201/E182, N202/E182, N201/E183, N201/E184, and N201/E185 (see Moss 2008 for more details). If Pacific cod were proportionately more abundant in these upper levels of Component III at 49-PET-067, perhaps this would indicate that the 3000–2000 cal BP period was one in which cod fishing was a dominant activity at both sites. This did not prove to be the case; within Levels 2A and 2B from these units, salmon made up 77 percent of the fish identified to family, Pacific cod made up 11 percent, and rockfish contributed 6 percent (Figure 13.1). While salmon are somewhat less abundant and cod somewhat more abundant in the upper part of Component III, the difference is not significant. At 49-PET-067, Pacific cod still comprise a relatively low percentage, and nothing approaching 49-PET-556 where they make up 54 percent of fish identified to family. The emphasis on cod seen at 49-PET-556 is not replicated at 49-PET-067 during the same time period. The two archaeological sites at Coffman Cove clearly represent activities focused on different resources—salmon in the case of the former and Pacific cod in the case of the latter. The taxonomic abundances do not vary significantly from the oldest to the most recent deposits at either site (with the exception of the very small Component IV at 49-PET-067). This suggests that the use of particular resources at these two sites did not substantially change over time. At neither site do we see evidence for evolutionary change, change in diet breadth, or resource depression.

Seasonality

Another alternative is that the difference between the two assemblages can be explained by seasonality. I have proposed that the Ferry Terminal Site was occupied during winter and spring (Moss 2007), whereas the Coffman Cove Site was occupied during summer and fall, and perhaps during other seasons (Moss 2008).

The idea that the Ferry Terminal Site was occupied during late winter–early spring is based on the dominance of Pacific cod in the assemblage and the inference that this species was most available to site residents during late winter–early spring, when it moves into the shallows to spawn (Bowers and Moss 2001; Newton and Moss 1984, 2005). The other fish represented in some abundance at the Ferry Terminal Site, particularly sculpin, rockfish, and spiny dogfish, are taxa often caught in the groundfish fishery today, that is, part of the Pacific cod bycatch (Thompson, Dorn, and Nichol 2006:169, 212). The salmon bones found at the Ferry Terminal site appear to have been from stored, not fresh, salmon. The bones of mammals and birds found at 49-PET-556 are relatively few, but the identified species—Canada goose, great blue heron, sandhill crane, marbled murrelet, sea otter, harbor seal, harbor porpoise, Sitka black-tailed deer, domestic dog, mink, beaver, and Keen's mouse—while not inconsistent with late winter–spring occupation, are not particularly informative. The availability of these taxa is not exclusive to the winter-spring season.

At the Coffman Cove Site, the presence of salmon cranial, pectoral, pelvic, and trunk elements shows that whole salmon were consumed and processed. The heavy salmon reliance and the presence of whole carcasses suggest fishing and processing during the peak of the salmon runs, which occur during the last week of August to mid-September in the Coffman Cove vicinity. Summer occupation is also indicated by harbor seal pup bones in Components I and II and a juvenile beaver bone in the upper part of Component II. Other species identified in the 49-PET-067 assemblage—Canada goose, great blue heron, northern fulmar, bald eagle, marbled murrelet, Steller's jay, harbor seal, sea otter, Dall's porpoise, harbor porpoise, Steller sea lion, domestic dog, Sitka black-tailed deer, beaver, black bear, Keen's mouse, land otter, mink, and marten—are not specific seasonal indicators. Although their relative abundance vis-à-vis salmon may be low, Pacific cod, rockfish, and sculpin are present in significant numbers at the Coffman Cove Site. Were these fish taken during the late winter–early spring season, as argued for the Ferry Terminal site? If so, the Coffman Cove Site was minimally occupied during summer, fall, and late winter–early spring, suggesting nearly year-round occupation.

The crux of these interpretations is the evidence for when Pacific cod was taken by the residents of both sites. As presented in Chapter 10 (this volume), Henry Katasse (Newton and Moss 1984, 2005), Oberg (1973), and de Laguna (1972) indicate that Pacific cod were obtained during March and April, when they were eaten fresh and/or smoked for future use. But were Pacific cod also taken during other seasons, including the summer when Cobb (1916) indicated they were found on the banks in Chatham and Icy straits and Lynn Canal? Even though Pacific cod move to deeper water in the fall in southeast Alaska, can fall and winter be ruled out as seasons of capture, even when we know Tlingit fishers were capable of taking deep-water fish such as halibut? Today, commercial fishers in southeast Alaska take Pacific cod in the winter using longlines or pots. Instead of relying on indirect evidence for the seasonality of Pacific cod fishing, ideally we would obtain direct evidence of season of capture from the fish bones themselves. Perhaps skeletal representation might be diagnostic of cod taken at different times of the year and processed for different purposes (see Partlow and Kopperl, this volume).

Looking at the cod bones from both assemblages, there is no apparent difference in skeletal representation. Of the 1,308 Pacific cod remains from 49-PET-556, 75 percent are vertebrae, 23 percent are cranial elements, and 2 percent are pectoral elements. Of the 528 Pacific cod remains from 49-PET-067, 79 percent are vertebrae, 19 percent are cranial elements, and 2 percent are pectoral elements. At this scale, the skeletal representation of Pacific cod at the two sites is quite similar.

Table 13.4. Length estimations of Pacific cod archaeological specimens from Coffman Cove

	N	Minimum length (cm)	Maximum length (cm)	Average length (cm)	Regression formula
49-PET-556					
Otoliths	30	33.4	63.6	52.0 ± 6.8	Harvey et al. 2000
Premaxillae	19	32.7	71.6	54.5 ± 10.8	Orchard 2003
49-PET-067					
Otoliths	2			43.8 ± 2.6	Harvey et al. 2000
Premaxillae	9	44.3	77.8	62.4 ± 11.5	Orchard 2003

One element that differs in representation is the otolith. While the Ferry Terminal site contained eighty Pacific cod otoliths/otolith fragments, 49-PET-067 yielded only four otoliths. Of the otoliths from the Ferry Terminal Site, thirty were measurable. Using the regression formula for Pacific cod from Harvey et al. (2000), fork lengths for the Ferry Terminal fish range from 33.4 to 63.6 cm (Table 13.4). Based on size, at least twenty-eight MNI are represented by these otoliths. The average length is 52.0 cm, with a standard deviation of 6.8 cm. For the two measurable otoliths from the Coffman Cove Site, the average length is 43.8 cm ± 2.6. Yet the number of otoliths from 49-PET-067 is too small to make much of this difference.

The sample of premaxillae from the Coffman Cove site is larger than that of otoliths. Using the regression formula (for the maximum height of the ascending process) developed by Orchard (2003), fork lengths from left premaxillae from the Ferry Terminal Site range from 32.7 to 71.6 cm (n = 19). The average length is 54.5 cm ± 10.8. This length estimate is larger, with a greater standard deviation, than that resulting from reconstructing body size (fork lengths) from otoliths. Since a larger sample of premaxillae than otoliths is available from 49-PET-067, this measure is useful for comparison between sites. From the 49-PET-067 assemblage, Pacific cod fork lengths range from 44.3 to 77.8 cm, with a mean of 62.4 cm ± 11.5 (n = 9). Using the two-sample t test, $t = 1.77$; the probability that the difference between the two samples is due to vagaries of sampling is less than 10 percent and greater than 5 percent. The Pacific cod taken from 49-PET-067 appear to have been somewhat larger than those from the Ferry Terminal Site.[*]

Cod size depends on age, sex, maturity, and environmental factors contributing to growth rate. Off the west coast of Canada, Pacific cod generally comprise only a few age classes; 84 to 95 percent of the cod are ages two to four years (Westrheim 1996:xiv). Assuming data from Hecate Strait cod populations as shown in Table 13.5 apply, and using the sizes reconstructed from premaxillae, it would appear that most of the cod caught by the Ferry Terminal Site residents were two to four years old, whereas those caught by the Coffman Cove Site residents ranged in age from two to six years old.[†] The absence of juvenile cod (less than one year old) from both assemblages might be taken to indicate that fishing at both sites did not occur during the summer. Yet juvenile Pacific cod bones are not routinely identified as such in archaeological sites on the Pacific coast, and this may be a result of low bone density and loss due to attrition or of recovery biases such as too coarse a screen size. Given these complications and in the absence of additional ecological data about Pacific cod in Clarence Strait, I cannot use these size data to isolate the season during which the fish were obtained—whether they were caught during spawning or the feeding season. Another problem is that the samples used to reconstruct fork length from both sites are fairly small.

---

[*] Since the relationship between Pacific cod length and weight varies both seasonally and geographically (Westrheim 1996:77), I will not generate weight estimates for the fish represented archaeologically.

[†] The data on cod size are more complicated than this. For example, for Hecate Strait, length frequencies of trawl-caught cod are multimodal. In the studies Westrheim (1996:43) synthesized, principal modes were at 47 cm and 56–62 cm during the January–March period. During the April–June period, principal modes were at 50–53 cm and 59–65 cm. Female cod are larger than males; in British Columbia waters, the length at which 50 percent of the cod are mature (L/50) is 41–53 cm for males and 47–56 cm for females. In places where cod grow to larger size, for example in Kamchatka, L/50 is 69 cm for both sexes (Westrheim 1996:78).

Table 13.5. Mean length at age for Pacific cod in northern Hecate Strait

Age	Length (cm)
1	27.7
2	42.8
3	54.6
4	63.7
5	70.9
6	76.5
7	80.9
8	84.4
9	87.0
10	89.1

Source: Data from Westrheim 1996:121.

It is possible to reconstruct the age distribution and season of capture of archaeological fish remains when baseline information on annuli formation in otoliths or vertebrae has been derived from monthly captures (e.g., Van Neer, Lõugas, and Rijnsdorp 1999). Such data do not currently exist, however, for Pacific cod. In addition, the edges of otoliths and vertebrae must be well preserved to ensure that the most recent episode of annulus formation is read. Unfortunately, the condition of the fish bones in the Coffman Cove sites would not be adequate for this kind of analysis. In the future, this type of study might allow comparison of the growth rates obtained from archaeological specimens to that from modern otoliths or vertebrae, to see how Pacific cod have adapted to modern conditions of industrialized fishing, climate change, etc. Yet the likelihood of being able to identify season of capture seems low when fisheries biologists have found that by ageing recaptured Pacific cod, their estimates are only "roughly correct" within one or two years (Roberson et al. 2005:158; West, Wischniowski, and Johnston, this volume; see also Andrews, Gobalet, and Jones 2003).

To sum up this section, direct evidence for the seasonality of Pacific cod fishing at either of the Coffman Cove sites is lacking. The inference that the Ferry Terminal Site was occupied during late winter–early spring primarily for cod fishing and processing remains viable, as does the inference that the Coffman Cove Site was occupied during the summer and fall, but probably also during other seasons. Site 49-PET-067 may have been a year-round settlement, but in the absence of architectural data, we cannot determine when such sedentary settlement may have begun. People occupying both Coffman Cove sites may have been members of a single social group, living in separate, possibly related, households. If so, perhaps activity at the Ferry Terminal Site intensified during the Pacific cod-fishing season, when some number of 49-PET-067 residents moved to 49-PET-556. The shorter occupation span at 49-PET-556 may be culturally meaningful—on the other hand, only a small portion of the Ferry Terminal Site was investigated, and portions of the site were destroyed during ferry terminal construction.

Resource Ownership

A final explanation to consider here involves the dynamics of social relationships and claims to territory and resources. One possibility is that during the 3000–2000 cal BP time period, the Coffman Cove area was occupied by two different social groups. Perhaps those occupying the Coffman Cove Site had lived in the area for a long time and enjoyed exclusive rights to the salmon fishery at Coffman Creek and nearby streams. Those who lived at the Ferry Terminal Site may have been relative newcomers to the area and/or seasonal visitors who took advantage of locally available Pacific cod.

The Coffman Cove Site has occupations dating to more than 5,000 years ago, currently the oldest in the Coffman Cove area. Admittedly, a comprehensive survey and dating program has never been conducted in this portion of Prince of Wales Island. Site 49-PET-067 is located closer to the mouth of Coffman Creek than is the Ferry Terminal Site. Its relative proximity to the salmon stream may have been one of the criteria contributing to its choice as the original settlement location at Coffman Cove. For Component I, the presence of harbor seal pups along with salmon supports summer occupation. In addition to the fish, the Component I faunal assemblage includes a large number of birds and mammals (Table 13.3), and though it lacks some of the taxa present in the more recent components (a few bird species, the porpoises, mink, marten), it does not lack many, especially considering its sample size. This might be interpreted as evidence for a group with wide-ranging knowledge of their environs and year-round settlement. The predominance of salmon in the 49-PET-067 assemblage is suggestive of a group with ownership rights to the local salmon stream over a long period of time. This contrasts markedly with the Ferry Terminal assemblage, with its low numbers of salmon vertebrae. Perhaps the residents of the Ferry Terminal site were seasonal visitors who came to the area to fish for Pacific cod at various times between 3000 and 2000 cal BP. Alternatively, they may have been relative newcomers trying to establish themselves at Coffman Cove, relying on local residents for winter provisions of dried salmon. These interpretations must remain tentative in the absence of direct evidence for the seasonality of Pacific cod fishing.

Conclusion

Although the explanation for the difference between the faunal assemblages from the Coffman Cove sites is not definitive, I have used a common approach, known as "inference to the best explanation" (Fogelin 2007; Kelley and Hanen 1988:360–368). Frequently, faunal analysts aim to examine change through time, but this is not always feasible with Northwest Coast shell middens where temporal resolution is poor. Both sites 49-PET-556 and 49-PET-067 are cumulative palimpsests and neither reveals significant change through time in the relative abundance of the major taxa. The two occupations overlap in time and are located only about 600 m apart from each other. While the Coffman Cove Ferry Terminal fish assemblage shows a focus on Pacific cod, rockfish and sculpin are also relatively abundant. Meanwhile, the Coffman Cove Site shows a heavy reliance on salmon. One clear result of this effort is that studying either of the Coffman Cove sites in isolation of the other would give a false impression of the resources used by the ancient residents of the area. This emphasizes how little we understand about long-term use of fisheries even within a small geographic area. In neither site do we see evidence of changing economic focus through time, although archaeologists seek out such evidence to support claims for technological or cultural evolutionary change, changes in diet breadth, or resource depression. In the case of the Coffman Cove sites, the best explanations probably incorporate differences in the seasonality of site occupations, coupled with differential access to resources. As described above, direct evidence for the seasonality of Pacific cod use has not been marshaled, and this is a major weakness of this study. Nevertheless, the Coffman Cove investigations tentatively suggest that ownership of resource territories such as salmon streams is a very ancient practice. Although more data on the ecology and spatial distribution of Pacific cod in southeast Alaska are desperately needed (see Chapter 10), clearly this species was of economic significance to the Tlingit and their ancestors. Resource territory ownership and stewardship may go a long way toward explaining why there is very little evidence for evolutionary change in subsistence or resource depression on the Northwest Coast over the last 5,000 years (Butler and Campbell 2004).

Acknowledgments

The 2006 investigations at 49-PET-067 were conducted under contract to Northern Land Use Research, Inc., with federal funds administered by the USDA Forest Service. I am grateful to Terry Fifield for all manner of support over the last decade to make data recovery from the site a reality. I also thank my co-principal investigators, Pete Bowers and Doug Reger, for their work on the project. Justin Hays and Catherine Williams have also been great colleagues, and I appreciate their many contributions. Data recovery at 49-PET-556 was conducted by NLUR with funds from the Alaska Department of Transportation and Public Facilities. Many fieldworkers and volunteers took part in these projects, and I thank them all. University of Oregon students who assisted in the lab include Eric Greenwood, Marissa Guenther, Carley Smith, Brendan Culleton, Erin Leonard, Nicki Dwyer, and Susan Morasci. Deidra Holum, Scott Forbes, and Grant Thompson generously shared their insights and experience. I am grateful to the City of Coffman Cove, Wrangell Cooperative Association, and the Sealaska Heritage Institute. Ross Smith provided a copy of his important master's thesis. Aubrey Cannon co-organized the 2008 SAA symposium where a short version of this analysis was presented, and Virginia Butler served as our superb discussant. I also thank Aubrey for his comments on the text, which greatly improved this chapter. Finally, I appreciate the continuing support of Jon and Erik Erlandson.

References Cited

Andrews, Allen H., Kenneth W. Gobalet, and Terry L. Jones. "Reliability Assessment of Season-of-Capture Determination from Archaeological Otoliths." *Bulletin, Southern California Academy of Sciences* 102 (2003): 66–78.

Bailey, Geoff. "Time Perspectives, Palimpsests and the Archaeology of Time." *Journal of Anthropological Archaeology* 26 (2007): 198–223.

Bowers, Peter M., and Madonna L. Moss. "The North Point Wet Site and the Subsistence Importance of Pacific Cod on the Northern Northwest Coast." In *People and Wildlife in Northern North America: Essays in Honor of R. Dale Guthrie*, S. Craig Gerlach and Maribeth S. Murray, eds., pp. 159–177. BAR British Archaeological Report International Series 944. Oxford: Archaeopress, 2001.

Butler, Virginia L., and Sarah K. Campbell. "Resource Intensification and Resource Depression in the Pacific Northwest of North America." *Journal of World Prehistory* 18 (2004): 327–405.

Butler, Virginia L., and James C. Chatters. "The Role of Bone Density in Structuring Prehistoric Salmon Bone Assemblages." *Journal of Archaeological Science* 21 (1994): 413–424.

Clark, Gerald H. Archaeological Testing at the Coffman Cove Site, Southeast Alaska. Paper presented at the 32nd Annual Northwest Anthropological Conference, Eugene, OR, 1979.

———. The Coffman Cove Site (49 PET 067): History of Investigations, Condition, and Significance. MS on file, Tongass National Forest, Juneau, 1981.

Cobb, John N. *Pacific Cod Fisheries*. Appendix IV to the Report of the US Commissioner of Fisheries for 1915. US Bureau of Fisheries Document 830, 1916.

Drennan, Robert D. *Statistics for Archaeologists: A Commonsense Approach*. New York: Plenum, 1996.

Fogelin, Lars. "Inference to the Best Explanation: A Common and Effective Form of Archaeological Reasoning." *American Antiquity* 72 (2007): 603–625.

Goldschmidt, Walter R., and Theodore H. Haas. *Haa Aani, Our Land: Tlingit and Haida Land Rights and Use*. Thomas F. Thornton, ed. Seattle: University of Washington Press; Juneau: Sealaska Heritage Foundation, 1998.

Harvey, J. T., T. R. Loughlin, M. A. Perez, and D. S. Oxman. Relationship between Fish Size and Otolith Length for 63 Species of Fishes from the Eastern North Pacific Ocean. US Department of Commerce NOAA Technical Report NMFS 150, 2000.

Kelley, Jane H., and Marsha P. Hanen. *Archaeology and the Methodology of Science*. Albuquerque: University of New Mexico Press, 1988.

Krebs, Charles J. *Ecological Methodology*, 2nd edition. Menlo Park, CA: Addison, Wesley, Longman, 1999.

de Laguna, Frederica. *Under Mount Saint Elias: The History and Culture of the Yakutat Tlingit*. Smithsonian Contributions to Anthropology, vol. 7. Washington, DC: Smithsonian Institution, 1972.

Moss, Madonna L. Coffman Cove Ferry Terminal Site (49-PET-556): Shell Midden Compositional Analyses and Analysis of the Vertebrate and Invertebrate Remains. Submitted to Northern Land Use Research, Fairbanks, AK, December, 2007.

———. Coffman Cove Village (49-PET-067): Analysis of the Vertebrate and Invertebrate Remains. Submitted to Northern Land Use Research, Fairbanks, AK, March, 2008.

Newton, Richard G., and Madonna L. Moss. *The Subsistence Lifeway of the Tlingit People: Excerpts of Oral Interviews*. Juneau: USDA Forest Service, Alaska Region Report No. 179, 1984.

———. *Haa Atxaayi Haa Kusteeyix Sitee, Our Food Is Our Tlingit Way of Life: Excerpts of Oral Interviews*. Juneau: USDA Forest Service, Alaska Region, R10-MR-30, 2005.

Oberg, Kalervo. *The Social Economy of the Tlingit Indians*. Seattle: University of Washington Press, 1973.

Orchard, Trevor. *An Application of the Linear Regression Technique for Determining Length and Weight of Six Fish Taxa: The Role of Selected Fish Species in Aleut Paleodiet*. BAR British Archaeological Reports, International Series 1172. Oxford: Archaeopress, 2003.

Reger, Douglas R. 1993 Investigations at the Coffman Cove Archaeological Site, PET-067: A Preliminary Review. Office of History and Archaeology Report Number 53. Anchorage: Alaska Division of Parks and Outdoor Recreation, Department of Natural Resources, 1995.

Reger, Douglas R., Madonna L. Moss, Peter M. Bowers, and Justin M. Hays. Recovery of Archaeological Data from the Ferry Terminal Site (PET-556), Coffman Cove, Alaska. Report prepared for Alaska Department of Transportation and Public Facilities, Southeast Region, Juneau, AK. Fairbanks: Northern Land Use Research, 2007.

Roberson, Nancy E., Daniel K. Kimura, Donald R. Gunderson, and Allen M. Shimada. "Indirect Validation of the Age-reading Method for Pacific Cod Using Otoliths from Marked and Recaptured Fish." *Fishery Bulletin* 103 (2005): 153–160.

Rushmore, Paul, Susan Colby, and James Bell. Final Report, Results of Archaeological Excavation and Monitoring at Site PET-067 "Parcel A," Coffman Cove, Alaska. MS on file, Alaska Office of History and Archaeology, Anchorage, 1998.

Smith, Ross E. "Structural Bone Density of Pacific Cod (*Gadus macrocephalus*) and Halibut (*Hippoglossus stenolepis*): Taphonomic and Archaeological Implications." Master's thesis, Portland State University, 2008.

Stein, Julie K., Jennie N. Deo, and Laura S. Phillips. "Big Sites—Short Time: Accumulation Rates in Archaeological Sites." *Journal of Archaeological Science* 30 (2003): 297–316.

Thompson, Grant G., Martin W. Dorn, and Daniel G. Nichol. "Assessment of the Pacific Cod Stock in the Gulf of Alaska." In *Stock Assessment and Fishery Evaluation Report for the Groundfish Resources of the Gulf of Alaska*. US Department of Commerce, National Oceanic and Atmospheric Administration National Marine Fisheries Service, and Alaska Fisheries Science Center, Seattle, 2006.

Van Neer, Wim, Lembi Lõugas, and Adriaan D. Rijnsdorp. "Reconstructing Age Distribution, Season of Capture and Growth Rate of Fish from Archaeological Sites based on Otoliths and Vertebrae." *International Journal of Osteoarchaeology* 9 (1999): 116–130.

Westrheim, S. J. On the Pacific Cod (*Gadus macrocephalus*) in British Columbia Waters, and a Comparison with Elsewhere, and Atlantic Cod (*G. morhua*). Canadian Technical Report Fisheries and Aquatic Sciences 2092, 1996.

Fish Traps and Shell Middens at Comox Harbour, British Columbia

Megan Caldwell, *Department of Anthropology, University of Alberta*

Intensive harvesting of marine resources on the Northwest Coast of North America has a long and important history prior to European contact. One means of mass harvesting aquatic resources is through use of large fishing structures, such as traps and weirs. Recent research into the presence, use, and antiquity of fishing structures has revealed not only that archaeological remains of these technologies are widespread along the Northwest Coast from Oregon to Alaska but also that they vary greatly in material, structure, and function (Betts 1998; Byram 1998, 2002; Chaney 1998; Eldridge and Acheson 1992; Greene 2005a, 2010; Losey 2008; Mobley and McCallum 2001; Moss, Erlandson, and Stuckenrath 1990; Moss and Erlandson 1998; Prince 2005; Tveskov and Erlandson 2003; White 2006, this volume). Fishing structures were constructed in various ways and included stone, wood, and basketry elements. Structures were built in varying forms, from straight alignments of wood stakes across rivers, to complex, multicomponent intertidal traps (Moss and Erlandson 1998:180), and they functioned to target one or many different species of fish, as well as to attract mammals and birds (Monks 1987). Only now is the importance and variability of Northwest Coast traps and weirs being appreciated. The research presented here is a summary of my master's research, which examined the use of a large complex of wooden stake fish traps in Comox Harbour, British Columbia, by looking at fish remains recovered from the Q'umu?xs Village site on the northern shore of the harbor (Caldwell 2008).

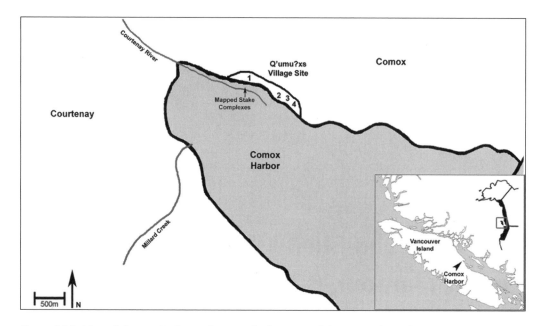

Figure 14.1. Map of Comox Harbour, showing the location of the Q'umuʔxs Village site and Areas 1–4 discussed in text.

The Archaeology of Comox Harbour

Comox Harbour is a large tidal bay situated near the center of the east side of Vancouver Island (Figure 14.1). The Tsolum and Puntledge rivers join to form the Courtenay River, which flows into the north end of the harbor (Clague 1976:2). Within the harbor are the remains of more than 200 wooden stake fish traps, with between 100,000 to 200,000 individual wooden stakes (Greene 2005a, 2010). This complex of wooden stakes represents over 1,200 years of human use of the intertidal zone to harvest fish (Cullon 2007:27–28; Greene 2005a, 2005b). Greene (2005a) has mapped eleven concentrations of wooden stakes and identified at least two structural types of traps. Both types consist of lead lines connected to holding pens. Additionally, Greene (2005b) has obtained twelve dates on wooden stakes from these eleven concentrations. Further mapping of fish traps was conducted in 2006 by the Hamatla Treaty Society, and an additional ten dates were obtained on stake complexes (Cullon 2007). Dates from the Comox Harbour fish traps range between 120 ± 40 BP and 1230 ± 60 BP (Cullon 2007; Greene 2005b).

The oldest traps in Comox Harbour have heart-shaped holding pens (Figure 14.2a), while more recent traps have holding pens with chevron-shaped wings (Figure 14.2b) (Greene 2005a). The Hamatla Treaty Society has found other types of trap structures along the shore of Courtenay River where it enters the harbor (Cullon 2007). While mapping two stake complexes along the Courtenay River, I found a heart-shaped trap similar to, but smaller than, those on the tidal flats on the south side of the harbor (Figure 14.2c). I also found a series of stake lines running perpendicular to the river (Figure 14.2d), which may be remnants of chevron-winged traps or possibly weir-type structures crossing the river channel. Unfortunately, recent dredging of the river channel has removed portions

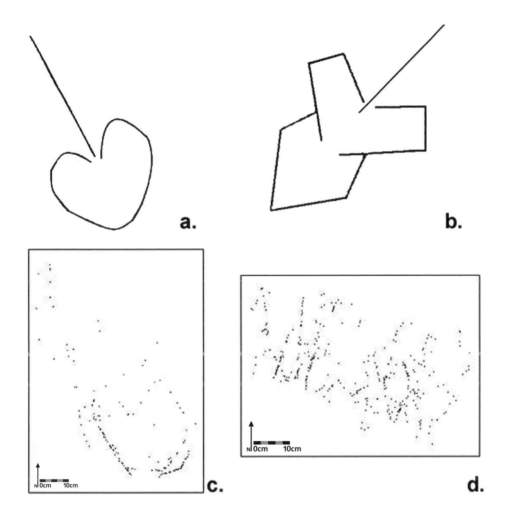

Figure 14.2. Fish traps in Comox Harbour: (a) heart-shaped fish trap, (b) chevron-winged fish trap (both after Greene 2005a); (c) heart-shaped fish trap along the north shore of Courtenay River; (d) stake complex along the north shore of Courtenay River with lines running perpendicular to the river.

of these complexes, and though more mapping is needed, it may be impossible to tell what structure(s) this wooden stake complex once formed.

Although little archaeological work has been done in the area, sites located near Comox Harbour indicate long use of the area by First Nations prior to contact. Previous work indicates human occupation from at least the Locarno Beach period (starting at ~3500 BP) at both the Millard Creek and Mission Hill sites located adjacent to the harbor (Capes 1964:18, 1977:82). A shell midden at Saltery Bay, across the Strait of Georgia from Comox Harbour, dates to 6700 ± 40 BP, indicating early occupation of the area (Pegg, Mason, and Wada 2007:43). Other fish traps and clam gardens are known throughout the area (Cullon 2007; Williams 2006).

Q'umuʔxs Village Site

This research examines the use of fish at Q'umuʔxs Village (DkSf-19), a large ethnographic winter village site at Comox Harbour. The site stretches approximately 1,500 m along the north shore of the harbor, adjacent to the Courtenay River. Comox Harbour is located at the boundary between ethnographically recorded Northern Coast Salish and Southern Kwakwaka'wakw territories. Prior to the seventeenth century, Northern Coast Salish groups occupied northeastern Vancouver Island, and the Comox Harbour area was inhabited by the Pentlatch, another Northern Coast Salish group (Kennedy and Bouchard 1990:442). Both European contact and warfare with Nuu-chah-nulth groups from the west coast of Vancouver Island led to loss of Pentlatch lives (Kennedy and Bouchard 1990:443). By 1886, only one Pentlatch family remained at Q'umuʔxs (Rohner 1969:59). The K'omoks were displaced southwards due to conflict with Southern Kwakwaka'wakw groups, eventually settling in Comox Harbour and intermarrying with the Kwakwala speakers to the north who had displaced them (Kennedy and Bouchard 1990:441; McMillan 2003:257).

Previous to my work, the age of Q'umuʔxs was unknown. Sandstone abraders and a quartz-crystal microblade, found at the site during earlier mitigation projects, led Lindberg (2000:13) to suggest that the site dated to the Locarno period (3500–2400 BP) onward. The oldest date I obtained confirms occupation from at least the Marpole period (2400–1500 BP) at 2370 ± 60 BP. Aboriginal occupation of Q'umuʔxs continued during European contact and into the present; part of the site sits within the bounds of the present-day K'omoks Reserve. Together with other sites in the Comox Harbour area, the antiquity of sites and ongoing habitation at Q'umuʔxs suggests continual use of Comox Harbour and surrounding areas.

I propose that Comox Harbour supported a long-term herring fishery prior to European contact and likely thereafter. Herring played an important part in the subsistence economy at Q'umuʔxs throughout its occupation, and it is likely that the introduction and expansion of wooden stake fish traps in Comox Harbour targeted the herring fishery.

Methods

I recovered fish remains from three arbitrarily selected areas of Q'umuʔxs using a bucket auger. Additional auger samples from a fourth area of the site excavated by Lindberg (2000) were obtained from the Courtenay District Museum and Archives. Site areas were chosen based on proximity to mapped traps and accessibility. Deposits in Area 1 and upper deposits in Area 2 have been disturbed by historic activity, while lower deposits in Area 2 and all deposits in Area 3 and Area 4 were undisturbed (Figure 14.1).

Cannon (2000) has shown how bucket auger sampling can be used to measure the density or abundance of fish remains, especially of herring and salmon. Bucket auger sampling is an ideal way of examining the main fish taxa targeted by the use of fish traps in Comox Harbour as it allows for collection of representative samples of fish remains at multiple locations across a site without accumulating large volumes of material to be processed. While the importance of salmon in Northwest Coast economies is often cited, recognition of the importance of herring as a subsistence resource has been hampered by the use of quarter-inch and even eighth-inch screens, which are often too large to recover elements from herring and other small fish species. The use of bucket augers to obtain samples that are later sieved through smaller mesh in controlled laboratory settings enables the recovery of relatively small-bodied taxa, such as herring, which would otherwise go unobserved.

Results

A total of seventeen auger samples and one column sample were analyzed. I analyzed materials larger than 2.36 mm from all eighteen samples and materials larger than 1.5 mm from one sample from each of the four site areas. Three auger samples are considered from Area 1, four from Area 2, and five each from Area 3 and Area 4; the column sample comes from Area 2. Five radiocarbon dates were obtained from Q'umu?xs. Three dates were obtained for Area 2 and two dates were obtained for Area 3 (Table 14.1). These dates place occupation of Q'umu?xs minimally between 2370 ± 60 BP and 1010 ± 50 BP, with the most recent date overlapping the known period of use of fish traps in Comox Harbour (Figure 14.3).

I recovered a total of 41,764 fish bones from these samples, and identified 34,577 elements; Table 14.2 shows the distribution of remains by taxon, site area, and mesh size. Pacific herring is the most abundant fish taxon recovered at Q'umu?xs, dominating the assemblage at 96.6 percent of all identified fish remains. Herring is present at all depths of deposits in all four areas of Q'umu?xs, and represents between 85.9 percent and 98 percent of identified fish remains in each area. The overall abundance of herring indicates that Comox Harbour sustained an important herring fishery for a continuous period of time from at least the Marpole period, and likely into the postcontact period. In contrast, salmon is represented in only small numbers in all four areas of Q'umu?xs at 2 percent of all identified fish remains. Six other fish taxa were recovered from Q'umu?xs: flatfish (Pleuronectiformes), greenling (*Hexagrammos* spp.), plainfin midshipman (*Porichthys notatus*), rockfish (*Sebastes* spp.), spiny dogfish shark (*Squalus acanthias*), and surfperch

Table 14.1. Radiocarbon dates from the Q'umu?xs Village site

Sample	Location	Depth below surface (cm)	Material	Measured radiocarbon age (BP)	¹³C/¹²C ratio (o/oo)	Conventional radiocarbon age (BP)	2 Sigma calibration (cal. BP)
Beta 240174	Column Sample	20–40	Shell	1810 ± 70	−1.9	2190 ± 70	1500–1240 Midpoint = 1370
Beta 240175	Column Sample	40–60	Shell	1790 ± 70	−0.5	2190 ± 70	1500–1240 Midpoint = 1370
Beta 240176	Column Sample	80–90	Shell	1960 ± 50	−0.4	2370 ± 60	1700–1370 Midpoint = 1535
Beta 241997	Auger 40	16–27	Shell	610 ± 50	−0.8	1010 ± 50	430–110 Midpoint = 270
Beta 241998	Auger 40	37–47	Shell	1440 ± 50	−1.1	1840 ± 60	1180–870 Midpoint = 1025

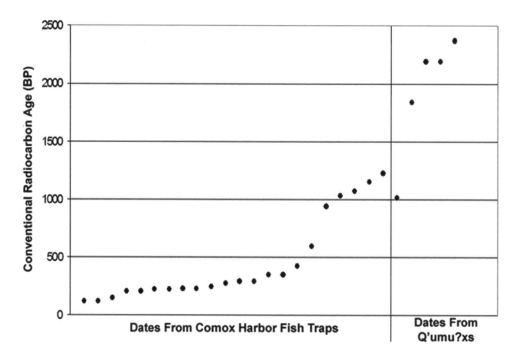

Figure 14.3. Radiocarbon dates from fish traps in Comox Harbour and Q'umuʔxs Village (trap dates obtained by Hamatla Treaty Society [Beta-221751-221760] and Greene 2005b [Beta-191580-191590; Beta-193293]).

Table 14.2. Frequency of fish remains at the Q'umuʔxs Village site

Area	Mesh size (mm)	Herring	%NISP	Salmon	%NISP	Other*	%NISP	NISP	%NSP	NSP
Area 1	2.36	1,069	85.5	136	10.9	46	3.7	1,251	80.5	1,554
	1.5	73	93.6	5	6.4	0	0.0	78	58.6	133
	Total	1,142	85.9	141	10.6	46	3.5	1,329	78.8	1,687
Area 2	2.36	3,312	95.9	57	1.7	83	2.4	3,452	83.1	4,155
	1.5	195	97.0	6	3.0	0	0.0	201	69.3	290
	Total	3,507	96.0	63	1.7	83	2.3	3,653	82.2	4,445
Area 3	2.36	23,523	97.1	411	1.7	290	1.2	24,224	84.8	28,582
	1.5	1,368	96.4	43	3.0	9	0.6	1,419	80.2	1,770
	Total	24,891	97.1	454	1.8	299	1.2	25,643	84.5	30,352
Area 4	2.36	3,556	98.1	30	0.8	40	1.1	3,626	79.4	4,564
	1.5	317	97.2	7	2.1	2	0.6	326	45.5	716
	Total	3,873	98.0	37	0.9	42	1.1	3,952	74.8	5,280
Site Total		33,413	96.6	695	2.0	470	1.4	34,577	82.8	41,764

* This category consists of six additional taxa that were recovered in low frequencies at the Q'umuʔxs Village site: flatfish (Pleuronecti-formes); greenling (*Hexagrammos* spp.); plainfin midshipman (*Porichthys notatus*); rockfish (*Sebastes* spp.); spiny dogfish shark (*Squalus acanthias*); surfperch (Embiotocidae).

Table 14.3. Density of fish remains from the Q'umuʔxs Village site

Area	Volume (l)[a]	Herring/l	Salmon/l	Other/l[b]	NISP/l	NSP/l
1	11.25	102	13	4	118	150
2	38.98	90	2	2	94	114
3	33.26	748	14	9	771	913
4	6.98	555	5	6	566	756
Total	90.47	369	8	5	382	462

a. Volume in liters of sediments smaller than 2.36 mm.
b. This category consists of six additional taxa that were recovered in low frequencies at the Q'umuʔxs Village site: flatfish (Pleuronecti-formes); greenling (*Hexagrammos* spp.); plainfin midshipman (*Porichthys notatus*); Rockfish (*Sebastes* spp.); spiny dogfish shark (*Squalus acanthias*); surfperch (Embiotocidae).

(Embiotocidae). These taxa are all found in relatively low numbers and represent 1.4 percent of identified fish remains. These fish may have been bycatch from use of the traps, or caught by other means including spear, line, or net. The contributions of these six taxa to the diet would have been minimal.

Table 14.3 shows the densities of herring, salmon, and overall fish remains by site area. Overall densities and herring densities are higher in Areas 3 and 4 than in Areas 1 and 2. Salmon densities are low across Q'umuʔxs, as are the densities of other taxa. Soil pH analysis performed on samples from all four areas of the site shows that soil chemistry would not have affected bone preservation at Q'umuʔxs, as all soil pH levels are neutral, ranging between 7.07 and 7.86.

Of particular interest is the pattern seen in herring remains recovered from Area 3. A lens of highly fragmented California mussel and sea urchin remains dominates the upper levels of the deposit in this area. The highest densities of herring remains are recovered from this lens. As Table 14.4 shows, significant increases in the density of herring in all five samples from this area occur between 20 and 56.5 cm below surface. Two dates from this area were obtained from Auger 40. The older date of 1840 ± 60 BP comes from 37–47 cm below the surface, before herring density increases in this sample, and the younger date of 1010 ± 50 BP comes from 16–27 cm below the surface, after herring density increases in this sample. These dates indicate that the increase in herring density occurs after fish traps appear in Comox Harbour. Although this pattern is not replicated in all areas of the site, the timing of the increase in herring densities at Area 3 suggests the possibility that the fish traps in Comox Harbour were constructed as a means of intensifying an already important herring fishery.

Discussion

In order to understand the relationship between the fish traps in Comox Harbour and the fish remains from Q'umuʔxs, the timing of fish trap use in relation to fish remains needs to be understood. As previously mentioned, twenty-two radiocarbon dates have been obtained from fish traps located in Comox Harbour (Cullon 2007; Greene 2005a). These dates range

Table 14.4. Increases in herring density in upper levels of Area 3

Auger sample	Depth below surface (cm)	Herring NISP/l
34	13–20	1,885
	20–34	233
35	22.5–35	2,040
	35–43	458
36	49–56.5	1,891
	56.5–68	342
39	39–41	1,213
	41–49.5	149
40	27–37	1,435
	37–47	297

between 120 ± 40 BP and 1230 ± 60 BP, with a gap between approximately 600 ± 40 BP and 940 ± 60 BP. When associated with the dates from Q'umu?xs, it appears that these dates may indicate some association between fish resource use in Area 3 and trap utilization.

Herring spawned in large numbers in Comox Harbour in late winter, until the run disappeared sometime in the twentieth century (E. Hardy, 2007, pers. comm.). Lindberg (2000:1) reported the presence of kelp and eelgrass on the mud flats in front of Area 4, which would be attractive to spawning herring. Herring fishing is well documented in many Northwest Coast ethnographies, with herring fishing recorded as occurring during mid- to late March as the herring gathered in large numbers to spawn. On the northern North-west Coast, herring is also exploited in the fall as a source of oil (Emmons 1991:118; Moss, Butler, and Elder, this volume). Comox Harbour is mentioned specifically as a destination for herring fishing in late winter by Northern Coast Salish and southern Kwakwaka'wakw peoples during the ethnographic period (Assu 1989:12; Curtis 1974[1915]:19). The abundance of herring remains recovered at Q'umu?xs corroborates the likelihood that herring was an important resource in Comox Harbour prior to European contact.

As herring gather to spawn, they are relatively easy to capture en masse using dip nets and herring rakes (Curtis 1974[1915]:19). How can the abundance of herring remains seen at Q'umu?xs be linked to use of fish traps in Comox Harbour if an already efficient means of procurement was practiced? Perhaps instead of replacing existing technologies, the fish traps represent a means of increasing or intensifying the herring harvest. K'omoks Elders speak about the use of herring in the past, and suggest the possibility that traps in Comox Harbour may have been used to catch herring. Boas (1909:468) depicts a canoe at the convergence of two wooden stake weir lines used to catch eulachon, and it is possible that the traps in Comox Harbour were employed in a similar manner. In a large tidal bay like Comox Harbour, herring would move out of the bay as the tide recedes. Traps may have prevented herring from leaving the harbor with receding tides, making those herring available for gathering during low tides.

The low abundance of salmon is surprising based on the expectation that the traps in Comox Harbour were constructed to exploit the Courtenay River salmon run. This does not mean, however, that salmon was unimportant at Q'umu?xs. Information provided by K'omoks Elders indicates that salmon was exploited further upriver using different devices such as gaff hooks, and that it was an important part of the subsistence base. Elders also suggest that salmon caught with fish traps may have been transported back to the site already processed, without bones.

Based on the abundance of herring, the large complex of fish traps, and the relative absence of other species at Q'umu?xs, it appears that the fish traps within Comox Harbour were used to trap herring. Ethnographic evidence and information provided by K'omoks Elders also suggests that these traps may have been used to procure salmon despite the paucity of salmon remains at Q'umu?xs. Additional work at Q'umu?xs and the numerous other sites in the vicinity of Comox Harbour will be needed before we fully understand the use of the traps.

Conclusion

The fish assemblage at the Q'umu?xs Village site suggests a focus on herring exploitation throughout the occupation, while salmon are only present in low numbers contrary to expectations. The use of wooden stake fish traps begins around 1200 BP, and it appears that these traps were used to target herring. An increase in the density of herring remains from Area 3 suggests that these traps may have been used to intensify an already important herring fishery. While these results do not match the expectation of an intensive salmon fishery in Comox Harbour, they are not inconsistent with known archaeological and ethnographic practices on the Northwest Coast. As more research examines the age, variability, and associated village and processing sites of both stone (White, this volume) and wooden fish traps on the Northwest Coast, our expectations will be challenged as our understanding improves.

Acknowledgments

I thank the K'omoks First Nation for allowing me to conduct research within their territory, especially the Elders who graciously shared their time and knowledge with me. I am thankful for the assistance of the Hamatla Treaty Society, and for the use of their data. I thank my thesis committee, Drs. Fikret Berkes, S. Brooke Milne, and Greg Monks, for their valuable assistance in this research. Catherine Siba and the Courtenay District Museum and Archives allowed access to materials from DkSf-19 curated at that institution. Initial analysis was undertaken at the Department of Archaeology, Simon Fraser University, and I thank them for allowing me space in the lab and use of their equipment. Val McKinley of the University of Winnipeg provided soil pH analysis for this project. The Department of Anthropology at the University of Manitoba provided financial support. I would like to acknowledge the many individuals who assisted both in and outside of the field: Amanda Blackburn, Ashleigh Czyrnyj, Emily Holland, Charlie Johnson, Lisa Lefever, Alan Mitchell, Jr., Andrea Onodi, and Rachel Ten Bruggencate. Finally, I thank Drs. Aubrey Cannon and Madonna Moss for their feedback on an earlier draft of this chapter.

References Cited

Assu, H., with J. Inglis. *Assu of Cape Mudge: Recollections of a Coastal Indian Chief*. Vancouver: UBC Press, 1989.

Betts, Robert C. "The Montana Creek Fish Trap I: Archaeological Investigations in Southeast Alaska." In *Hidden Dimensions: The Cultural Significance of Wetland Archaeology*, Kathryn Bernick, ed., pp. 239–251. Vancouver: UBC Press, 1998.

Boas, Franz. *The Kwakiutl of Vancouver Island*. New York: G. E. Stechert, 1909.

Byram, R. Scott. "Fishing Weirs in Oregon Coast Estuaries." In *Hidden Dimensions: The Cultural Significance of Wetland Archaeology*, Kathryn Bernick, ed., pp. 199–219. Vancouver: UBC Press, 1998.

———. "Brush Fences and Basket Traps: The Archaeology and Ethnohistory of Tidewater Weir Fishing on the Oregon Coast." PhD diss., University of Oregon, 2002.

Caldwell, Megan. "A View from the Shore: Interpreting Fish Trap Use in Comox Harbour through Zooarchaeological Analysis of Fish Remains from the Q'umu?xs Village Site (DkSf-19), Comox Harbour, British Columbia." Master's thesis, University of Manitoba, 2008.

Caldwell, Megan E., Dana Lepofsky, Georgia Combes, Michelle Washington, John R. Welch, and John R. Harper. "A Bird's Eye View of Northern Coast Salish Intertidal Resource Management Features, Southern British Columbia, Canada." *Journal of Island and Coastal Archaeology*, forthcoming.

Cannon, Aubrey. "Assessing Variability in Northwest Coast Salmon and Herring Fisheries: Bucket-Auger Sampling of Shell Midden Sites on the Central Coast of British Columbia." *Journal of Archaeological Science* 27 (2000): 725–737.

Capes, Katherine H. *Contributions to the Prehistory of Vancouver Island*. Pocatello: Occasional Papers of the Idaho State University Museum No. 15, 1964.

———. "Archaeological Investigations of the Millard Creek Site, Vancouver Island, British Columbia." *Syesis* 10 (1977): 57–84.

Chaney, Greg. "The Montana Creek Fish Trap II: Stratigraphic Interpretation in the Context of Southeastern Alaska Geomorphology." In *Hidden Dimensions: The Cultural Significance of Wetland Archaeology*, Kathryn Bernick, ed., pp. 252–266. Vancouver: UBC Press, 1998.

Clague, J. J. *Sedimentology and Geochemistry of Marine Sediments near Comox, British Columbia*. Ottawa: Geological Survey of Canada Paper 76–21, 1976.

Cullon, Deidre. HTS Foreshore and Archaeological Management Training and Research Project: Final Report. MS on file, Hamatla Treaty Society, Campbell River, BC, 2007.

Curtis, E. S. *The North American Indian, Being a Series of Volumes Picturing and Describing the Indians of the United States, the Dominion of Canada and Alaska*. London: Johnson Reprint Company, 1974 (1915).

Eldridge, M., and S. Acheson. "The Antiquity of Fish Weirs on the Southern Coast: A Response to Moss, Erlandson and Stuckenrath." *Canadian Journal of Archaeology* 16 (1992): 112–116.

Emmons, George T. *The Tlingit Indians*. Frederica de Laguna, ed. Seattle: University of Washington Press; Vancouver: Douglas and McIntyre; New York: American Museum of Natural History, 1991.

Greene, Nancy A. A New Angle on Northwest Coast Fish Trap Technologies: GIS Total Station Mapping of Intertidal Wood-Stake Features at Comox Harbour, BC. http://www.canadian archaeology.com/comox/Comox_Harbour.html (accessed March 8, 2008), 2005a.

———. Resource Management Reports for DkSf-43 and DkSf-44. MS on file, Heritage Resource Centre, Ministry of Small Business, Tourism and Culture, Victoria, BC, 2005b.

———. "Comox Harbour Fish Trap Site. WARP (Wetland Archaeological Research Project) Web Report." http://newswarp.info/wp-content/uploads/2010/03/WARP-web-report.pdf, accessed July 26, 2010.

Kennedy, Dorothy I., and Randy T. Bouchard. "Northern Coast Salish." In *Northwest Coast,* Handbook of North American Indians, vol. 7, Wayne Suttles, ed., pp. 441–452. Washington, DC: Smithsonian Institution, 1990.

Lindberg, Jennifer J. Building a House at Qʷumu?Xʷs: Archaeological Site DkSf-019. MS on file, Heritage Resource Centre, Ministry of Small Business, Tourism and Culture, Victoria, BC, 2000.

Losey, Robert. "Late Holocene Fish Traps in the Transforming Landscapes of Willapa Bay, Washington." Paper presented at the 73rd annual meeting of the Society for American Archaeology, March 2008, Vancouver, BC.

McMillan, Alan D. "Reviewing the Wakashan Migration." In *Emerging from the Mist: Studies in Northwest Coast Culture History*, R. G. Matson, Gary Coupland, and Quentin Mackie, eds., pp. 244–259. Vancouver: UBC Press, 2003.

Mobley, Charles M., and W. Mark McCallum. "Prehistoric Intertidal Fish Traps from Central Southeast Alaska." *Canadian Journal of Archaeology* 25 (2001): 28–52.

Monks, Gregory G. "Prey as Bait: The Deep Bay Example." *Canadian Journal of Archaeology* 11 (1987): 119–142.

Moss, Madonna L., and Jon M. Erlandson. "A Comparative Chronology of Northwest Coast Fishing Features." In *Hidden Dimensions, the Cultural Significance of Wetland Archaeology*, K. Bernick, ed., pp. 180–198. Vancouver: UBC Press, 1998.

Moss, Madonna L., Jon. M. Erlandson, and Robert Stuckenrath. "Wood Stake Weirs and Salmon Fishing on the Northwest Coast: Evidence from Southeast Alaska." *Canadian Journal of Archaeology* 14 (1990): 143–158.

Pegg, Brian, Andrew Mason, and Gail Wada. Final Report on Archaeological Mitigation of DkSb-30, Saltery Bay, B.C., Telus North Island Ring Project. MS on file, Heritage Resource Centre, Ministry of Small Business, Tourism and Culture, Victoria, BC, 2007.

Prince, Paul. "Fish Weirs, Salmon Productivity, and Village Settlement in an Upper Skeena River Tributary, British Columbia." *Canadian Journal of Archaeology* 29 (2005): 68–87.

Rohner, Ronald P. *The Ethnography of Franz Boas: Letters and Diaries of Franz Boas Written on the Northwest Coast from 1886 to 1931.* Chicago: University of Chicago Press, 1969.

Tveskov, Mark A., and Jon M. Erlandson. "The Haynes Inlet Weirs: Estuarine Fishing and Archaeological Site Visibility on the Southern Cascadia Coast." *Journal of Archaeological Science* 30 (2003): 1023–1035.

White, Elroy A. F. (Xanius). "Heiltsuk Stone Fish Traps: Products of my Ancestors' Labour." Master's thesis, Simon Fraser University, 2006.

Williams, Judith. *Clam Gardens: Aboriginal Mariculture on Canada's West Coast.* Vancouver: New Star Books, 2006.

CHAPTER 15

An Archaeological History of Holocene Fish Use in the Dundas Island Group, British Columbia

Natalie Brewster, *Department of Anthropology, McMaster University*
Andrew Martindale, *Department of Anthropology, University of British Columbia*

Archaeological evidence of a high density of large village sites in the Dundas Island Group suggests intensive occupation of this region from at least 5,000 to about 1,500 years ago. Surprisingly, fish remains are relatively minor constituents of the faunal assemblages from these shell middens, especially when compared to other assemblages in the region. The unusual character of the Dundas Island Group assemblages is in marked contrast to elsewhere on the Northwest Coast. This case raises interesting questions regarding the nature of the subsistence economy and settlement history on these islands and the reconstruction of subsistence from faunal data.

The Tsimshian are known from ethnographic and ethnohistoric sources for their reliance on fish as the basis for their diet, their delayed-return subsistence economy (storage), and the production of surpluses associated with a suite of developments after ca. 3500 cal BP. In the following pages we present the results of the analysis of Dundas Islands faunas and consider the evidence for (a) large populations dependent on intertidal resources and (b) long-distance staple food transport to sustain the Dundas communities.

Tsimshian Fish Use

General descriptions of Northwest Coast subsistence and economic development indicate a focus on marine resources, particularly fisheries (Ames and Maschner 1999; Matson 1992). Indeed fishing has been identified as the resource procurement activity of greatest importance for the Tsimshian, so much so that in the past it has influenced many of their seasonal moves and the organization of labor (Stewart 1975). This section summarizes ethnographic and archaeological data that pertain to Tsimshian use of fish. These descriptions were used to formulate the expectations for what would be found through the analysis of the Dundas faunas and provide the context in which they are interpreted.

The Tsimshian caught many different species of fish in the vicinity of their habitations and consumed them immediately during much of the year. At times of great abundance of certain species, they would catch and preserve large quantities of fish to store for use during times of low productivity (Stewart 1975). The timing and significance of food storage in the development of Tsimshian culture history remains unclear. Ethnographic accounts of Tsimshian subsistence list salmon as the single most important resource (Garfield 1966). Pink, chum, sockeye, chinook, and coho salmon were utilized, but coho, pink, and sockeye were the most heavily used.

Ethnographic sources suggest that houses controlled salmon fishing locations. In this context, a "house" is a co-resident, multigenerational extended-family lineage that shares a narrative of common ancestry. Each house controlled multiple fishing locations that would give them access to all salmon species, as well as provide some protection against the failure of any single run (Halpin and Seguin 1990). Salmon became available in the early summer and could be caught in open water with hook and line. In the late summer and early autumn salmon would begin to school offshore in preparation for their move into the rivers in large numbers, where they were caught using mass capture techniques such as fish traps.

As Martindale (2006a, 2006b) notes, the ethnographic evidence for Tsimshian fisheries is complex and likely includes reference to different time periods across a dynamic pre- to postcontact history. On the northern coast of British Columbia the Skeena River with its tributaries was the most important salmon fishery and was probably the focus of food procurement throughout history (Aro and Shepard 1967). Following Martindale's (2006b:175) reordering of the conflicting ethnographic data, Boas's (1916) description of the Tsimshian moving from coastal winter villages to their fishing territories on the Skeena River and its tributaries in the summer predates Garfield's (1939) description of permanent villages on both the coast and the Skeena. Regardless of settlement pattern, Martindale (2006a:146) has argued that if the limiting factor in storing salmon is processing, by removing the head, innards, and vertebrae, filleting them, and smoking them to preserve for the winter an extended family of thirty can store a year's supply of fish in three to five weeks.

Eulachon (an oil-rich smelt) has been identified as the second most important food species for the Tsimshian (Garfield 1966). Eulachon was the first fish available in the springtime, when stores of dried foods were running low (Garfield 1966; Miller 1997). The Tsimshian fished for eulachon in their fishing territories along the Nass River. Each family would collect large quantities and either dry them or render them for oil. Eulachon grease was a preservative useful for long-term storage of otherwise short-term resources such as berries. As a result, it was highly prized and traded widely (MacDonald and Cove 1987). Grease was also brought back to villages and stored for use during the winter (Halpin and Seguin 1990). Based on ethnohistoric reports of eulachon grease production, an extended family of thirty could have produced between 175 and 250 gallons (about 800 to 1,100 liters) of oil per week, if processing time was the limiting factor (Martindale 2006a:146).

In the late spring herring would move to nearshore waters to spawn, when they would be most intensively fished. Like eulachon, herring could be dried or rendered for oil (Garfield 1966). Other economically important fish used by the Tsimshian include halibut and cod (Garfield 1966). These fishes were available during the winter, between November and January (Stewart 1975), and would be welcome foods during times of low productivity. Ethnohistoric sources clearly show that the Tsimshian food economy was capable of producing subsistence resources sufficient for supporting high densities of people. The Metlakatla (Venn Pass) area of Prince Rupert Harbour is the best-known area of high population density (Halpin and Seguin 1990), but other areas may also have supported many people.

Archaeological sources reflect a similar pattern of important food resources, as well as extend the range of species represented to include many that are not mentioned in ethnographic sources. The importance of fish in general to the Tsimshian is made clear from the remains found in archaeological sites. Fish, especially salmon, clearly dominate all other categories of food.

There is not a wide range of Tsimshian sites from which reliable faunal data are available; for instance, many of the Prince Rupert sites excavated as part of the North Coast Prehistory Project have inherent problems with nonsystematic collection of fish remains (Ames 2005). This has resulted in gross underrepresentation of the role of fisheries in the economy. In contrast, at Ridley Island (GbTn-19), the careful collection and screening of bulk samples has provided reliable faunal data. This site is a small camp located in Prince Rupert Harbour excavated in 1978 by May (1979). Fish clearly dominate the assemblage, with salmon the most abundant resource by far, with more than 10,000 elements recovered. The next most abundant fish was herring; other fish represented included dogfish, ratfish, skate, cod, greenling, flounder, surfperch, rockfish, sculpin, and sablefish.

McNichol Creek (GcTo-6) is a winter village in Prince Rupert Harbour alongside the creek of the same name. Coupland and colleagues (1993, 2003) reported over 90 percent of the elements recovered from the site were fish. As with Ridley Island, salmon is the most abundant taxon. Coupland suggests these salmon were obtained in the adjacent creek and caught from the Skeena River and then processed and transported to the village as stored food.

At both of the abovementioned sites, salmon is the most abundant fish, followed by herring. Eulachon is only present in low quantities at Ridley Island, but this may be because of their small size, not an indication of their economic importance. Though modest, the archaeological data from this village and campsite reflect a pattern of fish use consistent with ethnographic data and expand upon a pattern that extends to four other villages in the Prince Rupert Harbour: GbTo-77, Boardwalk, Tremayne Bay and Phillips Point (Coupland et al. 2010).

It is from these ethnographic and archaeological sources that expectations for the faunal assemblages from the Dundas Islands sites were derived. We expected that a full

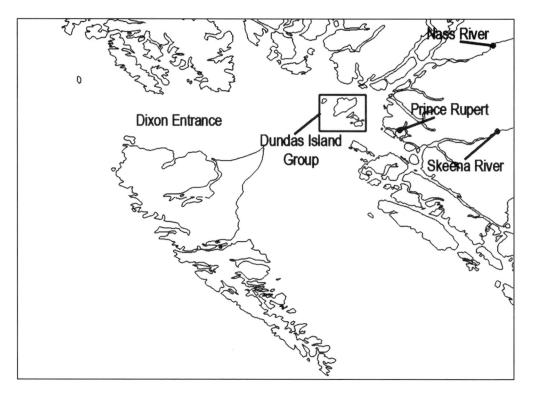

Figure 15.1. Regional map showing location of the Dundas Island Group.

suite of fish species would be recovered in relative abundances that reflect their value in the local economy, with salmon, herring, and eulachon as the most abundant. Other fish mentioned in ethnographies or recovered from previous excavations were expected to be found in lower numbers.

The Study Area

The Dundas Islands are located on the northern coast of British Columbia, just to the northwest of Prince Rupert, at the southern end of Dixon Entrance (Figure 15.1). The islands are situated in Tsimshian territory, and they are known to have been used historically for the collection of a variety of resources including deep-sea fish and intertidal shellfish. The group is made up of five main islands—Dundas, Zayas, Baron, Dunira, and Melville—and many islets.

The islands are located 12–15 km west of the mainland between the Nass River to the north and the Skeena River to the south. Both rivers support large fisheries that the Tsimshian are known to have relied upon. Salmon were taken from the Skeena and its tributaries (Garfield 1966); beginning in the nineteenth century, it became the center of one of the largest industrial fisheries on the west coast. The Nass River is important for the eulachon fishery, which in the past drew large numbers of Tsimshian, Tlingit, and Haida, although it was controlled by the Tsimshian (Garfield 1966; Mitchell and Donald 2001). The position of the islands therefore provided the inhabitants access to both fisheries, though they were not located in their immediate vicinity.

History of Investigation

In 1987, James Haggarty (1988) launched the Zayas Island Archaeological Project, in which systematic survey resulted in the identification of thirty-two sites. These included twelve historic habitation sites with house structures, seven shell middens, nine stone wall fish trap sites, and four canoe run sites. Having found a large village with thirty-five houses, Haggarty suggested people had lived on the Dundas Islands year-round.

In 1998, David Archer (2000) led the Dundas Islands Archaeological Project to investigate the large village identified by Haggarty and to assess the archaeological potential of the southernmost Melville Island. The survey located an additional twenty-nine sites, fifteen of which are historic camps. A winter village (GcTq-1) with twenty house features was mapped. Four surface samples were collected from the village site with dates ranging from 2400 ± 60 to 3280 ± 60 uncalibrated radiocarbon years and a basal date of 3210 ± 60 radiocarbon years (Archer 2000). The dates from the village roughly coincide with the abandonment of villages in Prince Rupert, a development potentially related to a period of increased conflict over territory between the Tsimshian and the Tlingit, which is recorded in oral histories (Archer 2000). Based upon the village dates, and the later presence of many historic camps, Archer (2000) suggested that the Tsimshian use of the islands changed during the historic period.

Between 2005 and 2007, Andrew Martindale and David Archer joined forces in the Dundas Island Project to conduct additional survey, increasing the total inventory to 121 sites (Table 15.1). The project initiated several site-specific studies, including detailed surface-contour mapping of five large villages, intensive subsurface coring to identify and map buried deposits, and combined coring and augering to establish a sequence of occupational patterns and obtain faunal samples. The main goals were to reveal the complex environmental and cultural history of the region with a specific focus on (a) finding evidence of Early Holocene occupation and (b) identifying points of agreement between the oral record (*adawx*) and the Late Holocene archaeological record. There is a special focus on the last 4,000 years because this was a period of increased migration and interaction among cultural groups in northern British Columbia. Analyses of oral histories (Marsden 2000, 2001; Martindale 2003; Martindale and Marsden 2003) describe a period of increased conflict between the Tsimshian and the Tlingit in which Dundas Island figured prominently as a location where key events occurred.

This paper is situated within this most recent research program; we present the results of the faunal analyses from remains recovered in the auger samples. Our objective is to investigate the fish resources used to support large winter villages, like those recorded by Haggarty (1988) and Archer (2000). Given that the Dundas Islands are somewhat removed from the mainland rivers where the Tsimshian typically collected fish for their winter stores, how were the villages on the islands sustained?

Table 15.1. Types of Dundas Island Group sites located during the 2005–2006 surveys

General site type	Sites	Total
Flat middens	34	34.6%
Ridge middens	5	5.1%
Villages	15	15.3%
Rockshelters	15 sites, 17 shelters	15.3%
CMT sites	10	9.8%
Campsites	2	2.0%
Lithic scatters	1	1.0%
Historic cabins	9	9.2%
Fish traps	6	6.1%
Canoe skids	1	1.0%
Total		98 (100%)

Source: Martindale and Letham (2008), Martindale et al. (2009)

Environment and Resource Availability

The Dundas Islands have been previously described as marginal (Ames 1998). This is true when the distributions of anadromous fish and terrestrial mammals are compared to mainland resources. As discussed below, many of the food resources used by the Tsimshian could be obtained from the islands. The islands are less than ideal for procurement and processing of those resources that occur in large quantities for short periods of time, for example, salmon and eulachon.

The Dundas Islands are removed from concentrations of anadromous fish, as they support no large runs. The north end of Dundas Island has five small runs of pink salmon, but these do not match the abundance found in mainland rivers. Haggarty (1988) found stone fish traps in this area as well as on the south side of Dundas Island, but none have been dated. Although these creeks and streams provide a source of salmon, only one species was available, and the small size of the runs would leave island inhabitants vulnerable to periodic failures. The offshore routes of migrating salmon follow the coast, but large numbers do not likely move throughout the islands (DFO 1999a, 1999b). One large run is known to pass north of Dundas Island and could be fished by trolling. Unlike in the Gulf of Georgia, there is no evidence for netting large numbers of salmon in open water (Ames and Maschner 1999).

Eulachon are not known to run in any streams on the islands. Eulachon generally remain deep in the water until they come into river mouths for spawning (Hay and McCarter 2000). To obtain eulachon, it is probable that people living on the islands went to the Nass River either to fish for eulachon themselves or to obtain them through trade.

Commercial fisheries data from the twentieth century show that herring are readily available from multiple locations on the islands, though the quantity and the location of their capture vary from year to year (Hay and McCarter 2007). Based upon this information, people on the islands would have had good access to herring. Halibut were also likely available, as the halibut fishery off the western side of Dundas Island was productive during the historic period (Garfield 1966).

Other food resources found in the Dundas region include sea mammals such as harbor seal and sea lion, which would be abundant in the spring when they congregate for breeding. The Tsimshian historically hunted sea mammals from offshore islands while they were engaged in collecting and processing seaweed (Garfield 1966).

Similarly, the Dundas Islands are known for their numerous beaches, habitat for shellfish and other intertidal resources such as sea urchins. Butter clams, littleneck clams, and barnacles are found on beaches along the channels between the larger islands and in the intertidal zones that connect small islands during low tide. These three taxa comprise the bulk of the midden contents at the Dundas sites, and Tsimshian continue to harvest these shellfish today.

No terrestrial mammals that might have been hunted by the Tsimshian occur on the islands. For deer, mountain goat, or bear, visits to the mainland were necessary. Varieties of berries and seaweed are found on the islands.

Methods

Faunal samples were obtained from a variety of shell middens in the study area. We designed a sampling strategy to include the widest range of midden types, defined on the basis of both local and regional variation (Martindale and Letham 2008). Our sample includes seven sites with elongate ridge or back middens and associated house features in front of the midden. Based on the presence of surface house depressions, these sites are thought to be villages. Twelve smaller middens with an irregular shape were also included. These middens reflect a wide variety of sizes, with six that are less than 15 m in diameter and five that are much larger. These sites are considered the remains of seasonal temporary habitations such as camps or resource procurement areas because of the absence of surface house depressions.

Even though it is an oversimplification, the typological distinction between large sites with evidence of long-term use and small sites with evidence of short-term, perhaps seasonal, occupation is useful. Martindale et al. (2009) have shown that large sites with house depressions can be villages in their upper, more recent components and smaller camps in their lower, older components. Similarly, variation exists among the large sites and among the small sites. For example, Martindale and Letham (2008) identify four types of villages based on different configurations of house depressions and shell ridges. Small sites vary in the shape and size of the shell midden area(s). Despite this variability, the distinction between large sites with more permanent residence and small sites exists in all time periods for the Dundas region. If, as McLaren (2008) argues, the mainland and island ecosystems developed in concert and the paleoenvironment of the Dundas Islands was relatively stable and characterized by a gradual increase in the availability of food resources, then the questions of how large sites were supplied with food and what role small sites played in this economic history are relevant.

A total of eighteen shell middens are represented in our sample (Figure 15.2, Table 15.2). Anthropogenic shell middens are characterized by large accumulations of a variety of shell

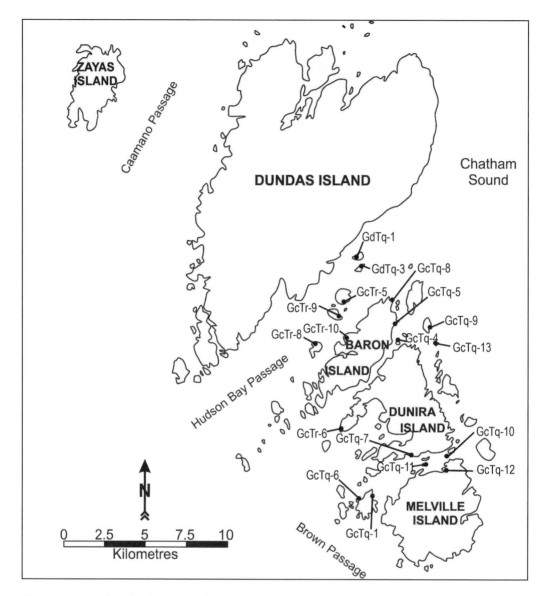

Figure 15.2. Dundas Island group site locations.

species, vertebrate faunal remains, and artifacts. Although it is possible that naturally occurring, biogenic shell "middens" may have been mined for material to use as construction fill, especially at large habitation sites, no evidence of this has been found. Thus shell middens, composed predominantly of large shell fragments, with butter clam, barnacle, and littleneck clam as the most abundant constituents, are primarily a product of the subsistence economy.

The shell middens in this sample also contain small amounts of charcoal and very few artifacts. These middens are distributed throughout the islands, though the focus was on three areas where sites cluster: Hudson Bay Passage, the far eastern end of the channel between Baron and Dunira islands, and the channel between Melville and Dunira islands. Each of these locales is a marine channel that links the inner Chatham Sound to the outer coast, and each has large intertidal areas with abundant shellfish habitat.

Table 15.2. Dundas Island Group shell midden data

Site	Augers	Depth (cm)	Levels	Site type (m)
GdTq-1	3	245	10	Village
GcTr-5	4	392	18	Village
GcTq-5	4	480	21	Village
GcTq-7	1	290	12	Village
GcTr-10	4	420	19	Village
GcTq-13	1	140	4	Camp (<15)
GcTr-8	2	440	20	Village
GcTq-8	1	220	9	Camp (<15)
GcTr-9	1	340	15	Camp (<15)
GcTq-1	2	440	18	Village
GcTq-6	2	320	15	Camp (>15)
GcTq-12	1	260	12	Camp (<15)
GcTq-9	1	160	5	Camp (<15)
GcTq-10	1	340	15	Camp (<15)
GcTq-4	2	368	17	Camp (>15)
GcTr-6	2	269	11	Camp (<15)
GdTq-3	2	410	16	Camp (>15)
GcTq-11	1	240	8	Camp (>15)

The shell middens vary in depth, with some village middens in excess of 5 m. Most middens are located on a 6 m high terrace, not very far back from the beach. Three middens (GdTq-3, GcTq-4, and GcTr-6) were sampled from the 13 m terrace. Both the 6 m and 13 m terraces are relict beaches representing still stands dated to about 8000–5000 cal BP and 15,000–12,000 cal BP, respectively (McLaren 2008). Since the sea level curve shows decline over the Holocene, these date ranges are *termini post quem*. The dating program thus far has focused on the large middens, and many smaller middens have not yet been dated (Table 15.3). The dated midden components are consistently more recent than the terrace dates.

Shell middens have been sampled following Cannon's (2000a, 2000b) methods. For each midden, a 7 cm diameter bucket auger was used to recover matrix in 20 cm increments. It was possible to collect the auger samples in such thick increments because the sampled middens consisted of large, loosely packed fragments of shell that easily moved up into the cylinder of the bucket auger. Shell matrix beginning from the uppermost shell-bearing level through to the base of the midden was bagged, labeled, and returned to the laboratory for analysis. Each auger sample was taken alongside an ESP (environmentalist's subsoil probe) percussion core sample. The ESP core gives an intact profile of the midden stratigraphy from the location where the auger is taken. It also indicates what the content of the auger will be, as well as the depth of the midden. Lastly, the core provides an undisturbed context from which shell or charcoal fragments can be removed to obtain datable samples.

Multiple auger samples were taken from the large shell middens. At village sites two to four samples were taken along the midden with an attempt to include samples from the

Table 15.3. Radiometric dates from Dundas Island Group sites

Site number[1]	Site type[2]	Sample elevation (m abl)[3]	Material[4]	Delta R[5](±100)	Delta 13C ‰[6]	C14 years BP (SD)	Calibrated age range (years BP)[7] (2 sigma) Lower	Upper	Lab. no.
GcTq-4	Village	11.1	char	n/a	−25.0	3650 ± 50	3842	4092	TO-13309
GcTq-4	Village	7.5	clam	200	−1.7	5290 ± 40	5195	5698	Beta 215179
GcTq-4	Village	12.2	clam	200	−0.7	6830 ± 70	6869	7404	Beta 215178
GcTq-5	Village	7.9	char	n/a	−25.0	1460 ± 50	1288	1417	TO-13310
GcTq-5	Village	5.4	clam	300	0	2180 ± 60	1209	1709	TO-13598
GcTq-5	Village	3.6	clam	300	0	2200 ± 60	1229	1729	TO-13597
GcTq-5	Village	7.5	clam	300	0	2780 ± 50	1851	2395	TO-13290
GcTq-5	Village	6.2	clam	200	0	3170 ± 50	2413	3027	TO-13291
GcTq-5[8]	Village	10.6	clam	400	−5.2	2765 ± 35	1885	2343	Poz 25881
GcTq-5[8]	Village	6.5	clam	200	2.8	3485 ± 35	2842	3365	Poz 25882
GcTq-5[9]	Village	3.5	clam	200	0	4620 ± 50	4280	4847	TO-13599
GcTq-5[9]	Village	4.9	clam	500	0	8829 ± 60	8566	9223	TO-13600
GcTq-6[8]	Village	6.7	clam	300	−1.5	2315 ± 35	1337	1822	Poz 25877
GcTq-6[8]	Village	4.9	clam	300	2.2	2355 ± 30	1380	1869	Poz 25878
GcTr-3	Other	5.7	clam	225	−1.2	3460 ± 40	2816	3350	Beta 215177
GcTr-3	Other	5.5	clam	200	−1.8	4440 ± 50	4001	4653	Beta 215176
GcTr-3[10]	Other	3.8	seed	n/a	−23.8	1815 ± 35	1690	1827	UCAIMS 21987[12]
GcTr-5	Village	5.8	clam	300	0	2140 ± 40	1173	1640	TO-13602
GcTr-5	Village	4.5	clam	200	−0.3	3070 ± 40	2334	2843	Beta 215175
GcTr-5	Village	2.3	clam	200	0	3000 ± 40	2147	2698	TO-13601
GcTr-5[11]	Village	10.0	clam	300	−3.1	2390 ± 40	1403	1914	Beta 215182
GcTr-6 (ex)	Other	8.3	char	n/a	−25.0	6800 ± 60	7566	7761	TO-13292
GcTr-6[10]	Other	8.5	muss	500	−4.8	7510 ± 20	7311	7678	UCIAMS 21881
GcTr-6[10]	Other	8.5	char	n/a	−25.0	6925 ± 50	7668	7860	UCIAMS 21984
GcTr-6[10] (ex)	Other	4.5	clam	200	−0.5	3145 ± 20	2401	2953	UCIAMS 21883
GcTr-6[10] (ex)	Other	3.2	char	n/a	−25.0	3645 ± 25	3888	4002	UCIAMS 21985[12]
GcTr-6[10] (ex)	Other	3.2	clam	200	−4.4	4200 ± 15	3718	4302	UCIAMS 21882
GcTr-6[10] (ex)	Other	10.4	char	n/a	−25.0	6185 ± 20	7007	7131	UCIAMS 30932
GcTr-6[10] (ex)	Other	8.9	char	n/a	−25.0	6490 ± 20	7412	7437	UCIAMS 30931

Source: Modified from Martindale et al. (2009).

1. All materials for dating were obtained from ESP core samples, except for samples from GcTr-6 (ex), obtained from excavation.

2. All samples for dating were obtained from shell middens. Villages are shell middens with associated surface depressions. The remaining samples are from shell middens without surface features. These sites are classified as "other."

3. abl = Elevation above barnacle line.

4. Abbreviations for material type: clam = clam shell, *Saxidomus gigantea*; muss = blue mussel shell, *Mytilus trossulus*; char = charcoal; seed = seed; whal = whale bone.

5. Delta-R and its error (± 70) is the local marine correction based on Eldridge and McKechnie's (2008) estimate of the local marine reservoir curve for the Dundas Islands area. These values are similar to those reported for the California coast (Kennett et al. 1997). A number of factors contribute to variation in eastern Pacific marine corrections (Deo, Stone, and Stein 2004; Southon and Fedje 2003).

6. Values without decimal places are the assumed values (0 for marine shell and −25 for charcoal) following Stuiver and Polach (1977).

7. Calibrated ranges were estimated using INTCAL04 for terrestrial carbon and MARINE04 for marine shell. MARINE04 includes the global marine correction (~405 years; Hughen et al. 2004).

8. Samples from cultural components within auger column adjacent to core; AMS assay.

9. Disturbed, stratigraphically reversed sample, older date was above younger.

10. McLaren (2008).

11. Samples taken from first contact of cultural component containing shell in auger column adjacent to core; radiometric rather than AMS assay.

12. These dates were incorrectly listed as shell and calibrated for marine reservoir effect in Martindale et al. (2009).

Table 15.3. *continued*

Site number[1]	Site type[2]	Sample elevation (m abl)[3]	Material[4]	Delta R[5](±100)	Delta 13C ‰[6]	C14 years BP (SD)	Calibrated age range (years BP)[7] (2 sigma) Lower	Upper	Lab. no.
GcTr-6[10] (ex)	Other	8.4	char	n/a	−25.0	6940 ± 20	7698	7829	UCIAMS 30930
GcTr-6[10] (ex)	Other	8.3	whal	300	−16.0	7300 ± 30	7299	7675	UCIAMS 31730
GcTr-6[10] (ex)	Other	8.3	char	n/a	−25.0	9690 ± 30	10883	10926	UCIAMS 28008
GcTr-7[10]	Other	16.0	char	n/a	−25.0	640 ± 60	537	678	UCIAMS 21983
GcTr-7[10]	Other	16.0	muss	400	−3.2	1395 ± 15	427	754	UCIAMS 21880
GcTr-8	Village	11.3	clam	300	0	2510 ± 50	1536	2083	TO-13288
GcTr-8	Village	10.2	clam	300	0	2960 ± 70	2065	2696	TO-13592
GcTr-8	Village	4.6	clam	300	0	2970 ± 70	2082	2704	TO-13591
GcTr-8	Village	6.7	clam	200	0	7000 ± 60	7083	7548	TO-13289
GdTq-1	Village	9.6	clam	200	0	4160 ± 70	4267	4903	TO-13596
GdTq-1	Village	8.5	clam	200	−0.2	4780 ± 40	4470	5079	Beta 215174
GdTq-1	Village	8.0	clam	200	0	4640 ± 70	4267	4903	TO-13595
GdTq-1	Village	12.5	clam	200	0	5140 ± 70	4917	5565	TO-13594
GdTq-1	Village	10.5	clam	200	0	6190 ± 70	6161	6719	TO-13593
GdTq-1[11]	Village	12.7	clam	300	−3.0	2440 ± 50	1448	1991	Beta 215181
GdTq-3	Village	8.2	clam	200	-1.2	6890 ± 50	6949	7422	Beta 215180
GdTq-3[8]	Village	11.6	clam	200	0.1	5540 ± 40	5473	5923	Poz 25879
GdTq-3[11]	Village	10.5	clam	200	-5.9	5230 ± 60	5037	5613	Beta 215183

center and the ends. For irregularly shaped middens, auger tests were placed toward the center and in areas where the midden had the greatest depth. This was done for intra-site comparison, as well as to compensate for the possibility of an auger test where the deposit was very shallow or the number of faunal remains unusually low. Many of the very small, patchy middens, however, had only one auger test.

Shell midden matrix was processed following Cannon's (2000a) methods. All materials from each level were wet screened through 2 mm mesh. The greater than 2 mm fraction was observed under 2x magnification with a stereoscopic microscope. All faunal remains were removed and identified to the lowest taxon possible. The identifiable elements are primarily fish vertebrae and vertebrae fragments. These were used to identify the economically important taxa and to assess variability.

Cannon (2000a) has shown that these collection and analytical methods are appropriate for sampling fish remains. Such samples are better suited to the recovery of ubiquitous remains that are more evenly distributed throughout midden deposits than they are to the recovery of less commonly occurring remains found in dispersed discrete deposits. The latter are more likely to be recovered in representative numbers using more traditional excavation and recovery methods. This sampling method also produces a much more manageable sample size than that recovered from a large-scale excavation, and small fish bones tend to be represented more accurately. Given the relatively short time required to obtain an auger sample, it is an efficient method for sampling multiple middens.

Table 15.4. Dundas Island Group faunal data

Site	NISP						Sample volume (l)		Density (NISP/l)		
	Fish	Salmon	Herring	Eulachon	Other	Unidentitfied	Total	<2mm	Fish	Salmon	Unidentified
GdTq-1	379	332	17	29	1	476	24.45	15.3	24.77	21.7	31.11
GcTr-5	671	552	46	72	1	1056	50.8	28.05	23.92	19.68	37.65
GcTq-5	896	703	103	82	8	1713	80.6	39.35	22.77	17.86	45.53
GcTq-7	204	143	24	33	4	1032	13.05	8.3	24.59	17.23	124.34
GcTr-10	500	415	25	59	1	676	53.55	26.85	18.62	15.46	25.18
GcTq-13	35	16	2	17	0	40	4.25	1.45	24.14	11.03	27.59
GcTr-8	347	181	89	53	24	1150	44.7	18	19.28	10.06	63.9
GcTq-8	49	39	4	5	1	86	8.25	4.45	11.01	8.76	19.33
GcTr-9	75	59	14	2	0	176	16.5	7.9	9.49	7.47	22.28
GcTq-1	382	187	51	126	18	715	48.8	26.45	14.44	7.07	27.03
GcTq-6	82	12	13	54	3	94	13.5	6.6	12.42	6.6	14.2
GcTq-12	27	23	0	4	0	22	10.65	4.85	5.57	4.7	4.54
GcTq-9	33	7	22	4	0	63	9.8	2.25	14.7	3.11	28.0
GcTq-10	36	18	5	5	8	161	13.35	5.8	6.21	3.1	27.76
GcTq-4	64	36	23	0	5	169	23.55	14.15	4.52	2.54	11.94
GcTr-6	14	7	4	0	3	22	17	6.05	2.31	1.16	17.85
GdTq-3	18	8	7	0	3	89	25.5	13.05	1.38	0.61	6.82
GcTq-11	11	0	8	0	3	14	9.05	3.25	3.38	0	4.31

The data resulting from this analysis are presented in Table 15.4, as NISP for the three most abundant taxa (salmon, herring, and eulachon) as well as a category for all other fish, and one for unidentifiable fish. Following Cannon (2000a), the total volume of each auger sample and the volume of the fraction less than 2 mm are presented for each site. The densities of identified and unidentified fish bone are calculated against the less than 2 mm fraction for each site. The reason for this is due to the tendency for the extremely large size of shell fragments in some samples to inflate the matrix volume and thereby depress the relative density of fish. The density is calculated as NISP/liter of the less than 2 mm fraction.

Results

Data from eighteen sites provide evidence for patterns common to the region as well as variation within the Dundas area itself. Given what is known about the importance of fisheries to the Tsimshian and the range of species in other archaeological assemblages, we expected high densities of vertebrae and a wide range of species. Yet the Dundas faunal assemblages are characterized by a low density of fish and low taxonomic diversity.

Salmon, herring, and eulachon are the three most abundant taxa, with salmon the most abundant. Other taxa present in the middens include greenling, cod, northern anchovy, and gunnel, although their low numbers suggest they were not of great importance. Notably absent from the assemblages are flatfish such as halibut, which the Tsimshian are known to have used (Garfield 1966).

Two patterns of fish use have been identified, which roughly correlate with the two different types of sites: villages and camps. The larger middens associated with house depressions have the highest proportion of salmon. At these sites the difference in abundance between salmon and other fish species is greatest. The second- and third-ranked fish at these sites are herring and eulachon. This pattern is similar to that found at the McNichol Creek site, though the total numbers are much lower in the Dundas sites. The smaller middens produced much smaller assemblages, with a more even distribution of major taxa, though salmon are the most abundant.

The density measures suggest that some villages had greater access to salmon than others. For instance, GdTq-1 and GcTr-5 have higher total fish and salmon densities than the others, particularly when compared to GcTr-4 and GcTq-1. At some sites, salmon is not the most abundant taxon. At GcTq-13 and GcTq-6, eulachon is the most abundant fish, while at GcTq-11 and Gctq-9, herring is the most abundant. These may be seasonal indicators of specialized use in spring. Since both species of fish could be stored, it is difficult to assess whether this interpretation is correct.

Discussion

The differences between the Dundas assemblages are interesting, but of greater interest is comparison with other sites in the larger region. Expected taxa were found, but the extremely low density and diversity of fish remains were unexpected. Faunal remains were recovered in good condition, so preservation does not explain the low density.

The sampling strategy used here is the same employed in Cannon's study of sites from the central coast of British Columbia. Central coast sites from the Namu area have produced high densities of fish remains, as illustrated in Table 15.5. The densities of total fish and salmon from the Dundas villages do not approach anything near the central coast village densities. Even sites GdTq-1 and GcTr-5, located relatively close to the salmon runs on north Dundas Island, do not yield densities that fall within the range of central coast villages. The Dundas village assemblages more closely resemble the small camps of the central coast. Densities from the Dundas camps are very low, suggesting fish may not have been caught or prepared there. The sample of eleven sites representing a similar range of villages, residential bases, and small encampments from Rivers Inlet showed comparable densities of fish bone to those from the Namu area (Cannon, Yang, and Speller, this volume). These comparisons suggest that the Dundas assemblages are representative of the Dundas middens and distinct from other areas of the coast. This has broader implications for understanding fish resource use and the importance of fish in the region.

The low densities of fish remains in the Dundas middens is particularly notable because (a) the Tsimshian are known to have relied heavily on fish and (b) the sampling strategy used was specifically targeted toward recovering fish remains. We can conclude that fewer fish remains were deposited in middens on the Dundas Islands than in other Northwest Coast sites. Several possible behavioral explanations can be proposed.

One possibility is that the occupants of the Dundas Islands were using fewer fish than in other regions. Salmon and herring could have been obtained locally, and eulachon was

Table 15.5. Central Coast faunal data

Site	NISP fish		Sample volume (l)		Density NISP/<2mm vol	
	Total	Salmon	Total	<2mm	Total	Salmon
Winter village						
ElSx-1	5870	3222	62.7	51.1	114.9	63.1
ElSx-3	3356	1555	30.7	22.9	146.5	67.9
EkSx-12	1992	1599	26.2	23.0	86.6	69.4
Spring/summer village						
ElTb-1	4802	423	29.8	24.3	197.6	17.4
Base camp						
ElSx-5	989	363	21.4	14.4	68.7	25.2
ElSx-10	1433	481	27.6	20.0	71.7	24.0
ElSx-18	567	274	13.2	8.5	66.7	32.3
Specific-purpose camp						
ElTa-25	105	50	16.6	6.0	17.5	8.3
ElTb-2	97	5	3.0	2.5	32.0	2.0
ElTa-21	19	14	6.2	5.3	3.6	2.7
ElSx-8	7	0	4.3	2.8	2.5	0.0
ElTa-3	137	86	2.4	2.0	69.7	43.8
Small multipurpose camp						
ElTa-18	31	27	12.1	10.7	2.9	2.5
ElSx-4	46	27	6.0	4.5	10.2	6.0
ElSx-16	170	16	4.1	2.5	68.0	6.3
Rocky Islet camp						
ElSx-6	26	9	0.5	0.4	65.0	23.7
ElSx-17	1	1	0.8	0.7	1.4	1.5

Source: Cannon (2002); Cannon, Yang, and Speller (this volume)

likely obtained from the mainland. Salmon could have been caught by hook and line in open water or collected from the streams that support runs on the north end of Dundas Island. It is here that the remains of nine precontact stone wall fish traps have been identified, providing evidence of past fishing. The differential access to salmon suggested by the greater abundance at GcTr-5 and GdTq-1 could be due to their proximity to these runs. The inhabitants of these two villages may have controlled the fish traps and thus had greater access to salmon. A locally based economy, even with the runs on north Dundas, would require greater reliance on other food sources, possibly shellfish or sea mammals. Preliminary analysis of shellfish does suggest they were intensively collected (Burchell and Brewster 2007). Yet it seems unlikely that large dense occupations could have been supported over the long term on an economy based primarily on intertidal species because of upper limits in collection capacities and susceptibility to overharvest. Similarly, a sustained and intensive harvest of sea mammals can quickly lead to resource depression (Lyman 2003). Thus while it is possible that these villages may have been supported using the local resource base, it seems unlikely that such an economy could have been sustained for the long occupation periods on the Dundas Islands.

Alternatively, direct historical analogy to ethnographic sources suggests that the faunal remains may underrepresent the importance of fish, specifically salmon, because of methods of fish processing and deposition of fish bones elsewhere on the landscape. People traveled up the Skeena to fish and returned to coastal villages with stored food. If the majority of salmon consumed in the study area represented a stored resource collected from the Skeena River, it could explain the low densities of fish remains in the Dundas Islands middens. While such a pattern is known from recent times, our data suggest that the antiquity of this regional movement of stored foods extended back over 2,000 and possibly as long as 5,000 years ago. Similar subsistence processing has been observed by Matsui (1996) in Japan for transported trout and salmon. Village sites where the inhabitants primarily subsisted on stored foods were located in coastal areas, away from the riverine settings where they were collected and processed. Like the Dundas sites, middens in these villages contained high numbers of vertebral fragments, while whole vertebrae and other elements were rare.

Given the distance and difficulty of obtaining fresh salmon in large quantities in the study area, it is also possible that stored salmon was more heavily relied upon here than in other areas. This suggests that for the island inhabitants, their subsistence economy was regional in scope, requiring a wide-ranging local resource base but focused on a distant food supply. Such a pattern has been proposed for the Prince Rupert Harbour area dating back to 5000 BP (MacDonald and Inglis 1981).

This interpretation raises other questions. First, low fish bone densities on the Dundas Islands characterize the Early Holocene. Does this mean that the regional salmon fishery extends back to this time, with implications for community leadership, economic management, and land tenure? This seems unlikely, because the Dundas evidence is much older than that from sites located along the Skeena River itself (Martindale 2006a). Yet, further research may indeed reveal that so-called complex traits of the more recent period existed in the Early Holocene.

This hypothesis has other regional and historical political implications. If people were fishing the Skeena River and living on the coast, why did they travel to the outer islands to build their habitations? Would they not simply have lived on the mainland coast? The simplest explanation might be that people were already living on the coast and the Dundas area represents an expansion westward of a population centered on Prince Rupert Harbour that was fishing the Skeena before 5000 cal BP. The low density of salmon bones in the Dundas sites contrasts with high densities in Prince Rupert Harbour sites; presumably people in both areas were obtaining salmon from the Skeena River. A few Dundas sites predate the earliest sites in Prince Rupert Harbour, which raises other questions. We can conclude only that the low density and diversity of fish remains that characterize the Dundas sites make them distinctive, and that the long history of the Skeena River fishery will require regional comparisons, such as those attempted here.

Another possibility under consideration is that the settlement history of the Dundas Island sites was less consistent over time than that of the central coast sites Cannon investigated. Cannon (2010, pers. comm.) reports little significant variation in the densities of fish remains throughout the depths of shell midden deposits at central coast sites.

In contrast, intra-site level-by-level analysis of the Dundas assemblages is beginning to reveal a complex history of site settlement, with some levels from village sites with fish bone densities comparable to those from central coast villages. When complete, this analysis of the Dundas faunal assemblages may show a history of site use with periods of less intensive occupation, in which shellfish constituted the primary focus of subsistence at the smaller campsites, punctuated by periods of more intensive occupation, exhibited by high densities of fish remains at villages. If these patterns are borne out in further analyses, they indicate that Dundas Islands settlement was quite variable over time, as suggested by the volatile settlement history evident in Tsimshian oral history.

Although analysis is still under way, the Dundas Islands villages exhibit an intense level of occupation dependent on resources other than fisheries or on fisheries far removed from the sites themselves. The Dundas record may resemble patterns of settlement-subsistence on the central coast in some ways, but with more short-term variability. Clearly, a better understanding of the focus and intensity of fishing is key to understanding the nature and unique history of the Dundas Island settlements.

Acknowledgments

We would like to thank Madonna Moss and Aubrey Cannon for inviting us to contribute to this volume. We acknowledge the permission and support of the Tsimshian people, their elected and hereditary leaders, the Lax Kw'alaams Band Council, and the Allied Tsimshian Tribes Association. Thank you to the 2005–2006 field crews of the Dundas Islands Project, in particular Steve Dennis, Taylor Ryan, and Brian Pritchard for assistance in collecting the core and auger samples. Sue Formosa and Trevor Orchard provided the base maps used in this paper. We owe a big thanks to Jon Ahrens for volunteering to wash all the auger samples, as well as to Aubrey Cannon, Madonna Moss, and Meghan Burchell for reading and commenting on earlier drafts of this chapter. Funding for this research was provided by the Social Sciences and Humanities Research Council of Canada.

References Cited

Ames, K. M. "Economic Prehistory of the Northern British Columbia Coast." *Arctic Anthropology* 35 (1998): 68–87.

——. *The North Coast Prehistory Project Excavations in Prince Rupert Harbour, British Columbia: The Artifacts.* BAR International Series 1342. Oxford: British Archaeological Reports, 2005.

Ames, K. M., and H. D. G. Maschner. *Peoples of the Northwest Coast: Their Archaeology and Prehistory.* London: Thames and Hudson, 1999.

Archer, D. The Dundas Islands Archaeological Project: A Report on the 1998 Field Season. MS on file (Permit 1998–182), Resource Information Centre, British Columbia Heritage Conservation Branch, Victoria, 2000.

Aro, K. V., and M. P. Shepard. "Salmon of the North Pacific Ocean. Part IV. Spawning Populations of North Pacific Salmon." *International North Pacific Fisheries Commission Bulletin* 23 (1967): 225–229.

Boas, F. *Tsimshian Mythology.* Washington, DC: 31st Annual Report of the Bureau of American Ethnology for the years 1909–1910, pp. 29–1037, 1916.

Burchell, M., and N. Brewster. "Faunal Indicators of Site Occupation and Marine Resource Use at the Dundas Islands Group, British Columbia." Poster presented at the 72nd annual meeting of the Society for American Archaeology, Austin, TX, 2007.

Cannon, A. "Assessing Variability in Northwest Coast Salmon and Herring Fisheries: Bucket-Auger Sampling of Shell Midden Sites on the Central Coast of British Columbia." *Journal of Archaeological Science* 27 (2000a): 725–737.

———. "Settlement and Sea-Levels on the Central Coast of British Columbia: Evidence from Shell Midden Cores." *American Antiquity* 65 (2000b): 67–77.

Coupland, G., C. Bissell, and S. King. "Prehistoric Subsistence and Seasonality at Prince Rupert Harbour: Evidence from the McNichol Creek Site." *Canadian Journal of Archaeology* 17 (1993): 59–73.

Coupland, G., R. H. Colten, and R. Case. "Preliminary Analysis of Socioeconomic Organization at the McNichol Creek Site, British Columbia." In *Emerging from the Mist: Studies in Northwest Coast Culture History*, R. G. Matson, Gary Coupland, and Quentin Mackie, eds., pp. 152–169. Vancouver: UBC Press, 2003.

Coupland, G., K. Stewart, and K. Patton. "Do You Never Get Tired of Salmon? Evidence for Extreme Salmon Specialization at Prince Rupert Harbour, British Columbia." *Journal of Anthropological Archaeology* 29 (2010): 189-207.

Deo, J. N., J. O. Stone, and J. K. Stein. "Building Confidence in Shell: Variations in the Marine Radiocarbon Reservoir Correction for the Northwest Coast over the Past 3,000 Years." *American Antiquity* 69 (2004): 771–786.

Department of Fisheries and Oceans Canada (DFO). Stock Status of the Skeena River Coho Salmon. DFO Science Stock Status Report D6–02, 1999a.

———. Skeena River Sockeye Salmon. DFO Science Stock Status Report D6–10, 1999b.

Eldridge, M., and I. McKechnie. "(Re)calibrating Regional Chronologies on the North Coast: Data from Prince Rupert Harbour." Paper presented at the 73rd annual meeting of the Society for American Archaeology, Vancouver, BC, March 2008.

Garfield, V. E. *Tsimshian Clan and Society*. University of Washington Publications in Anthropology 7(3): 169–339. Seattle: University of Washington, 1939.

———. "The Tsimshian and Their Neighbors." In *The Tsimshian and Their Arts*, V. E. Garfield and P. S. Wingert, eds., pp. 1–70. Seattle: University of Washington Press, 1966.

Haggarty, J. "Settlement Pattern and Culture Change in the Dundas Group, North Coastal British Columbia." Paper presented at the 41st annual Northwest Anthropological Conference, Tacoma, WA, 1988.

Halpin, M., and M. Seguin. "Tsimshian Peoples: Southern Tsimshian, Coast Tsimshian, Nishga, and Gitksan." In *Northwest Coast*, Handbook of North American Indians, vol. 7: Wayne Suttles, ed., pp. 267–284. Washington, DC: Smithsonian Institution, 1990.

Hay, D., and P. B. McCarter. Status of the Eulachon *Thaleichthys pacificus* in Canada. Canadian Stock Assessment Secretariat, Research Document 2000/145. Ottawa: Fisheries and Oceans Science, 2000.

———. Herring Spawning Areas of British Columbia: A Review, Geographical Analysis and Classification. Fisheries and Oceans Canada, Revised Edition, http://www.pac.dfo-mpo.gc.ca/sci/herring/herspawn/pages/project.e.htm, 2007.

Hughen, K. A., M. G. L. Baillie, E. Bard, J. W. Beck, C. J. H. Bertrand, P. G. Blackwell, C. E. Buck, G. S. Burr, K. B. Cutler, P. E. Damon, R. L. Edwards, R. G. Fairbanks, M. Friedrich, T. P. Guilderson, B. Kromer, G. McCormac, S. Manning, C. Bronk Ramsey, P. J. Reimer, R. W. Reimer, S. Remmele, J. R. Southon, M. Stuiver, S. Talamo, F. W. Taylor, J. van der Plicht, and C. E. Weyhenmeyer. "MARINE 04 Marine Radiocarbon Age Calibration 0–26 cal KYBP." *Radiocarbon* 46 (2004): 1059–1086.

Kennett, D. J., B. L. Ingram, J. M. Erlandson, and P. Walker. "Evidence for Temporal Fluctuations in Marine Radiocarbon Reservoir Ages in the Santa Barbara Channel, Southern California." *Journal of Archaeological Science* 24 (1997): 1051–1059.

Lyman, R. L. "Pinniped Behavior, Foraging Theory, and the Depression of Metapopulations and Non-depression of a Local Population on the Southern Northwest Coast of North America." *Journal of Anthropological Archaeology* 22 (2003): 376–388.

MacDonald, G. F., and J. J. Cove. *Tsimshian Narratives 2: Trade and Warfare.* G. F. MacDonald and J. J. Cove, eds. Mercury Series Paper 3. Hull: Canadian Museum of Civilization, 1987.

MacDonald, G. F., and R. I. Inglis. "An Overview of the North Coast Prehistory Project (1966–1980)." *BC Studies* 48 (1981): 37–63.

Marsden, S. *Defending the Mouth of the Skeena: Perspectives on Tsimshian Tlingit Relations.* Prince Rupert, BC: Tin Ear Press, 2000.

———. "Defending the Mouth of the Skeena: Perspectives on Tsimshian Tlingit Relations." In *Perspectives on Northern Northwest Coast Prehistory,* J. S. Cybulski, ed., pp. 61–106. Archaeological Survey of Canada, Mercury Series Paper 160. Hull: Canadian Museum of Civilization, 2001.

Martindale, A. "A Hunter-Gatherer Paramount Chiefdom: Tsimshian Developments through the Contact Period." In *Emerging from the Mist: Studies in Northwest Coast Culture History,* R. G. Matson, G. Coupland, and Q. Mackie, eds., pp. 12–50. Vancouver: UBC Press, 2003.

———. "Methodological Issues in the Use of Tsimshian Oral Traditions (Adawx) in Archaeology." *Canadian Journal of Archaeology* 30 (2006a): 159–193.

———. "Tsimshian Houses and Households through the Contact Period." In *Household Archaeology on the Northwest Coast,* E. Sobel, A. Trieu Gahr, and K. M. Ames, eds., pp. 140–158. Ann Arbor, MI: International Monographs in Prehistory, 2006b.

Martindale, A., and B. Letham. "Regional Patterning and Cultural History on the Dundas Islands." Paper presented at the 73rd annual meeting of the Society for American Archaeology, Vancouver, BC, 2008.

Martindale A., B. Letham, D. McLaren, D. Archer, M. Burchell, B. Schöne. "Mapping of Subsurface Shell Midden Components through Percussion Coring: Examples from the Dundas Islands." *Journal of Archaeological Science 36 (2009): 1565–1575.*

Martindale, A., and S. Marsden. "Defining the Middle Period (3500 BP to 1500 BP) in Tsimshian History through a Comparison of Archaeological and Oral Records." *BC Studies* 138 (2003): 13–45.

Matson, R. G. "The Evolution of Northwest Coast Subsistence." In *Long-Term Subsistence Change in Prehistoric North America,* D. R. Croes, R. Hawkins, and B. L. Isaac, eds., pp. 367–428. Research in Economic Anthropology Supplement 6. Greenwich, CT: JAI Press, 1992.

Matsui, A. "Archaeological Investigations of Anadromous Salmonid Fishing in Japan." *World Archaeology* 27 (1996): 440–460.

May, J. Archaeological Investigations at GbTn-19, Ridley Island: A Shell Midden in the Prince Rupert Area, British Columbia. MS. No. 1530 on file, Information Management Services (Archaeological Records), Canadian Museum of Civilization, Hull, 1979.

McLaren, D. "Sea Level Change and Archaeological Site Locations on the Dundas Island Archipelago of North Coastal British Columbia." PhD diss., University of Victoria, 2008.

Miller, J. *Tsimshian Culture: A Light Through the Ages.* Lincoln: University of Nebraska Press, 1997.

Mitchell, D., and L. Donald. "Sharing Resources on the North Pacific Coast of North America: The Case of the Eulachon Fishery." *Anthropologica* 43 (2001): 19–35.

Southon, J., and D. W. Fedje. "A Post-glacial Record of ^{14}C Reservoir Ages for the British Columbia Coast." *Canadian Journal of Archaeology* 27 (2003): 95–111.

Stewart, F. L. "The Seasonal Availability of Fish Species Used by the Coast Tsimshians of Northern British Columbia." *Syesis* 8 (1975): 375–388.

Stuiver, M., and H. A. Polach. "Discussion: Reporting of ^{14}C Data." *Radiocarbon* 19 (1977): 355–363.

CHAPTER 16

Patterns of Fish Usage at a Late Prehistoric Northern Puget Sound Shell Midden

Teresa Trost, *Cascadia Archaeology, Seattle*
Randall Schalk, *Cascadia Archaeology, Seattle*
Mike Wolverton, *Cascadia Archaeology, Seattle*
Margaret A. Nelson, *Cascadia Archaeology, Seattle*

Ethnographic sources describe a distinctive settlement system for the Whidbey and Camano island areas of Puget Sound (Gibbs 1855:432; Osmundson 1961). That system involved fall salmon fisheries located along lower courses of mainland rivers and winter villages located on islands. Results of recent archaeological excavations at Cama Beach State Park on Camano Island suggest that this settlement system developed here about 1100 cal BP out of an earlier, more widespread system in which both villages and fall salmon fisheries were riverine (Schalk and Nelson 2008). In this chapter we examine how this inferred change in land use is expressed in the fish remains recovered at Cama Beach. We begin by describing the environmental and archaeological context of Cama Beach and then detail the methods and results of the fish remains analysis. These data shed light on how the Cama Beach midden was nested within a broader land use system and provide important evidence documenting the appearance of the island village/riverine salmon fisheries settlement strategy.

Environment

Camano Island is an irregularly shaped island that encompasses approximately 65 km² (Figure 16.1). To the east, the mouth of the Stillaguamish River has built its delta out almost to

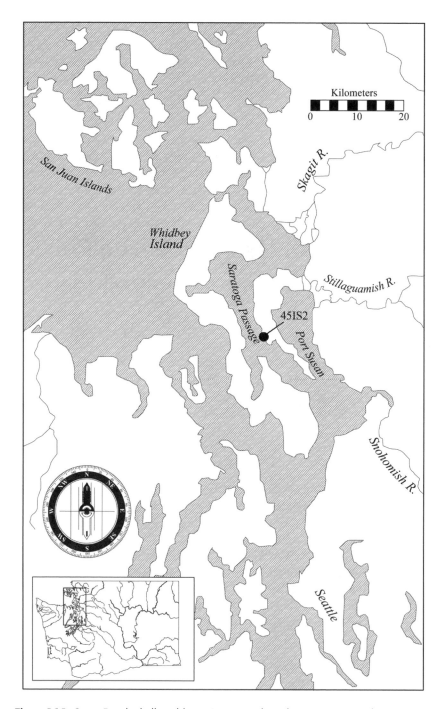

Figure 16.1. Cama Beach shell midden (45-IS-2) and northern Puget Sound region.

the island, which has become essentially attached to the mainland. Whidbey Island is about 3 km west across the deep marine waters of Saratoga Passage. Topography in the Cama Beach vicinity is characterized by rolling uplands associated with a 9 km long north-south ridge that constitutes the highest part of the island and rises to about 120 m above sea level. The island's natural vegetation is within the Puget Sound subtype of the Western Hemlock (*Tsuga heterophylla*) Zone described by Franklin and Dyrness (1973). This area lies within the rain shadow of the Olympic Mountains and has a predominance of excessively drained, nutrient-poor glacial soils and sediments. Camano Island has no perennial rivers or streams, although a number of springs and small intermittent streams drain surface run-off, including a small stream at the south end of Cama Beach (Island County 2005).

On the western shore of Camano Island lies Cama Beach, a 350 m long, 40 to 85 m wide sand and gravel cuspate foreland fronting steep bluffs composed of glacial drift. North and south of the archaeological site are narrow gravel and sand beaches at the base of the bluffs. Sometime before 1675 cal BP (Beta-154888), northward longshore drift deposited sediment from the eroding bluffs, resulting in the formation of a wide backshore area with a lagoon open to the tides. Circa 1300 cal BP, the lagoon was closed off by a sandbar. It is possible the sandbar did not completely close the lagoon or was breached during higher high tides for a period of time after 1300 cal BP. From 1300 to 1100 cal BP, the beach ridge continued to grow north, curving toward the base of the bluffs, while the foreshore continued to expand westward and the lagoon filled in with sediment eroded from the bluff and anthropogenic deposits. Circa 1100 cal BP, the beach had assumed its basic modern appearance. A single beach ridge forms the western margin of the landform from the south end of the cuspate bar to its north end. The backshore deposits dating ca. 1700–1100 cal BP are below modern high tides at about 1.1 m above sea level. Tectonic subsidence has not been entirely refuted as a possible explanation for this. Sea level reconstructions for Whidbey Island and Kitsap Peninsula, however, indicate that 1,600 years ago sea levels were likely 80 to 160 cm lower than at present (Schalk et al. 2008:16.1–16.3).

Cama Beach is exposed to the prevailing southwesterly winds and lacks protected water for beaching watercraft, although conditions would have been more favorable before the lagoon was closed off. Nevertheless, for much of the past the beach was poorly suited for winter residence. During data recovery excavations in the depths of winter, it became obvious that there are substantial differences in microclimate along the length of the site. On very cold, windy days, the northern portion of Cama Beach was notably more protected from the wind and, therefore, a much more comfortable place to work. In the past, some protection from the wind on the backshore would have been provided by the beach ridge and this effect would have been augmented by any accumulation of drift logs along the crest of the ridge.

Although many animals can be found at or near Cama Beach today, in terms of regional distribution, the location has no particularly salient food resource concentrations. Camano Island is near two of Puget Sound's largest rivers: the Skagit and Snohomish, as well as the smaller Stillaguamish (Figure 16.1). Chinook, chum, coho, pink, and sockeye salmon, along with cutthroat trout, migrate through or are resident in small numbers in Saratoga Passage (Williams, Laramie, and Ames 1975, Stillaguamish 03–04). Surf smelt and herring spawn in Saratoga Passage and Port Susan (Bargmann 1988:57, 64; Stick 2005). In addition, the waters surrounding Camano Island are habitat to more than 200 marine fishes including cods, perches, rockfishes, greenlings, sculpins, and at least fourteen species of flatfishes (DeLacy, Miller, and Borton 1972; Washington Surveys and Marine Land Management Division [SMLMD] 1974). The subtidal zone fronting Cama Beach has eelgrass beds that are partially exposed on the lowest tides. Such beds are highly productive and provide feeding, rearing, and resting areas for a variety of waterfowl; marine fish including salmon (Salmonidae), flatfish (Pleuronectidae), and sculpin (Cottidae); and invertebrates including green sea urchins (*Strongylocentrotus droebachiensis*), crabs (Brachyura), horse clams (*Tresus* spp.), butter clams (*Saxidomus giganteus*), littleneck clams (*Protothaca staminea*), moonsnails (*Polinices lewisii*), and dogwinkles (*Nucella* spp.; SMLMD 1974).

A variety of land and marine mammals is present along the shores and in the waters of Puget Sound. The largest common native mammal on Camano Island is the black-tailed deer (*Odocoileus hemionus columbianus*), which is smaller than its counterpart on the mainland. Smaller mammalian taxa include coyote (*Canis latrans*), raccoon (*Procyon lotor*), river otter (*Lontra canadensis*), beaver (*Castor canadensis*), porcupine (*Erethizon dorsatum*), muskrat (*Ondatra zibethicus*), weasel and mink (*Mustela* spp.), Townsend's chipmunk (*Tamias townsendii*), and Douglas squirrel (*Tamiasciurus douglasii*; Burt 1952; Osmundson 1964). Species that were probably once on Camano Island include elk (*Cervus elaphus*), black bear (*Ursus americanus*), and gray wolf (*Canis lupus*; Dalquest 1948; Johnson and Cassidy 1997). Most marine mammals are not found in high concentrations in Saratoga Passage. Some sea mammals can be spotted at various times of the year including sea lions, orcas, porpoises, and gray whales. Today harbor seals (*Phoca vitulina*) haul out overnight on Camano Island's western shore.

Site Archaeology

For archaeological investigation, Cama Beach was divided into five areas based on geomorphology: the north foreshore (Area A), splayed beach ridges at reattachment to shore (Area B), main beach ridge (Area C1), backshore (Area C2), and south shore (Area D; Figure 16.2). Cascadia Archaeology was contracted by the Washington State Parks and Recreation Commission (Parks) in 1999 to mitigate disturbance of archaeological deposits resulting from the planned rehabilitation of a historic resort that is now Cama Beach State Park. The mitigation included hand excavation of 138 m³ of archaeological deposits and mechanical excavation of trenches 2,000 m long, 0.4 to 1.75 m wide, and up to 2 m deep. Analytical units, or "Periods," were defined based primarily on the distribution of conventional radiocarbon dates and portion of the landform on which the archaeological deposits occur, as well as some distinctions in the artifact, faunal, feature, and stratigraphic data (Table 16.1). The bracketing conventional radiocarbon dates for a period are rounded off for simplicity. Dates are presented here as calibrated dates (cal BP). After the conversion to calibrated dates, four dates fall outside the upper and lower limits for Periods as defined by conventional radiocarbon dating. These deviations are small, being 5, 15, 20, and 55 years, and thus for our purposes, the definition of Periods is unchanged. When the calibrated dates returned two intercepts, the midpoint was used. Our discussion focuses

Table 16.1. Periods defined for Cama Beach prehistory

Period	Time span (ca. cal BP)	Area of landform
T0	na	Across site, below midden
T1	1700–1400	Southern third of Area C2
T2	1400–1100	Northern two-thirds of Area C2
T3	1100–600	Areas A, B (foreshore), C1, and D
T4	600–250	Areas A, B (backshore), and C1
T5	na	Across site, disturbed midden

Source: Refer to Nelson (2008:8.20).

on Periods 2, 3, and 4. Period 1 is excluded from the discussion here because of the paucity of faunal remains recovered. Periods 0, 5, and 6 are either historic or mixed prehistoric, historic, and noncultural deposits.

Each landform area (A, B, C, D) has its own complex stratigraphy. Here, we describe the stratigraphy characteristic of each area and the age of cultural deposits. Area A is composed of stratified, shell-rich midden deposited on underlying beach foreshore sands beginning around 1000 cal BP. Differences in densities of faunal remains and artifacts and in matrix composition below and above a mass wasting event(s) with a terminal date of 595 cal BP (Beta-249645) warranted a division of the last 850 years of Cama Beach prehistory at ca. 600 cal BP. Either (1) occupation of Area A ended ca. 300 cal BP or (2) post-300 cal BP prehistoric deposits were removed during logging or resort construction. Area B cultural deposits are bedded within splayed beach ridge gravels and date to between ca. 1100 and 300 cal BP. Deposits in Area C1 reflect development of the primary beach ridge. Radiocarbon dates from Area C1 span ca. 1100 to 250 cal BP. The stratigraphy in Area C2 is different than elsewhere, probably due to a combination of cultural and natural factors. These strata are distinguished by their black color, lower proportion of shell, and greater density of cultural lenses and features compared to the cultural strata elsewhere at the site. In most cases, the matrix is gravelly sand and silt. Cultural strata closest to the bluff and toward the south in Area C2 were deposited ca. 1700 to 1400 cal BP; closer to the eastern row of cabins, they date ca. 1400 to 1100 cal BP.

Very few tools clearly associated with fishing were recovered. The use of nets is represented by only three lithic artifacts *tentatively* described as net weights, one from Area A and two from Area C1. Groundstone (n = 3) and flaked stone (n = 7) triangular points recovered could have been used for arming toggling harpoons to spear large fish or hunt sea mammals. Bone tools likely used for fishing include small bipoints, unipoints, and toggling harpoon valves (n = 48). Twelve such artifacts were recovered from Area A and eighteen from Area B, spanning Periods 3 and 4. Areas C1, C2, and D each have seven or fewer such artifacts.

In deposits dated before 1100 cal BP, tools recovered are primarily large chopper-like utilized flakes produced from locally available cobbles and boulders. A substantial number of these items also had cortical anvils, while some had discrete grinding facets on their ventral surfaces. All had heavily battered and flattened edges. The abundance and

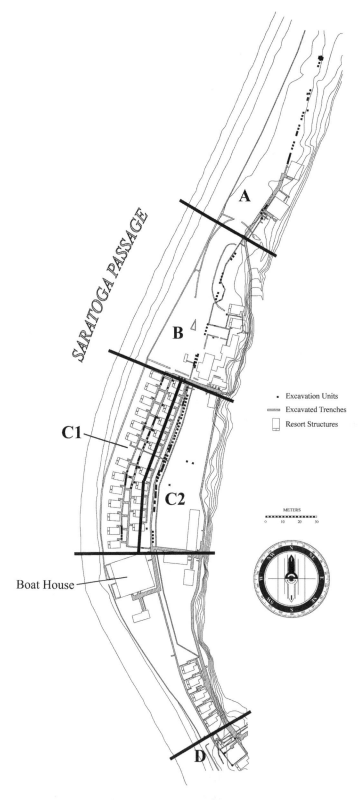

Figure 16.2. Cama Beach Site (45-IS-2) showing excavations, resort features, and landform area designations A–D. Lagoon was in the south part of C2 and unlabeled segment to its south.

consistency of these tools suggest they were processing tools of some type, though for what specific purpose is not clear. Cooking stones used to boil liquids were also relatively common. After 1100 cal BP, woodworking is well represented in Area B by stone, bone, and antler adze/chisel blades and antler wedges (n = 96). Hunting equipment such as stone projectile points declines in abundance and is altogether absent from Area A. With the exception of Area B, the use of cooking stones to boil liquids declines as well.

Regarding seasonality, the Cama Beach site was used during spring and summer throughout its history, but after 1100 cal BP, evidence for winter use increases in some site areas. Winter use is evidenced in greater relative abundance of mussel and sea urchin remains in the north and south shore areas between ca. 600 and 300 cal BP. Those same areas have low frequencies of moon snail and ratfish, which are interpreted as spring/summer indicators. Although moon snail is not usually considered a food source, the concentrations and context of moon snail shell strongly suggests Cama Beach residents were collecting multiple individuals. Moon snails would be most available to harvest in large quantity during spring and summer when they are at peak abundance in shallower waters and can be collected during lower low tides; in winter they seem to prefer deeper waters (Harbo 1997:207; Kozloff 1976:216, 1983:284; Schaefer 1997). In the field, pockets of moon snail shell were noted within cultural deposits in the backshore units, including one comprising a feature and another associated with a fire-modified rock feature. Lab analysis revealed two additional concentrations of moon snail shell within cultural deposits, one in north shore deposits dating to Period 3 and one in backshore deposits containing mixed prehistoric and historic sediments. Moon snail was also found throughout two backshore units that contained multiple features. Field observations and lab analysis indicate moon snail shell was absent or exceedingly rare in other deposits. Ratfish, used for its oil, skin, and as a marginal food source, is more abundant nearshore in spring (April–June) in Puget Sound but can be caught year-round in deeper waters (about 75 m; Lamb and Edgell 1986:60–66; Quinn, Miller, and Wingert 1980). As discussed earlier, the north shore is protected from prevailing winds and, consequently, the warmest portion of the landform during the winter.

In the backshore, a number of postholes were identified, but they were too small to have been posts for the cedar plank houses described ethnographically at winter villages. They could have been used for mat shelters during warmer periods of the year, although they were not associated with definitive interior living surfaces. Based upon the argument of Cannon, Burchell, and Bathurst (2008) regarding the relationship between shellfish maturity (age) and proximity to winter villages, the Cama Beach shellfish size data suggest that this site was likely located in proximity to a winter village after 1100 cal BP (Trost 2008a:10.16–10.23). Further support of this interpretation is that Cama Beach is near known historic winter village sites on Camano Island (Osmundson 1964).

Methods of Fish Analysis

The fish remains discussed here were recovered from 12.62 m³ of sediment in twenty-eight excavation units dispersed across Cama Beach. The units, typically 1 m x 1 m, were excavated by natural stratigraphic layers, with arbitrary 10 cm levels used for vertical control within thicker strata. All sediments were wet-screened in the field, and later dried and screened in the lab over quarter-inch mesh with a subsample (typically 25 liters per level) screened over a nested eighth-inch mesh. We discuss the quarter-inch sample, comprised of specimens caught in quarter-inch mesh as well as in situ finds. We also present analysis of the eighth-inch sample, comprised of two size fractions: specimens caught in quarter-inch mesh and specimens that fell through the quarter-inch mesh (i.e., the eighth-inch size fraction). The eighth-inch sample has remains from eight of the twenty-eight units and all areas of the site but Area C1. Fish remains were identified by Teresa Trost and Rebecca Wigen, Pacific Identifications. Reference collections housed at the National Marine Mammal Lab–Alaska Fisheries Science Center and University of Washington Fish Collection, both in Seattle, and at the Department of Anthropology, University of Victoria, were used. Salmon vertebrae widths were measured to discern if different salmon species or populations were harvested (Cannon and Yang 2006:132–133; Casteel 1976:83–87; Schalk 1991:73–76; see also Orchard and Szpak, this volume). Specimens were quantified by number of specimens (NSP) and number of specimens identified to family or more precise taxonomy (NISP). A subset of seventeen elements of the fish skeleton was identified to taxon (following Leach and Webb 1997:5–8). Analysis by Trost has shown that this approach does not significantly affect the representation of taxa.

Results

The quarter-inch sample of fish remains contains 12,528 NISP. After adjustments for subsampling and differences in sample volumes, the eighth-inch sample contains 16,147 NISP (Table 16.2). When comparing the quarter-inch sample to the eighth-inch sample, it must be kept in mind that in the former, the frequencies of larger elements are augmented, and in the latter, smaller elements are emphasized. All identified specimens are from marine or anadromous fish. Prior to describing the details of the fish analysis, we discuss the relative abundance of fish in relation to other animals.

Table 16.2. Composition of fish remains in quarter-inch and eighth-inch samples

Taxa	Quarter-inch sample		Eighth-inch sample	
	NISP	%NISP	NISP	%NISP
Flatfish	3,954	32	3,531	22
Sculpin	3,402	27	5,212	32
Salmon	2,211	18	2,654	16
Perch	1,801	14	2,848	18
Ratfish	530	4	199	1
Cod	111	1	145	1
Other	519	4	1,558	10
Totals	12,528	–	16,147	–

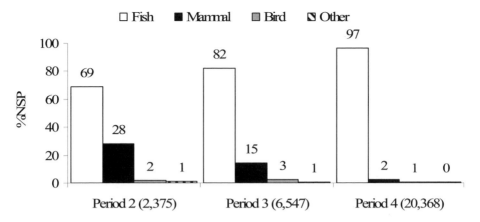

Figure 16.3. Relative abundance of vertebrate classes by period, quarter-inch sample. Percentages are shown atop bars. NSP is shown in parentheses on *x*-axis.

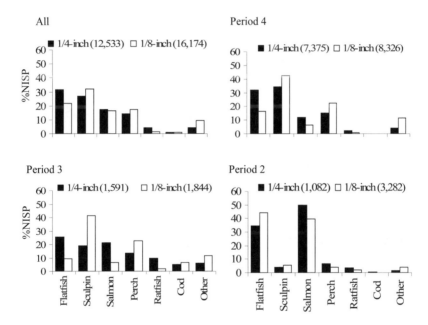

Figure 16.4. Relative abundance of fish taxa, quarter-inch and eighth-inch samples. NISP is in parentheses.

Changes in the relative abundance of vertebrate classes across time suggest the contribution of fish increased (Figure 16.3). Mammal remains comprise 28 percent of the quarter-inch sample in Period 2 but by Period 4 they are a mere 2 percent of the vertebrate remains. Birds occur in low numbers. It is possible that the higher frequency of mammal remains in the older deposit is because fish bone decomposes at a faster rate than mammal bone. We cannot control for this possibility, but the fact that excavation unit C82, dating to Period 2, has one of the largest samples of fish remains suggests this is unlikely.

Flatfish, sculpin, salmon, and perch are the most frequently recovered taxa in both the quarter-inch and eighth-inch samples (Figure 16.4). Ratfish, being a cartilaginous fish, has fewer elements that are likely to preserve compared to a bony fish, but based on the

Table 16.3. Relative abundance (percentage) of fish size classes, eighth-inch sample

Total length (mm)	Period 2 (n = 458)	Period 3 (n = 535)	Period 4 (n = 1,442)
<15	12	24	22
16–24	32	29	30
25–39	47	41	41
40–59	7	5	6
60>	2	1	1

Note: Salmon are excluded from analysis.

frequency and ubiquity of ratfish specimens, this fish was an important prey species at Cama Beach (Wigen 2002:5.3). The other aforementioned taxa also occur in every excavation unit analyzed. Although identifying specimens beyond family was not always a priority, based on specific identifications and observations, remains of starry flounder, pile perch, Pacific staghorn sculpin, and ratfish are the best represented species in the collection (Kopperl 2001; Trost 2008b; Wigen 2002).

Number of taxa, frequencies, and size distributions by period indicate an expansion in the number of fish taxa exploited over time and in the types of fish that were most heavily exploited. In Period 2, salmon and flatfish remains are much more abundant than those of any other taxa (40–50 percent and 34–40 percent NISP, quarter-inch and eighth-inch samples, respectively). Twelve fish taxa were identified in this period. We will show that most of the salmon remains are from small-size salmonids, so in terms of dietary input, the flatfish and salmon are approximately equal, assuming the flatfish were average-size adults (Emmett et al. 1991; Eschmeyer and Herald 1983; Ralston 2005). Excluding salmon, there are fewer remains from fish estimated to be less than 15 cm total length in Period 2 than in Periods 3 and 4 (Table 16.3).

The species of salmon or targeted population changes soon after the end of Period 2 (Figure 16.5). As mentioned, salmon species were inferred using size classifications derived from the width of the transverse axis of vertebrae. Three vertebrae of very large size (>14.45 mm), one from Area D and two from Area A deposits dating ca. 600–250 cal BP, are attributed to chinook salmon. A hypural identified as chum salmon was recovered from Area A in deposits dating to 600–250 cal BP. This identification is tentative, however, due to the small number of reference specimens available. The smaller specimens (pink or sockeye) in Period 3 are from deposits dating to 985 cal BP (Beta-230315), indicating the shift to large-size salmon occurred after 1000 cal BP but before 600 cal BP. Given the ecological requirements of pink and sockeye salmon, the nature of salmon spawning habitats in the locality, and the strength of pink salmon in rivers such as the Skagit and Stillaguamish historically, we suggest that the small salmon are likely to be pinks rather than sockeye. They also could possibly be immature chinook ("blackmouth") or coho taken in saltwater (Miller and Borton 1980; Snyder n.d.:46); however, based on identifications, that seems unlikely. After 600 cal BP, large-size salmon are prevalent among the salmonid remains.

The fish used in Period 2 suggests reliance on salmon and flatfish. The lagoon probably acted as a natural tidal trap, which would have provided easy access to a variety of

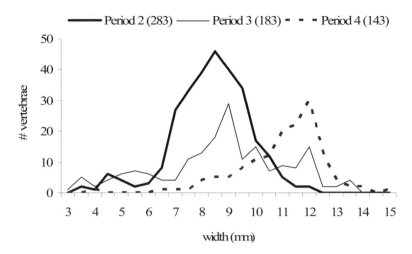

Figure 16.5. Salmon species inferred from vertebral widths in quarter-inch sample. NISP shown in parentheses. Widths <8.0 mm are considered sockeye or pink, 8.0–10.5 mm are considered any species, and >10.5 mm are considered chum or chinook.

taxa and sizes of fish, yet the archaeology indicates people did not take advantage of this circumstance. Although we cannot definitively state that salmon were not caught off Cama Beach, Schalk and Wolverton (2008:13.27–13.28) infer from the quantity of boiling stones in Period 2 that people brought dried pink salmon to the site and reconstituted it by boiling. Their inference is based on the co-occurrence of cooking stones used for boiling and salmon remains.

During Period 3, the quarter-inch sample reveals a broadening of the resource mix with twenty-one fish taxa represented. Sculpin, perch, and ratfish remains are present in relatively large numbers but no single family contributes more than 25 percent of the collection, indicating a fairly diverse, or unspecialized, subsistence strategy. The eighth-inch sample indicates an increased emphasis on sculpin and perch and a relatively large increase in the numbers of ratfish and cod from Period 2. The true input of each taxon to subsistence probably lies somewhere between the relative abundances calculated from the quarter-inch and eighth-inch samples. As mentioned, salmon remains appear to be from a mixture of both small and large species, with smaller vertebrae primarily from older deposits in the south end of Area B. When importance in regards to subsistence is calculated based not just on NISP but also on the size of the individual represented, the contribution of flatfish and salmon increases relative to sculpin.

During Period 4, the mix of fish resources expands to twenty-six taxa. Unlike in Period 3, few ratfish or cod specimens are found in either the quarter-inch or eighth-inch samples. Salmon vertebrae in Period 4 are primarily from larger salmon, including chinook and chum. The diversification seen during Period 3 may have decreased, with more of a focus on flatfish and sculpin relative to other taxa (32 percent and 34 percent NISP in the quarter-inch sample, respectively; 16 percent and 43 percent NISP in the eighth-inch sample, respectively). The variety of taxa and inclusion of small fish suggests nonselective harvest technologies were used (Table 16.2; refer to Wigen 2002). Today, as was likely the case in Periods 3 and 4, the shore is not well suited for a tidal trap. Due to the low slope of the beach, only on a lower low tide is a relatively large expanse of the intertidal zone exposed. Seining for this suite of fish does not seem particularly practical from Cama Beach because many of these taxa lie on the bottom, and seines would have been easily damaged by rocks and debris. Yet people living at Cama Beach could easily travel to trap or netting locations elsewhere in Sarasota Passage within a single day. It would seem that Cama Beach may have been selected for features other than immediate proximity to fish (see Monks, this volume).

The variety of shoreline types (sand and gravel beach, rocky shore) along Saratoga Passage does not seem to have changed over time, so we do not suspect fish or marine mammal habitat changed. Earlier we noted the lack of preservation bias (see also Trost 2008b). The number of identified taxa is related to sample size for our Period samples (cf. Grayson 1984). However, the number of taxa is not simply an artifact of sample size. Periods 2 and 4 represent approximately equivalent time spans, and volumes of excavation differ by 150 percent (2.34 m^3 and 3.52 m^3, respectively). If an equivalent number of fish were caught during both periods, the NISP for Period 4 should increase by 150 percent. A much greater increase is present, indicating a true change in the amount and diversity of fish harvested. Calculations based on all fish remains (NSP) further support this claim (Trost 2008b:9.10).

Conclusion

Several patterns observed in the fish remains recovered at Cama Beach are consistent with a change in settlement strategy ca. 1100 cal BP. Marine fish became a significantly greater proportion of the subsistence economy during the span of occupation Cama Beach. Between ca. 1400 and 1100 cal BP, food resources seem to have consisted principally of black-tailed deer and elk along with a few fish such as pink salmon and flatfish (Trost 2008b). In the six centuries that followed, the relative frequencies of marine fishes increased compared to mammals and birds. During these later periods, prey niche width expanded, as evidenced by an increase in the number of fish taxa and more equal representation across taxa and size class.

Many of these marine fish, such as sculpin, ratfish, and perch, are relatively small, are difficult to harvest in quantity, and, in the case of ratfish, are relatively marginal resources (Cannon 1995). These fish taxa could be characterized as low-return resources, and increased dependence on them seems to represent a form of subsistence intensification. What is possibly most significant about the fish represented in Periods 3 and 4 (1100–250 cal BP) is that many of them can be harvested at nearly any time of the year and thus could be exploited for consumption during those annual lows in subsistence productivity when people depended heavily on stored foods. In view of the logistical advantages of wintering in proximity to locations where staples of stored salmon are amassed, the capacity to exploit a variety of marine resources for immediate consumption as a supplement to stored foods must have been a primary incentive to the establishment of winter villages on saltwater in this area. The island village/riverine salmon fishery positioning strategy seems to make sense in the context of a system responding to rising consumer demand by emphasizing resources exploited for immediate consumption rather than for storage.

Reliance on marine fish at Cama Beach was a local expression of a wider trend to exploit a greater quantity and variety of marine resources, including a number of lower-valued ones, across Puget Sound. In their regional assessment of data from sites in the Puget Sound–Gulf of Georgia region spanning the past 10,000 years of prehistory, Butler and Campbell (2004:364–365, 374, 392) found no evidence of increasing specialization on salmon. Instead, a widening of the diet niche width is suggested based on increased representation of herring. The reef-net fisheries of the Straits Salish in the southern Gulf of Georgia can also be seen as an elaboration of technologies developed on mainland rivers for saltwater environments (Burley 1979; Snyder n.d.). Although the exact timing for the development of the reef-net fishery in the San Juan Islands is unknown, it is thought to have occurred within the past 2,000 years. Recent archaeological investigations in the San Juans documented a major increase in the number of shell middens occupied between about 1000 and 500 cal BP (Taylor and Stein 2007). This expansion coincides with an increase in population for the Northwest Coast (Ames and Maschner 1999:53–55) and evidence for expanding marine resource dependence at Cama Beach. In conjunction with these trends, defensive sites begin to appear in the Gulf and San Juan islands ca. 1100–1000 BP (Keddie n.d.). Such sites have been recorded on Whidbey and Camano islands, although they remain undated. The appearance of such sites is likely symptomatic of increased competition for resources that may have driven the kind of subsistence intensification observed at Cama Beach.

Acknowledgments

The Cama Beach archaeological testing and data recovery efforts were supported by the Washington State Parks and Recreation Commission in consultation with the Swinomish, Lower Skagit, and Tulalip tribes. We are especially appreciative of the efforts of Dan Meatte, Lisa Kelley, Jeanne Wahler, and David Newcombe of Washington State Parks. We are also indebted to Damon McAlister, Bob Kugen, and Mark Van Vliet of Parametrix, Inc., for their help and advice in the field and afterward. For their support and camaraderie, a heartfelt thanks is owed to the staff at Cama Beach and Camano Island state parks: Jeff Wheeler, Melanie Ford-Bissey, Peg Hayes Tipton, Richard Donovan, Tom Riggs, and Pete and Sue Green. The size of our crews prevents us from recognizing each individual, but to all the field personnel and archaeological technicians, especially those who endured months of midwinter working conditions and subfreezing temperatures, we are extremely grateful. Our appreciation also goes to the Tribal and local volunteers who assisted in the field lab and water-screening area. For maintaining remarkable focus over long periods of meticulous and tedious work, we are thankful to our lab technicians. For graciously allowing access to their collections and for willingness to answer questions, Teresa Trost acknowledges Jim Thomason, Bill Walker, Sara Finneseth, and Erin Siek at the National Marine Mammal Lab, Alaska Fisheries Science Center, Seattle, and Katherine Pearson Maslenikov, University of Washington Fish Collection.

References Cited

Ames, Kenneth M., and Herbert D. G. Maschner. *Peoples of the Northwest Coast: Their Archaeology and Prehistory*. London: Thames and Hudson, 1999.

Bargmann, Greg. A Plan for Managing the Forage Fish Resources and Fisheries of Washington. Washington Department of Fish and Wildlife, Olympia, WA, 1988.

Burley, David. "Marpole: Anthropological Reconstruction of a Prehistoric Northwest Coast Culture Type." PhD diss., Simon Fraser University, 1979.

Burt, William H. *A Field Guide to the Mammals*. Boston: Houghton Mifflin, 1952.

Butler, Virginia L., and Sarah Campbell. "Resource Intensification and Resource Depression in the Pacific Northwest of North America: A Zooarchaeological Review." *Journal of World Prehistory* 18 (2004): 327–405.

Cannon, Aubrey. "The Ratfish and Marine Resource Deficiencies on the Northwest Coast." *Canadian Journal of Archaeology* 19 (1995): 49–60.

Cannon, Aubrey, Meghan Burchell, and Rhonda Bathurst. "Trends and Strategies in Shellfish Gathering on the Pacific Northwest Coast of North America." In *Early Human Impact on Megamolluscs*, Andzej T. Antezak and Roberto Cipriani, eds., pp. 7–22. Oxford: BAR British Archaeological Reports International Series 1865, 2008.

Cannon, Aubrey, and Dongya Yang. "Early Storage and Sedentism on the Pacific Northwest Coast: Ancient DNA Analysis of Salmon Remains from Namu, British Columbia." *American Antiquity* 71 (2006): 123–140.

Casteel, Richard. *Fish Remains in Archaeology and Paleo-environmental Studies*. New York: Academic Press, 1976.

Dahlquest, Walter W. *Mammals of Washington*. Lawrence: University of Kansas, 1948.

DeLacy, Allan C., Bruce S. Miller, and Steven F. Borton. Checklist of Puget Sound Fishes. Washington Sea Grant Publication, WSG-72-3. Seattle: Division of Marine Resources, University of Washington, 1972.

Emmett, Robert L., Susan Hinton, Steven L. Stone, and Mark E. Monaco. *Distribution and Abundance of Fishes and Invertebrates in West Coast Estuaries*, vol. II: *Species Life History Summaries*. Rockville, MD: ELMR Report No. 8, NOAA/NOS Strategic Environmental Assessments Division, 1991.

Eschmeyer, William N., and Earl S. Herald. *A Field Guide to Pacific Coast Fishes of North America: From the Gulf of Alaska to Baja, California*. Peterson Field Guide Series. Boston: Houghton Mifflin, 1983.

Franklin, Jerry F., and C. T. Dyrness. *Natural Vegetation of Oregon and Washington*. Portland: US Forest Service, Pacific Northwest Forest and Range Experiment Station, 1973.

Gibbs, George. "Report to Captain McClellan on the Indian Tribes of Washington Territory." In *Report of Explorations for a Route…from St. Paul to Puget Sound*, by I. I. Stevens, pp. 402–434. Vol. 1 of Reports of Explorations and Surveys…from the Mississippi River to the Pacific Ocean….1853–4. 33rd Congress, 2nd Sess. Senate Executive Document No. 78 (Serial No. 758). Washington, DC: Beverly Tucker, Printer, 1885.

Grayson, Donald K. *Quantitative Zooarchaeology*. San Diego: Academic Press, 1984.

Harbo, Rick M. *Shells and Shellfish of the Pacific Northwest*. Madeira Park, BC: Harbour Publishing, 1997.

Island County. Camano Island Nonpoint Pollution Prevention Action Plan. MS on file, Island County Surface Water Management, Public Works, Coupeville, WA, 2005.

Johnson, R. E., and K. M. Cassidy. Terrestrial Mammals of Washington State: Location Data and Predicted Distributions. *Washington State Gap Analysis—Final Report*, vol. 3. Washington Cooperative Fish and Wildlife Research Unit. Seattle: University of Washington, 1997.

Keddie, Grant. Aboriginal Defensive Sites. MS on file, Royal British Columbia Museum, Victoria, BC, n.d.

Kopperl, Robert E. "Fish Remains." In *Archaeological Testing at Cama Beach State Park*. Cascadia Archaeology, pp. 7.1–7.12. Submitted to the Washington State Parks and Recreation Commission. MS on file, Department of Archaeology and Historic Preservation, Olympia, WA, 2001.

Kozloff, Eugene N. *Seashore Life of Puget Sound, the Strait of Georgia, and the San Juan Archipelago*. Seattle: University of Washington Press, 1976.

———. *Seashore Life of the Northern Pacific Coast*. Seattle: University of Washington Press, 1983.

Lamb, Andy, and Phil Edgell. *Coastal Fishes of the Pacific Northwest*. Madeira Park, BC: Harbour Publishing, 1986.

Leach, Foss, and Murray Webb. *A Guide to the Identification of Fish Remains from New Zealand Archaeological Sites*. Wellington: Archaeozoology Laboratory Museum of New Zealand Te Papa Tongarewa, 1997.

Miller, B. S., and S. F. Borton. *Geographical Distribution of Puget Sound Fishes: Maps and Data Source Sheets*. Seattle: Washington Sea Grant Publication, 1980.

Nelson, Margaret A. "Site Structure: Stratigraphy and Horizontal Distributions." In *The Archaeology of the Cama Beach Shell Midden (45IS2) Camano Island, Washington*, R. F. Schalk and M. A. Nelson, eds., pp. 8.1–8.27. Draft report submitted to Washington State Parks and Recreation Commission. MS on file, Department of Archaeology and Historic Preservation, Olympia, WA, 2008.

Osmundson, John S. "Camano Island—A Succession of Occupations from Prehistoric to Present Time." *Washington Archeologist* 5(4) (1961): 2–18.

———. "Man and His Natural Environment on Camano Island, Washington." Master's thesis. Pullman: Washington State University, 1964.

Quinn, T. P., B. S. Miller, and R. A. Wingert. "Depth Distribution and Seasonal and Diel Movements of Ratfish, *Hydrolagus colliei*, in Puget Sound." National Marine Fisheries Service, National Oceanic and Atmospheric Administration *Fishery Bulletin* 78 (1980): 816–821.

Ralston, Stephen. An Assessment of Starry Flounder off California, Oregon, and Washington. National Oceanic and Atmospheric Administration. Santa Cruz, CA: Southwest Fisheries Science Center, 2005.

Schaefer, John. "Lewis' Moon Snail–*Polinices lewisii*." http://www.dfw.state.or.us/MRP/shellfish/other/featured_snails.asp, 1997.

Schalk, Randall F. "Fish Remains." In *A Content Based Study of the West Sound Shell Midden Disturbance (45SJ133), San Juan County, Northwestern Washington* by S. Kenady, pp. 73–76. Contributions in Cultural Resource Management No. 33. Pullman: Center for Anthropology, Washington State University, 1991.

Schalk, Randall, and Margaret A. Nelson, eds. *The Archaeology of the Cama Beach Shell Midden (45IS2) Camano Island, Washington*. Draft report submitted to Washington State Parks and Recreation Commission. MS on file, Department of Archaeology and Historic Preservation, Olympia, WA, 2008.

Schalk, Randall, Margaret A. Nelson, Teresa Trost, Jana Boersema, and Mike Wolverton. "Summary and Conclusions." In *The Archaeology of the Cama Beach Shell Midden (45IS2) Camano Island, Washington*, R. F. Schalk and M. A. Nelson, eds., pp. 16.1–16.24. Draft report submitted to Washington State Parks and Recreation Commission. MS on file, Department of Archaeology and Historic Preservation, Olympia, WA, 2008.

Schalk, Randall, and Mike Wolverton. "The Cooking Stone." In *The Archaeology of the Cama Beach Shell Midden (45IS2) Camano Island, Washington*, R. F. Schalk and M. A. Nelson, eds., pp. 13.1–13.31. Draft report submitted to Washington State Parks and Recreation Commission. MS on file, Department of Archaeology and Historic Preservation, Olympia, WA, 2008.

Snyder, Sally. "Aboriginal Saltwater Fisheries: Swinomish, Lower Skagit, Kikialus, and Samish Tribes of Indians." MS on file, Department of Sociology and Anthropology, University of Windsor, Ontario, n.d.

Stick, Kurt. 2004 Washington State Herring Stock Status Report. Washington Department of Fish and Wildlife, Fish Management Division, Olympia, WA, 2005.

Taylor, Amanda, and Julie Stein. "The San Juan Islands Archaeological Project, 2006." MS on file, Burke Museum of Natural History and Culture, University of Washington, Seattle, 2007.

Trost, Teresa. "Invertebrate Remains." In *The Archaeology of the Cama Beach Shell Midden (45IS2) Camano Island, Washington*, R. F. Schalk and M. A. Nelson, eds., pp. 10.1–10.24. Draft report submitted to Washington State Parks and Recreation Commission. MS on file, Department of Archaeology and Historic Preservation, Olympia, WA, 2008a.

———. "Vertebrate Remains." In *The Archaeology of the Cama Beach Shell Midden (45IS2) Camano Island, Washington*, R. F. Schalk and M. A. Nelson, eds., pp. 9.1–9.36. Draft report submitted to Washington State Parks and Recreation Commission. MS on file, Department of Archaeology and Historic Preservation, Olympia, 2008b.

Washington Surveys and Marine Land Management Division (SMLMD). *Washington Marine Atlas*, vol. 2, *South Inland Waters*. State of Washington, Department of Natural Resources, Olympia, 1974.

Wigen, Rebecca J. "Vertebrate Faunal Remains." In *Supplemental Testing at 45IS2 Cama Beach State Park, Camano Island, Washington*, Cascadia Archaeology, pp. 5.1–5.7. Submitted to the Washington State Parks and Recreation Commission. MS on file, Department of Archaeology and Historic Preservation, Olympia, WA, 2002.

Williams, R. Walter, Richard M. Laramie, and James J. Ames. A Catalog of Washington Streams and Salmon Utilization: vol. 1, Puget Sound. Washington Department of Fisheries, Olympia, WA, 1975.

CHAPTER 17

Herring Bones in Southeast Alaska Archaeological Sites

The Record of Tlingit Use of Yaaw (Pacific Herring, Clupea pallasii)

Madonna L. Moss, *Department of Anthropology, University of Oregon*
Virginia L. Butler, *Department of Anthropology, Portland State University*
J. Tait Elder, *Department of Anthropology, Portland State University*

The Pacific herring of southeast Alaska are best known today as the source of sac roe prized by Asian markets. In March of every year in Sitka, under the watchful eye of the Alaska Department of Fish and Game, herring egg development is monitored closely before the fleet of herring seiners and gillnetters rush to catch as much as they can before the quota is reached. The 2009 herring quota for Sitka Sound was more than 14,500 tons (ADFG 2009). The purse seines of herring are too heavy to be lifted onto the boats, so the herring are vacuumed on board from the nets. The herring are then transferred to tenders, and then onto Japanese freighters. In Japan, the roe is removed from the females, and the rest of the fish carcasses, along with all the males, are ground into fish meal (ADFG 2005). In this way, the commercial sac roe industry removes herring from the ecosystem before they can consummate the most important milestone in their life cycle: reproduction.

Pacific herring is the foundation of North Pacific marine ecosystems, and herring are key prey species for a variety of animals. Because the sac roe fishery is among the most lucrative in the region, its success has an enormous impact on the economy of Sitka, and southeast Alaska more generally. Yet contemporary harvest levels are set based on abundances of the late twentieth century, when herring were already depleted. At the same time, subsistence users of herring eggs have seen reductions in the quantities of eggs available for their harvest (Roby Littlefield, 2009, pers. comm.; Schroeder and Kookesh 1990:14; Thornton et al. 2010). Long-term use of herring by the Tlingit (and the Haida, also of southeast Alaska) is not well documented over space or through time, yet this information can inform us about preindustrial patterns of herring abundance and distribution.* In this chapter, we briefly describe Tlingit use of herring known from ethnographic sources and contemporary practice. Then we describe our work compiling data from zooarchaeological records, as part of cultural anthropologist Thomas Thornton's Herring Synthesis Project. We then review some of the challenges of recovering herring bones from archaeological sites. The results of our literature review and data compilation set the stage for future research on the long-term Tlingit use of herring in southeast Alaska.

Tlingit Use of Herring—Not Just Eggs, Not Just Spring

Previous research provides insight into the Indigenous herring fishery of southeast Alaska—the value of eggs, meat, and oil; capture methods; and season of use. Herring were eaten as fresh food, dried and smoked, rendered for their oil, and used as bait. Ethnographic and traditional uses are described below tracing the seasons of use.

The spring arrival of herring was (and is) a critically important time for the Tlingit, when eggs were (and are) harvested from subtidal eelgrass beds or from hemlock branches and kelp strategically placed in the intertidal zone (Emmons 1991:147; Newton and Moss 2005:15; Niblack 1890:299; Oberg 1973:69; Schroeder and Kookesh 1990; Swanton 1905). Herring spawning times vary by location. In Craig, herring spawn in mid-March; in Sitka, late March or early April; in Auke Bay, late April; and in Yakutat, May (Hay et al. 2000:419; de Laguna 1972:403; Skud 1959:4). Herring eggs were eaten fresh but were also dried and mixed with fat for winter use and trade (Emmons 1991:147; Krause 1979:123; Niblack 1890:299; Schroeder and Kookesh 1990:10). Besides eggs, schooling herring were also taken in the spring (Krause 1979:123).

The Tlingit and other Northwest Coast groups did not focus exclusively on herring during the spring spawn. The spectacular phenomenon of herring spawning also attracted a host of other animals, including birds (gulls, ducks, geese, eagles, and crows), fish (chinook and coho salmon, lingcod, halibut, Pacific cod, hake, black cod, dogfish), and marine mammals (harbor seals, Steller sea lions, porpoises, whales; de Laguna 1960:29; Newton and Moss 2005:15; Niblack 1890:299; Schroeder and Kookesh 1990:16; Skud 1959). While spawning occurs fairly quickly, it is an event that concentrates a variety of species, all animals of interest to Tlingit fishers. In some places along the Northwest Coast, people took advantage of this massing behavior, extending it in time, to permit a wide range of animals to be harvested. Monks (1987) called this strategy "prey as bait" and proposed that the stone fish trap at Deep Bay on Vancouver Island was used in this way, based on the faunal remains nearby. By retaining herring in the stone trap, the Aboriginal people were able to

---

* Although this chapter focuses on the Tlingit, the Kaigani Haida also occupied southeast Alaska prior to European contact (Emmons 1991; Langdon 1979; Moss 2008a). The Kaigani used herring, and the community of Hydaburg, in particular, continues to collect herring eggs in the spring (Schroeder and Kookesh 1990).

take not just herring but all the other species that prey upon herring. Caldwell (this volume) has presented a strong case that the fish traps in Comox Harbour, also on Vancouver Island, were used to take herring with a specialized focus, although her samples are too small to reliably gauge the use of mammals and birds. Mobley and McCallum (2001:43) have documented the double-lead-and-enclosure type of fish trap in four sites near Petersburg, Alaska, which they suggested might have been used to catch herring. Whether or not the Tlingit used traps to focus on herring or take herring along with other species attracted by the spring spawn is a topic for future research.

Besides the early spring fishery, the Tlingit (and birds, fish, and mammals) also took herring later in the spring, during summer, and into the fall. Oberg (1973:69) states that herring oil was processed during May on the islands. Schools of herring congregate into what some observers call "hot spots," attracting humpback whales and other species mentioned above (Nahmens 2008). Although herring abundance can be temporary, dragnetting, dip-netting with baskets, or using herring rakes during these times could be very effective (Krause 1979:123; de Laguna 1960:28–29, 116–117; Niblack 1890:292, 299; Oberg 1973:69). Herring were also dried on strings or alder sticks and cured in the smokehouse for future use (Emmons 1991:145–147; Krause 1979:123).

During the fall, the fat content of herring is at its highest in its annual cycle (Emmons 1991:145). Herring rakes were used to gather thousands of herring from which oil was rendered. The herring were boiled in boxes or a small canoe to extract the oil, which was skimmed off the top and placed in seal bladders or boxes for storage (Emmons 1991:143–145). The herring mash was also eaten (Newton and Moss 2005:15). Favorite Bay near Angoon is a famous locality for harvesting herring for oil production, and this is where many Tlingit were staying in late October 1882, when the U.S. Navy bombarded Angoon (de Laguna 1960:42, 168, 170). An eyewitness to these events, Billy Jones, described how Angoon residents were putting up the fall run of herring and rendering the oil at the time the gunboats came to Angoon. In Jones's words, "they left us homeless on the beach" (de Laguna 1960; Reckley 1982:11). This helps to illustrate how critical herring oil produced in the fall was to the winter food supply of the Angoon Tlingit.*

For Tlingit and Haida—past and present—herring were important not just for eggs but also for meat, oil, and bait. Herring were an important resource used throughout much of the seasonal cycle, for as many as eight months of the year—from March through the spring and summer and into late October.

The Role of Zooarchaeology in the Herring Synthesis Project

Thornton's Herring Synthesis Project brought together historical, ethnographic, archaeological, and biological information along with local and traditional ecological knowledge to understand the history of herring use in southeast Alaska. Individuals in the small communities of southeast Alaska have witnessed the reduction of herring populations and of spawning areas during their lifetimes, due to a variety of factors that are still being studied,

---

* For more on the bombing of Angoon, see de Laguna (1960). This event is still remembered in Angoon today (Bowers and Moss 2006), partly because the U.S. Navy has never apologized for it.

including overharvesting of sac roe and fish, increase in sea mammal predation, disease, habitat destruction, and climate change. We assessed how humans and herring have interacted as part of a dynamic ecosystem over the long term. Ultimately, we would like to provide information that can be used to manage herring more effectively, which may involve protecting and potentially restoring areas where herring spawned and schooled in the past.

Our role in the Herring Synthesis Project was to study past herring distribution and abundance using zooarchaeology (Moss, Butler, and Elder 2010). We created a database summarizing current knowledge of zooarchaeological records from southeast Alaska. Taxonomic information for herring and other vertebrate and invertebrate fauna was compiled from all archaeological site reports, including USDA Forest Service and contract reports not widely distributed. We targeted sites that received some subsurface testing—from shovel probes to substantial excavation. All faunal data were included to provide estimates of relative abundance. Direct measurement of absolute prehistoric animal population levels using zooarchaeology is difficult; our approach examined trends by comparing herring abundance relative to other taxa. Reviewing the entire faunal record allowed us to consider ways the Tlingit relied on broader food webs and how these strategies may have varied over time and space. The database lists published and unpublished site reports; excavation methods, including screen size; volume excavated; condition; age estimates; and site location. The screen size used during excavations impacts measures of taxonomic abundance. Given their small size, herring remains are especially prone to loss and are numerically underrepresented unless fine mesh sieving and laboratory analysis of bulk samples are undertaken. We considered variation in analytic decisions and sampling approaches across the site records to ensure comparability. We analyzed faunal records to identify temporal and spatial trends in herring and other animal records. Given constraints in sampling, we documented patterns in herring use over time and space, in relation to trends in settlement pattern, social organization, technology, and changing climate or sea levels. Here, we present first-order results of the zooarchaeological component of the project.

Herring Bones in Archaeological Sites—Challenges to Recovery

The main challenge to archaeological recovery of herring bones is that they are quite small. In southeast Alaska, adult herring can reach a length of 38 cm (O'Clair and O'Clair 1998:343), but most herring are smaller, usually less than 30 cm long (Hay et al. 2000:420). The jaws (dentaries and premaxillae) lack teeth, but the vomer does have fine teeth. Herring cranial and pectoral bones are light and thin, and they typically fragment into small pieces. Exceptions are the prootic and pterotic, which are quite robust and distinctive and can be found in abundance in archaeological deposits. The most numerous elements in any fish are the vertebrae, rays, and spines. Rays and spines are generally not diagnostic to species, however, so the herring vertebra is the most abundant and readily identified element. While the haemal and neural spines usually break off, vertebral centra are often found intact (Figure 17.1). The number of vertebrae in a herring varies by geographic region but ranges from forty-six to fifty-five per individual (Lassuy 1989).

Figure 17.1. Herring vertebrae from 49-PET-067, Coffman Cove, Alaska. Photograph by Gyoung-Ah Lee, University of Oregon.

When the herring skeleton is disarticulated, and the elements break down into fragments, the individual bones are very small and can fall through the screens archaeologists use while excavating if the mesh size is too large. Although herring bones can be recovered by sharp-eyed screeners using quarter-inch mesh, most herring bones will fall through screens with openings this size. Screens with mesh sizes of one-eighth inch or smaller are needed to ensure recovery of herring bones (Cannon 2000; Moss 2007).

In the history of southeast Alaska archaeology, early excavations (and into the 1980s; Rachel Myron, 2009, pers. comm.) were done without any screening of site sediments, so that if herring bones are not reported, it does not mean that they were not present. The same statement is true of any excavation that has not used screens at least as fine as one-eighth inch. Most excavations, even in the twenty-first century, employ quarter-inch screens even though these are inadequate for herring bone recovery. To compensate for this, some of us take bulk samples (typically 2 liters or more) in which all of the archaeological deposit is recovered, and then this material is sorted and analyzed in the laboratory under controlled conditions. Such bulk samples then provide a more adequate indication of herring abundance.

Results of the Literature Review

As of March 2009, 2,846 archaeological sites were recorded in southeast Alaska (McCallum 2009).* We reviewed sixty-six excavation reports: twenty-two were published in peer-reviewed journals, theses, dissertations, and museum publications, and forty-four were from less accessible contract and Forest Service reports. Based on these reports, we found

---

* According to Mark McCallum (2009, pers. comm.), the 2,846 sites listed on the Alaska Heritage Resource Survey for the Tongass National Forest do not include sites on Alaska Native corporation lands, other federal lands, or private lands, except for a few hundred weirs and traps found on state tide lands. This total also includes historical mines, fox farms, etc., including those of primarily Euro-American affiliation.

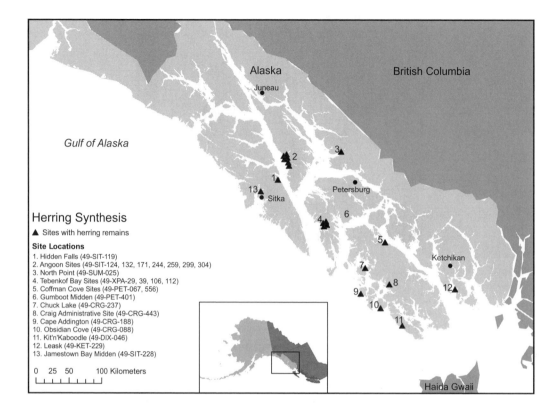

Figure 17.2. Map of southeast Alaska showing sites with herring remains.

that 181 sites have undergone some subsurface testing or excavation; most of these are shell midden sites where the shell helps preserve bone. Of these, faunal remains were recovered from ninety-three sites. Of these ninety-three sites, only sixty-four were reported and described in sufficient detail for us to understand the recovery and analytical methods used. There were twenty-eight sites at which eighth-inch or finer mesh screens were used, producing results we judge as reliable for quantitative analyses based on the comparative collections employed and/or analyst expertise. Of this subset of twenty-eight sites, twenty-one (75 percent) contained herring bones (Figure 17.2), which indicates that when adequate recovery methods are used, the archaeological record s hows widespread use of herring in the past. Two other sites yielded herring bones despite the methods used (49-CRG-237, 49-CRG-443); these are also included on the map, with a total of twenty-three sites with herring.

Of the twenty-three sites with herring bones, thirteen contained in excess of fifteen herring remains. Seven of the twenty-three sites were studied as part of Moss's (1989) dissertation research on Admiralty Island where 2 mm mesh screens were used to analyze bulk samples. All but one of the sites excavated in that study produced herring bones. Another seven sites are sites Moss has been involved in excavating and analyzing at some point, and all included at least some fine mesh screening. To reiterate, if fine mesh screens are used, herring bones will likely be found.

Table 17.1. Results of analyzing a small bulk sample from 49-PET-067, Unit N 201/E 182 (Level 4D)

Taxon	>quarter inch			<quarter inch scanned			<quarter inch 200 ml subsample		
	Total	Vert.	Cran.	Total	Vert.	Cran.	Total	Vert.	Cran.
Pacific herring				190	167	23	152	136	5
Salmonid	73	58	15				11	9	2
Unidentified fish	20								
Unidentified mammal	2								

Note: "Vert." stands for vertebrae and vertebral fragments; "Cran." stands for cranial, pectoral, and pelvic elements.

Some Sites with Herring

The earliest herring bones in a southeast Alaska archaeological site are from Chuck Lake on Heceta Island (Ackerman, Reid, and Gallison 1989) dated to ca. 8200 BP. Although the site is now located away from the ocean shoreline, at the time it was first occupied, the Chuck Lake site was positioned along the shore of an estuary that has since receded with isostatic rebound. This is the oldest vertebrate assemblage from all of southeast Alaska, indicating use of herring during the Early Holocene. It is also an example of a site that produced herring bones even without using fine mesh screens. The shellfish found in association, in addition to the limestone bedrock in the area (which increases soil pH), undoubtedly aided long-term preservation.

Most records of herring date from the last 4,000 years. Many long-standing Tlingit towns and villages are associated with important historical herring stocks (Skud 1959). Herring bones are found in both Coffman Coves sites, at 49-PET-067 as well as the Ferry Terminal Site, 49-PET-556 (Figure 17.2; see Moss, Chapter 13). Coffman Cove is located on the northeastern shore of Prince of Wales Island, adjacent to Clarence Strait. At 49-PET-067, herring bones were not evenly distributed throughout the shell midden (Moss 2008b). For example, during excavation of Level 4D in unit N 201/E 182, a concentration of bone was observed in the field and consequently collected as a small bulk sample (~250 ml). This sample was not screened in the field but was transported intact to the University of Oregon laboratory. In the lab analysis of this sample, the first step was to screen the sample over quarter-inch mesh and identify the bones. The salmon,* unidentified fish, and mammal bones found here are presented in Table 17.1, in the three columns labeled ">quarter-inch screen." Since Moss observed many bones falling through the quarter-inch screen, this material was scanned using a magnifier to identify herring bones, as presented in the middle columns of the table. To facilitate sorting, this material was screened over 2 mm mesh, but even then, herring vertebrae were lost through the screen. Due to the high density of bones still remaining in the fine matrix, Moss spent an additional four hours

* This sample yielded a relatively large number of salmon cranial elements. Because this sample was removed intact in the field, it underwent less damage through field recovery and initial processing than most other samples. Field screening (including water screening) can be tough on bones, but wet salmon cranial bones may be particularly vulnerable to breakage during field screening. The identifiability of salmon cranial bones may be reduced through both "dry" and water screening in the field, a phenomenon that is probably not limited to Coffman Cove.

picking small identifiable bones from the 200 ml subsample until she exceeded the number of herring vertebrae needed to provide a round number for bone density. As shown in the third set of columns in Table 17.1, this intensive sorting resulted in both salmon and herring remains. From the 303 herring vertebrae pulled from the 200 ml subsample, Moss estimated a minimum density of 150 herring vertebrae per 100 ml of fine matrix.

This is a remarkably high density, and a minimum number of six individual herring are represented. Because so many cranial bones were found, this may represent a place where all parts of the herring skeleton were deposited—whether this was from processing fresh herring for immediate consumption or from the disposal of bones from rendering oil is not clear. This density is more than seven times higher than that represented at Katit in Rivers Inlet (see Cannon, Yang, and Speller, this volume), although the analytical methods used are not the same. The concentration at 49-PET-067 may not be typical; without more sampling from the site, we cannot know. The study does highlight the localized nature of fish bone distribution in archaeological sites (see also Partlow 2006) and the need to develop sampling methods that adequately capture this variation. This example also emphasizes that sole reliance on quarter-inch mesh screens during excavation biases the recovery of herring.

The results from Coffman Cove also help illustrate the degree to which screen size affects the relative abundance of species found in archaeological sites. From the Ferry Terminal Site (49-PET-556), in the samples screened in the field with quarter-inch mesh, only 3 percent of the fish bones were herring. In the bulk samples processed in the lab, 65 percent of the fish bones were herring (Moss 2008c). Such bulk samples then provide a more adequate indication of herring abundance.

At a number of sites with relatively abundant herring bones, Pacific cod are also well represented. These include Cape Addington on Noyes Island, the Coffman Cove Ferry Terminal Site, the Leask site on Annette Island, Hidden Falls on Baranof Island, Killisnoo Picnicground Midden on Admiralty Island (49-SIT-124), and two sites in Tebenkof Bay on Kuiu Island (Figure 17.2; see Moss, Chapter 10). Since Pacific cod is thought to have been obtained in the early spring, these sites may indicate use of herring in March and/or April. In contrast, sites 49-PET-067 at Coffman Cove and Garnes Point (49-SIT-304) on Admiralty Island are thought to have been occupied in the fall, and one of the site activities may have been processing of herring oil (Moss 1989, 2008b).

Concluding Thoughts

We know Pacific herring is a bellwether species for North Pacific marine ecosystems. Herring were once abundant in Japan, and their importance to the Ainu is encapsulated in the phrase "herring is god fish" (Hamada 2009). Japan's herring fishery collapsed in the mid-twentieth century after more than 250 years of heavy fishing, although restoration efforts are now under way (Hamada 2009). It is the collapse of the Japanese herring fishery that has led to the high demand for sac roe from southeast Alaska. Can we learn from the Japanese experience?

Today, it appears that biologists and managers in Alaska have only a limited understanding of the longer-term population dynamics and ecology of herring. Since herring were overfished in southeast Alaska relatively early in the twentieth century (Funk 2009), archaeology can provide a more complete picture of the spatial distribution and abundance of herring, especially when the archaeological data are integrated with the traditional environmental knowledge and historical harvest levels. As is the case for many species, we suspect herring are being managed in a depleted status that represents a mere fraction of their historical abundance and distribution.

O'Clair and O'Clair (1998:345) have described five main stocks of herring in southeast Alaska, based on where the fish overwinter. These include Sitka, Craig, Deer Island/Etolin Island near Wrangell, Ketchikan, and Auke Bay. How representative are these of past herring stocks in the area? Today, some biologists avoid definition of stocks based on biological criteria, and use the concept of stock only as a management tool. One way to investigate this question is study of DNA in contemporary herring and ancient DNA in archaeological herring bones. By assessing the genetic variation in zooarchaeological samples of herring and comparing this with that of contemporary herring, we could develop better information as to the condition and status of today's herring. This is just one way that archaeology can potentially contribute new knowledge to sustainable herring management. Until then, we urge all archaeologists working along the Northwest Coast to use fine mesh screens and sample at sufficient intensity to obtain representative samples of herring and other small fish bones from archaeological assemblages. These small fish may be the key to understanding the marine ecosystems of the North Pacific.

Acknowledgments

A version of this chapter was first presented at Sharing Our Knowledge: A Conference of Tlingit Tribes and Clans, in Juneau, Alaska, in March 2009. We acknowledge the support of the North Pacific Research Board and Tom Thornton for including us in the Herring Synthesis Project. USDA Forest Service archaeologists Terry Fifield, Jane Smith, Martin Stanford, Myra Gilliam, Mark McCallum, and others have provided copies of unpublished reports. Moss would like to thank Risa Carlson for excavating the concentration of herring bones from 49-PET-067 and recovering it as a special sample. She is grateful to Pete Bowers, Catherine Williams, and Justin Hays of Northern Land Use Research. We thank Joan Dale, Office of History and Archaeology, for providing updates of the AHRS site records, and David Percy, Portland State University, for assistance with GIS. We also thank the City of Coffman Cove for supporting archaeological work in their town. Finally, we thank Gyoung-Ah Lee, University of Oregon, for photographing the herring bones in Figure 17.1.

References Cited

Ackerman, R. E., K. C. Reid, and J. D. Gallison. "Heceta Island: An Early Maritime Adaptation." In *Development of Hunting-Fishing-Gathering Maritime Societies along the West Coast of North America*, A. Blukis Onat, ed., pp. 1–29. Proceedings of the Circum-Pacific Prehistory Conference, Seattle. Pullman: Washington State University, 1989.

Alaska Department of Fish and Game, Division of Commercial Fisheries. Salmon and Herring Harvesting. http://www.cf.adfg.state.ak.us/geninfo/geninfo.php, 2005.

———. Sitka Sound Sac Roe Herring Fishery Announcement, February 17, 2009. http://documents .cf1.adfg.state.ak.us/AdfgDocument.po?DOCUMENT = 20576 (accessed April 17, 2009).

Bowers, Peter M., and Madonna L. Moss. "A Giant in the Rainforest: Frederica de Laguna's Contributions to the Anthropology of Southeast Alaska." *Arctic Anthropology* 43 (2006): 63–77.

Cannon, Aubrey. "Assessing Variability in Northwest Coast Salmon and Herring Fisheries: Bucket-Auger Sampling of Shell Midden Sites on the Central Coast of British Columbia." *Journal of Archaeological Science* 27 (2000): 725–737.

Emmons, George T. *The Tlingit Indians*. Frederica de Laguna, ed. Seattle: University of Washington Press; Vancouver: Douglas and McIntyre; New York: American Museum of Natural History, 1991.

Funk, Fritz. "Assessing the Impacts of Historical Commercial Fisheries on Herring Stocks." Paper presented at Sharing Our Knowledge: A Conference of Tlingit Tribes and Clans, Juneau, AK, March 2009.

Hamada, Shingo. "Herring Is God Fish: Historical and Cultural Importance of Herring in Hokkaido." Paper presented at Sharing Our Knowledge: A Conference of Tlingit Tribes and Clans, Juneau, AK, March 2009.

Hay, D. E., R. Toreson, R. Stephenson, M. Thompson, R. Claytor, F. Funk, E. Ivshina, J. Jakobsson, T. Kobayaski, I. McQuinn, G. Melvin, J. Molloy, N. Naumenko, K. T. Oda, R. Parmanne, M. Power, V. Radchenko, J. Schweigert, J. Simmonds, B. Sjostrund, D. K. Stevenson, R. Tanasichuk, Q. Tang, D. L. Watters, and J. Wheeler. "Taking Stock: An Inventory and Review of World Herring Stocks in 2000." In *Herring: Expectations for a New Millennium*, F. Funk, J. Blackburn, D Hay, A. J. Paul, R. Stephenson, R. Toreson, and D. Witherell, eds., pp. 381–454. Fairbanks: University of Alaska Sea Grant, 2000.

Krause, Aurel. *The Tlingit Indians* [5th printing]. Seattle: University of Washington Press, 1979.

de Laguna, Frederica. *The Story of a Tlingit Community: A Problem in the Relationship between Archeological, Ethnological and Historical Methods*. Washington, DC: Smithsonian Institution Bureau of American Ethnology Bulletin 172, 1960.

———. *Under Mount Saint Elias: The History and Culture of the Yakutat Tlingit*. Washington, DC: Smithsonian Contributions to Anthropology vol. 7, 1972.

Langdon, Stephen J. "Comparative Tlingit and Kaigani Adaptation to the West Coast of the Prince of Wales Archipelago." *Ethnology* 18 (1979): 101–119.

Lassuy, D. R. Species Profiles: Life Histories and Environmental Requirements of Coastal Fishes and Invertebrates (Pacific Northwest)—Pacific Herring. US Fish and Wildlife Service Biological Report 82(11.126). US Army Corps of Engineers, TR-EL-82–4, 1989.

McCallum, W. Mark. "What's It All Mean?: A Review of over 50 Years of Radiocarbon Dating on the Tongass National Forest." Paper presented at the annual meeting of the Alaska Anthropological Association, Juneau, March 11–14, 2009.

Mobley, Charles M., and W. Mark McCallum. "Prehistoric Intertidal Fish Traps from Central Southeast Alaska." *Canadian Journal of Archaeology* 25 (2001): 28–52.

Monks, Greg G. "Prey as Bait: The Deep Bay Example." *Canadian Journal of Archaeology* 11 (1987): 119–142.

Moss, Madonna L. "Archaeology and Cultural Ecology of the Prehistoric Angoon Tlingit." PhD diss., University of California, Santa Barbara. Ann Arbor, MI: University Microfilms, 1989.

———. "The Killisnoo Picnicground Midden (49-SIT-124) Revisited: Assessing Archaeological Recovery of Vertebrate Faunal Remains from Northwest Coast Shell Middens." *Journal of Northwest Anthropology* 41 (2007): 1–17.

———. "Outer Coast Maritime Adaptations in Southern Southeast Alaska: Tlingit or Haida?" *Arctic Anthropology* 45 (2008a): 41–60.

———. Coffman Cove Village (49-PET-067): Analysis of the Vertebrate and Invertebrate Remains. Submitted to Northern Land Use Research, Fairbanks, AK, March 2008b.

———. Coffman Cove Ferry Terminal Site (49-PET-556): Shell Midden Compositional Analyses and Analysis of the Vertebrate and Invertebrate Remains. Submitted to Northern Land Use Research, Fairbanks, AK, January 2008c.

Moss, Madonna L., Virginia Butler, and James Tait Elder. "Archaeological Synthesis." In *Herring Synthesis: Documenting and Modeling Herring Spawning Areas within Socio-Ecological Systems Over Time in the Southeastern Gulf of Alaska,* by Thomas Thornton, Virginia Butler, Fritz Funk, Madonna Moss, Jamie Hebert, and Tait Elder, pp. 161–233. North Pacific Research Board Project #728. http://herringsynthesis.research.pdx.edu/ (accessed October 1, 2010).

Nahmens, Jim. "For the Love of Herring," Alaska Sea Adventures newsletter. http://www.yacht alaska.com/pubs/2008springnews/SpringASANews2008.pdf, 2008.

Newton, Richard G., and Madonna L. Moss. *Haa Atxaayi Haa Kusteeyix Sitee, Our Food Is Our Tlingit Way of Life: Excerpts of Oral Interviews.* Juneau: USDA Forest Service, Alaska Region, R10-MR-30, 2005.

Niblack, Albert P. *The Coast Indians of Southern Alaska and Northern British Columbia.* Washington, DC: Annual Report of the US National Museum for 1888, pp. 225–386, 1890.

Oberg, Kalervo. *The Social Economy of the Tlingit Indians.* Seattle: University of Washington Press, 1973.

O'Clair, Rita M., and Charles E. O'Clair. *Southeast Alaska's Rocky Shores: Animals.* Auke Bay, AK: Plant Press, 1998.

Partlow, Megan A. "Sampling Fish Bones: A Consideration of the Importance of Screen Size and Disposal Context in the Pacific Northwest." *Arctic Anthropology* 43 (2006): 67–79.

Reckley, G. Eve. In Commemoration: The 100th Anniversary Commemoration, October 24, 25, 26, 1982, of the Bombardment and Burning of Angoon, October 26, 1882. Angoon, AK: Kootznoowoo Heritage Foundation, 1982.

Schroeder, Robert F., and Matthew Kookesh. *The Subsistence Harvest of Herring Eggs in Sitka Sound, 1989.* Alaska Department of Fish and Game, Division of Subsistence, Technical Report No. 173, 1990.

Skud, Bernard Einar. *Herring Spawning Surveys in Southeast Alaska.* Washington, DC: US Fish and Wildlife Service Special Scientific Report—Fisheries No. 321, 1959.

Swanton, J. R. "Tlingit Method of Collecting Herring-Eggs." *American Anthropologist* N.S. 7 (1905): 172.

Thornton, Thomas, Virginia Butler, Fritz Funk, Madonna Moss, Jamie Hebert, and Tait Elder. *Herring Synthesis: Documenting and Modeling Herring Spawning Areas within Socio-Ecological Systems Over Time in the Southeastern Gulf of Alaska.* North Pacific Research Board Project #728. http://herringsynthesis.research.pdx.edu/ (accessed October 1, 2010).

CHAPTER 18

Conclusion
The Archaeology of North Pacific Fisheries

Aubrey Cannon, *Department of Anthropology, McMaster University*
Madonna L. Moss, *Department of Anthropology, University of Oregon*

The eclectic chapters in this volume represent the fisheries archaeology research that has developed from the pioneering proposals and early research efforts of Casteel, Butler, and others. The chapters synthesize much of what we now know about the histories of Indigenous fisheries prior to European contact and colonization. The studies resolve some apparent contradictions in regional long-term trends in the focus and strategies of fisheries, such as the timing and importance of salmon fishing, and highlight state-of-the-art methods currently applied to the analysis of fish remains from archaeological sites. They also break new ground in knowledge and understanding of the organization of fisheries, and they set new directions for the questions to be asked and issues to be resolved in understanding the patterns and histories of fisheries on the north Pacific coast.

Focus, Structure, and Timing of Fisheries

Fisheries were the mainstay of north Pacific subsistence economies throughout the archaeological history of this region. When the appropriate recovery techniques are applied, archaeological sites in the region are shown to contain massive quantities of fish remains, numbering well into the tens if not hundreds of thousands and typically 85–95 percent of recovered vertebrate fauna. Local and regional fisheries also exhibit considerable variability over time and between locations. As important as salmon is shown to be in many locations, in other places taxa such as cod or herring surpassed salmon in abundance and significance. In some areas salmon appear not to emerge as a focal resource until very late in the period before European contact.

Beyond the ubiquitous importance of fisheries and the scale of regional and temporal variability now evident, a number of studies in this volume are beginning to reveal structures in the organization and management of fisheries. In their overview of fishing strategies, Cannon, Yang, and Speller outline major and minor regional systems of salmon acquisition and transport to residential sites. The unexpected fish bone densities in sites of the Dundas Islands group, described by Brewster and Martindale, suggest the possibility of residential groups on the islands participating in such a regional system focused on the mainland fisheries of the Nass and Skeena rivers. At the same time, this study underlines the difficulty of investigating such regional-scale fisheries. The subsistence economy of the Cama Beach site, described by Trost et al., may also be better understood by considering it within a larger regional system of fisheries organization and management.

Cannon, Yang, and Speller describe a widespread strategy of local acquisition of fish from the immediate vicinity of residential sites and regional systems of harvest involving small-scale dispersed facilities such as fish traps and weirs at the locations of small streams. Continuity in patterns of fisheries harvest also hints at particular structures of resource control and managed access, which in some sense can be termed ownership. This is evident at the two Coffman Cove sites described by Moss, where distinctive fisheries focused on salmon and cod, respectively, over a millennium. The millennia-long pattern of local stream salmon fisheries in the Namu region on the central coast of British Columbia, documented by Cannon, Yang, and Speller, also suggests a high degree of residential group control. From the available evidence, these patterns persist over very long periods of time and indicate a high level of sustainability in fisheries adaptations.

Long-term consistency in the pattern and intensity of fishing noted by Cannon, Yang, and Speller and Prince should put to rest any lingering notions that large-scale, storage-based fisheries only developed as the relatively late culmination of a long evolutionary process. At the same time, evidence for relatively recent transitions in fisheries from rockfish to salmon in Haida Gwaii (Orchard, this volume) and on the west coast of Vancouver Island (Monks, this volume) indicates the extent of variability among fisheries within the region and the effects of specific events and developments on local fisheries. In both these cases it will require further study to ascertain the precise nature of these contingencies.

Along with these particular historical patterns, north Pacific fishing seems to exhibit a general trend toward more intensive focus on a few mainstay taxa, such as salmon, cod, and herring, supplemented by a diverse array of other fish. This is evident in an increase in the number of remote harvest facilities such as stone wall fish traps (Cannon, Yang, and Speller; White, this volume) and wooden stake traps and weirs (Caldwell; Prince, this volume) dating to the last 2,000 years (see also Byram 2002; Moss and Erlandson 1998). If this relatively well-established pattern is confirmed through further investigation and more extensive dating programs, it will still have to be understood within a broader regional matrix of long-standing fishing intensity and a variety of more specific transitions to intensive fishing throughout the past 2,000 years.

Methods: Generative and Evaluative

Perhaps the most striking aspect of the chapters in this volume is their adoption and development of innovative new methods, both for the generation of new types of data and for evaluating knowledge claims based on traditional forms of data, such as the numbers of identified fish remains or the representation of skeletal elements. Several papers are notable for developing multisite regional-scale investigation in place of the individual site–centric perspective that once prevailed in Pacific coast archaeology. In discussing the Coffman Cove sites, Moss notes how different the understanding of local adaptation would have been had it been based on one site (cod) or the other (salmon), but how much more complex and interesting the possibilities for the organization and structure of fisheries become when evidence from both sites is available. Monks is also able to draw a broader basis for understanding the reasons for site locations through detailed comparisons of the faunal remains from two sites located near one another but in different environmental zones. Working with tribal oral historians, White learned that different fish traps dispersed over Heiltsuk territory were used at different times during an extended period ranging from August through early November.

Cannon, Yang, and Speller demonstrate the importance of multisite investigation for identifying the hyperlocal orientation of central coast salmon fisheries. Brewster and Martindale make the case for a different and potentially broadly regional pattern of fisheries-based subsistence from the consistent paucity of fish remains in sites on the Dundas Islands. The location of numerous sites at prime fishing locations on the upper Skeena River and the uniformity of associated artifact assemblages support Prince's contention that the artifacts are fish-processing tools and indicate a long-standing focus on salmon fisheries in the region. Orchard uses evidence from sites on Gwaii Haanas to document a clear transition from rockfish- to salmon-based subsistence economies, a pattern that might have seemed much less clear or much more contentious if based on a smaller number of sites. Moss also draws on data from a large number of sites to demonstrate the widespread importance of cod within the subsistence economies of southeast Alaska, and Partlow and Kopperl are able to make a case for the equivocal nature of evidence for cod storage based on the representation of skeletal elements at a wide range of different sites.

Equally important to the regional focus taken by most studies in this volume is the application of new methods to document and investigate the nature of fisheries. Prince makes a convincing case for using artifacts suited to fish processing as a surrogate measure of the important role of fishing at sites in which fish remains are not preserved. With respect to the recovery of fish bones themselves, three papers in this volume (Brewster and Martindale; Caldwell; Cannon, Yang, and Speller) rely on study of fish remains recovered through auger sampling of archaeological sites. This method proves especially useful for time-effective programs of multisite investigation and for intensive recovery and quantification of faunal remains, especially the ubiquitous remains of salmon and herring that constitute the majority of fish in many locations.

The analytical techniques applied to fish bones after their recovery and identification cover the gamut of physics, chemistry, and biology and serve an equally wide range of analytical purposes. Orchard and Szpak use a combination of advanced radiography, osteometrics, and stable isotope analysis to develop a cost-effective method for identifying the species focus of salmon fisheries, which in application to the fisheries of Gwaii Haanas shows a primary emphasis on chum and pink salmon. Cannon, Yang, and Speller apply the costlier but more precise method of ancient DNA analysis to samples of salmon vertebrae from the central coast of British Columbia to show variability in the species focus of fisheries at particular sites and change over time at Namu. They make a broader case for residential site fisheries focused on the streams and rivers in their immediate vicinities.

Betts, Maschner, and Clark use osteometric analysis of cod bones from sites in the Gulf of Alaska and the Gulf of Maine to document long-term fluctuations in the individual sizes and population fecundity of cod. They tie these trends to variations in climate and show that cod populations experienced considerable fluctuations prior to industrial fishing. West, Wischniowski, and Johnston apply stable isotope analysis of cod otoliths to monitor local variations in temperature associated with broader regional fluctuations of the Little Ice Age. These studies effectively demonstrate the value of archaeological fish bone assemblages for monitoring long-term fisheries biology and climate, both of which in turn have implications for human populations and their patterns of dependence on fish-based subsistence.

Other studies in this volume turn to issues concerned with human behavior in relation to fisheries and address whether skeletal element representation can be used to identify cod processing for storage. The issue is important for understanding the presence and basis of sedentary winter village settlement in areas where salmon either were not available or were not used. Although there is still no consensus as to how to determine whether salmon were processed for storage, it is an issue that until now had not been fully explored for cod. Smith et al. apply X-ray absorptiometry to assess the density of cod elements and find they are overall three to five times as dense as salmon bones, though there is little variation in the density of particular elements. This indicates that taphonomy alone cannot explain the uneven representation of elements often found in archaeological sites, suggesting a behavioral basis for these patterns, such as butchering and processing for storage.

In their study, Partlow and Kopperl find that cod cranial elements are overrepresented compared to vertebrae in most archaeological sites on Kodiak Island and from throughout the north Pacific. They suggest that none of these assemblages unequivocally indicates cod storage, though they also note the complexities introduced by combinations of fresh consumption and processing for storage at individual sites. Elemental representation alone, at least from individual sites, is an insufficient basis for inferences of cod storage. As with similar efforts to define salmon storage, it may be time to put this question aside until multiple lines of evidence and clear patterns from multiple sites within regions are available for comparison. The signature of stored Pacific cod may or may not be the same as that of the Atlantic cod that Partlow and Kopperl use as a base for comparison.

The overriding theme in the application of methods in most chapters in this volume is the need to apply multiple independent lines of investigation to build up corroborating evidence in favor of one interpretation or another. Where definitive answers are not forthcoming on the basis of available evidence, the lesson is to use the patterns that are derived to set new directions for investigation and to develop new hypotheses for testing through the application of existing and new methods.

Applied Fisheries Archaeology

Many of the studies in this volume focus on local archaeological questions or the application of new methods to generate and evaluate data from fish remains, but to varying degrees and especially in combination they also yield new information relevant to the contemporary management and future viability of fisheries. The study by Betts and colleagues uses data derived from archaeologically recovered cod remains to set a baseline for assessing the health of contemporary cod stocks. This type of baseline has proved impossible to establish from historically documented patterns that were themselves the products of the impacts of commercial fishing.

The Betts, Maschner, and Clark study goes the furthest of all the chapters in realizing the potential for fisheries archaeology to contribute to the growing field of historical ecology (Balée 2006; Lyman and Cannon 2004; Rick and Erlandson 2008). They suggest that the contemporary Gulf of Alaska cod fishery is sustainable, in contrast to that of the Gulf of Maine. Moss, Butler, and Elder suggest that today's commercial herring sac roe fishery may or may not be sustainable. In their review of the archaeological record of herring use in southeast Alaska, Moss, Butler, and Elder point out how preindustrial Tlingit use of herring was geographically widespread but of much smaller scale than that of today. They suggest that study of the aDNA of archaeological herring bones may help to understand the current status of herring in southeast Alaska, and allow managers to address contemporary conflicts over fisheries resources with more reliable data. As David Montgomery (2003) has pointed out, Northwest salmon are threatened with extinction today, even though we should have learned the lessons of overfishing from Great Britain and New England. The collapse of Japanese herring fisheries in the last century (Hamada 2009) should also be considered. The decline of herring in the eastern Pacific is noted by Caldwell, who mentions that herring have disappeared from Comox Harbour. As future studies begin to build portraits of the variable size, composition, and distribution of fish populations throughout the region, it should be possible to understand more fully the propensity for fisheries to follow long-term fluctuations that transcend the scale of recent impacts. Whether these patterns will provide hope or further despair regarding the potential for contemporary stocks to recover, especially with the growing impact of global warming, remains to be seen.

From another perspective, West and colleagues show the value of fish remains from archaeological contexts for constructing fine-grained temporal and regional maps of local variations within broader patterns of long-term climate change. To date, these types of studies have not moved much beyond refinement of methods and the demonstration of broader potential, but they are an important indicator of the vastness of the untapped record available in archaeological sites, where indicators of climate are often found in readily datable contexts.

Although fish remains from archaeological sites undoubtedly have far more potential as sources of information about past environments than has thus far been realized, perhaps the more immediate applied value of fisheries archaeology on the north Pacific is its affirmation of the pervasive importance of fisheries to Aboriginal populations. This is highlighted in White's chapter, in which forty-two stone fish traps were used primarily to catch salmon. White learned that when in use, some traps remained filled with water, producing a holding pen for live salmon. Holding live salmon in such a pool allowed Heiltsuk to select lean fish when they reached the optimal stage for smoke-drying. In this way, they avoided fish getting moldy or going rancid if they were taken when too oily or (alternatively) too spawned-out. When fishing season was over, "gates" in the trap were opened to prevent the entrapment of fish. These were just a few of the conservation techniques practiced by the Heiltsuk (Jones 2000). All of the studies show the importance of fisheries to Native subsistence economies, but many go beyond to begin to outline the structures of relationships between social and political groups and the fisheries-based landscape in which they derived their living.

A number of papers demonstrate long-standing, finely structured patterns of fish use, even within restricted regions, such that we can begin to speak, as Virginia Butler noted in her comments on the conference session on which this volume is based, about peoples' relationships not just with salmon but with pink, chum, and sockeye as individual species. We can see evidence of structured organization of access to particular species, even within the limited area of a location such as Coffman Cove in southeast Alaska (Moss, this volume). Archaeologists can also provide support for long-term structured access to and reliance on the salmon of the upper Skeena River watershed on the basis of artifact assemblages, even when direct evidence of salmon fishing is not available (Prince, this volume).

Archaeology has only just begun to document the history and full complexity of fishing strategies represented by the development of traps and weirs discussed by White (2006, this volume) and Langdon (2007) on the basis of oral histories and ethnographies. The full complexity of the herring fisheries represented by such remarkable wooden stake fish trap sites as that at Comox Harbour is only beginning to be realized with the mapping and dating of these complexes (Greene 2010) and now with the analysis of fish remains from associated sites (Caldwell, this volume).

The complexity of Native relationships with fisheries has probably surprised many archaeologists, if not the Native people with whom they work, but it is clear that even the extensive relationships with fish described ethnographically are at best a partial and fragmentary account of systems of fisheries organization that were vastly changed with European contact (Ford 1989). Although oral histories and traditional environmental knowledge still have enormous untapped potential for informing society about the forms and management of fisheries (White, this volume), archaeology, as shown by many of the studies in this volume, can provide a deep historical dimension to that knowledge as well as a level of detail that might not otherwise be forthcoming. Arguably, archaeology has made little contribution to the advancement of Native interests, but with regard to fisheries, we may be poised for greater relevance than at any time in the past.

Cultures and Histories

Appreciation of the richness of archaeological histories on the north Pacific coast and the role of fisheries in those histories is just emerging with the achievement of a critical mass of regional studies and accumulated data. It has taken a research program carried out over the course of thirty years, for example, to pursue the implications and meanings of responses to the historical contingency of periodic salmon fishery failure at Namu (Cannon, Yang, and Speller, this volume), and even in this case, with the availability of new data from the application of new methods, parts of the story remain unclear. Further research may fill in some of the gaps in this local story, but ultimately any locally focused study must also be fit within larger-scale regional historical frameworks.

Not surprisingly, most of the chapters focused on local and regional histories have left questions unanswered. All will require new research, more data, and additional archaeological insight to resolve. This applies to understanding the relationship between local settlements and the distinctive fishing orientations at Coffman Cove (Moss, this volume) and Cama Beach (Trost et al., this volume). It also applies to the explanation of the transition to salmon fisheries in Gwaii Haanas (Orchard, this volume) and on the west coast of Vancouver Island (Monks, this volume). As Orchard notes for Gwaii Haanas, it is impossible to explain this as either the result of the spread of technological innovations, such as stone wall fish traps, or demographic trends. For now, it seems sufficiently remarkable to have been able to document the trend at all, though of course the next step will be to devise research strategies that will lead to resolution of its cause. The results of the Brewster and Martindale study leave open many new questions about the organization of regional-scale subsistence economies and fisheries in particular, and the methods and evidence we have available to discern these archaeologically.

Integrating data from environmentally oriented studies with those focused more on historical transitions in local and regional cultures and settlements will be a difficult challenge but is now possible. Studies such as those by Betts, West, and colleagues show a way toward developing these types of syntheses as sufficient data become available. Additional challenges exist in efforts to define and document histories of intensification of fisheries, and again this will depend on the synthesis of broadly based regional data. Moss and Erlandson (1998) had earlier taken major strides in this process by documenting and synthesizing the dates of wooden stakes from fish traps and weirs. Application of methods such as optically stimulated luminescence (OSL) dating to stone wall fish traps might advance this type of information still further.

Regardless of the sophistication of new and existing methods applied to fisheries archaeology and the productive questions these studies pose, the way forward toward recognizing new patterns and understanding their basis in human histories and environmental contexts is with additional data analyzed at multiple scales. Fortunately, those data exist in abundance in the tens of thousands of archaeological sites that pepper the vast coastlines of the north Pacific. With the types of research efforts exemplified by papers in this volume, that potential is well on its way to being realized now and into the foreseeable future.

References Cited

Balée, William. "The Research Program of Historical Ecology." *Annual Review of Anthropology* 35 (2006): 75–98.

Byram, R. Scott. "Brush Fences and Basket Traps: The Archaeology and Ethnohistory of Tidewater Weir Fishing in Oregon." PhD diss., University of Oregon, 2002.

Ford, Pamela J. "Archaeological and Ethnographic Correlates of Seasonality: Problems and Solutions on the Northwest Coast." *Canadian Journal of Archaeology* 13 (1989): 133–150.

Greene, Nancy. A. "Comox Harbour Fish Trap Site. WARP (Wetland Archaeological Research Project) Web Report." http://newswarp.info/wp-content/uploads/2010/03/WARP-web-report.pdf (accessed July 26, 2010).

Hamada, Shingo. "Herring Is God Fish: Historical and Cultural Importance of Herring in Hokkaido." Paper presented at Sharing Our Knowledge: A Conference of Tlingit Tribes and Clans, Juneau, AK, March 25–28, 2009.

Jones, Jim. "'We Looked After All the Salmon Streams': Traditional Heiltsuk Cultural Stewardship of Salmon and Salmon Streams: A Preliminary Analysis." Master's thesis, University of Victoria, 2000.

Langdon, Stephen J. "Tidal Pulse Fishing: Selective Traditional Tlingit Fishing Techniques on the West Coast of the Prince of Wales Archipelago." In *Traditional Ecological Knowledge and Natural Resource Management*, Charles R. Menzies, ed., pp. 21–46. Lincoln: University of Nebraska Press, 2007.

Lyman, R. Lee, and Kenneth P. Cannon. *Zooarchaeology and Conservation Biology*. Salt Lake City: University of Utah Press, 2004.

Montgomery, David R. *King of Fish: The Thousand-Year Run of Salmon*. Boulder, CO: Westview Press, 2003.

Moss, Madonna L., and Jon M. Erlandson. "A Comparative Chronology of Northwest Coast Fishing Features." In *Hidden Dimensions: The Cultural Significance of Wetland Archaeology*, Kathryn Bernick, ed., pp. 180–198. Vancouver: UBC Press, 1998.

Rick, Torben D., and Jon M. Erlandson, eds. *Human Impacts on Ancient Marine Ecosystems: A Global Perspective*. Berkeley: University of California Press, 2008.

White, Elroy A. F. "Heiltsuk Stone Fish Traps: Products of My Ancestors' Labour." Master's thesis, Simon Fraser University, 2006.

Matthew W. Betts is curator of Atlantic provinces in the archaeology and history division of the Canadian Museum of Civilization in Gatineau, Quebec.

Natalie Brewster is a doctoral student in the Department of Anthropology at McMaster University in Hamilton, Ontario.

Virginia L. Butler is a professor of anthropology at Portland State University in Oregon.

Megan Caldwell is a doctoral student in the Department of Anthropology at the University of Alberta in Edmonton.

Aubrey Cannon is a professor of anthropology at McMaster University in Hamilton, Ontario.

Donald S. Clark is a fisheries biologist with the Department of Fisheries and Oceans Canada at St. Andrews Biological Station in New Brunswick.

J. Tait Elder earned his master's degree in anthropology at Portland State University in Oregon in 2010 and is employed as a consulting archaeologist at ICF International.

Christopher Johnston is a fisheries biologist with the Alaska Fisheries Science Center, Resource Ecology Fisheries Management, Age Determination Program in Seattle, Washington.

Robert E. Kopperl is an archaeologist with SWCA Environmental Consultants in Seattle, Washington.

Andrew Martindale is an assistant professor of anthropology at University of British Columbia in Vancouver.

Herbert D.G. Maschner is a professor of anthropology at Idaho State University in Pocatello, Idaho.

Gregory G. Monks is a professor of anthropology at the University of Manitoba in Winnipeg.

Madonna L. Moss is a professor of anthropology at the University of Oregon in Eugene.

Margaret A. Nelson is a senior archaeologist and project manager for Cascadia Archaeology in Seattle, Washington.

Trevor J. Orchard is a sessional lecturer in the Department of Anthropology at the University of Toronto in Ontario.

Shelia Orwoll is a research coordinator in the Bone Mineral Unit, Oregon Health and Sciences University in Portland.

Megan A. Partlow is an archaeologist in the Department of Anthropology, Central Washington University in Cheney, Washington.

Paul Prince is a faculty member in anthropology at Grant MacEwan University in Edmonton, Alberta.

Randall Schalk is owner and manager of Cascadia Archaeology in Seattle, Washington.

Ross E. Smith is an archaeologist formerly with Northwest Archaeological Associates and now with SWCA Environmental Consultants in Seattle, Washington.

Camilla Speller received her PhD in archaeology from Simon Fraser University in Burnaby, British Columbia, in 2009 and is currently a postdoctoral fellow at the University of Calgary.

Paul Szpak is a doctoral student in the Department of Anthropology, University of Western Ontario in London, Ontario.

Teresa Trost is project manager and laboratory manager for Cascadia Archaeology in Seattle, Washington.

Catherine F. West is a postdoctoral fellow at the National Museum of Natural History, Smithsonian Institution, in Washington, D.C.

Elroy White, whose Heiltsuk name is Xanius, is a contract archaeologist in Bella Bella, British Columbia.

Catherine Wilson-Skogen is a senior research assistant in the Bone Mineral Unit, Oregon Health and Sciences University in Portland.

Stephen Wischniowski is a fisheries biologist with the International Pacific Halibut Commission in Seattle, Washington.

Mike Wolverton is a lithics analyst and scientific illustrator for Cascadia Archaeology in Seattle, Washington.

Dongya Yang is an associate professor of archaeology, Simon Fraser University in Burnaby, British Columbia.

Note: Boldface italicized page numbers refer to maps. All other illustrations, tables, and figures are indicated by italicized page numbers.